Simulation-Driven Design by Knowledge-Based Response Correction Techniques

Slawomir Koziel • Leifur Leifsson

Simulation-Driven Design by Knowledge-Based Response Correction Techniques

Springer

Slawomir Koziel
Engineering Optimization
& Modeling Center
Reykjavik University
Reykjavik, Iceland

Leifur Leifsson
Department of Aerospace Engineering
Iowa State University
Ames, Iowa
USA

ISBN 978-3-319-30113-6 ISBN 978-3-319-30115-0 (eBook)
DOI 10.1007/978-3-319-30115-0

Library of Congress Control Number: 2016937931

Printed on acid-free paper

This Springer imprint is published by Springer Nature
The registered company is Springer International Publishing AG Switzerland

Preface

Computer simulation models have become the most important design tools in the majority of engineering disciplines. They are used for verification purposes but also directly in the design process, in particular, for design optimization. Although desirable, and in many situations indispensable, simulation-driven design may be quite challenging. A fundamental bottleneck is the high computational cost of the simulation models. This often makes the use of conventional optimization algorithms prohibitive as these techniques require—especially for higher-dimensional parameter spaces—a large number of objective function evaluations. The high computational cost issue can be alleviated to some extent by utilization of adjoint sensitivities or automated differentiation. In general, however, robust, automated, and efficient optimization of expensive computational models remains an open problem.

Probably the most promising approach to computationally efficient handling of expensive computational models is surrogate-based optimization (SBO) where direct optimization of the expensive (high-fidelity) simulation model is replaced by iterative construction, enhancement, and re-optimization of its cheap representation, referred to as a surrogate model. A number of SBO techniques have evolved over the recent years. The most important difference between them lies in the surrogate model construction. The two major types of surrogate models can be distinguished: data-driven models obtained by approximating sampled high-fidelity simulation data, and physics-based ones constructed by suitable correction of the underlying low-fidelity model. Because the low-fidelity models embed system-specific knowledge, usually a small number of high-fidelity simulations are sufficient to configure a reliable surrogate. The most straightforward way of constructing the surrogate model from the underlying low-fidelity model is to correct its response. Various response correction techniques and related SBO algorithms have been proposed in the literature. Some of these methods are based on simple analytical formulas; others aim at tracking the response changes of the low-fidelity model, thus ensuring better generalization capability, although at the cost of increased formulation complexity and more restrictive assumptions regarding applicability of a given method.

This book focuses on surrogate-based optimization techniques exploiting response-corrected physics-based surrogates. For the sake of keeping the text self-contained, relevant background material is also covered to some extent, including the basics of classical and surrogate-assisted optimization, both based on data-driven and physics-based surrogates. The main part of the book provides an exposition of parametric and non-parametric response correction techniques as well as discusses practical issues on the construction and trade-offs for low-fidelity models. Furthermore, we consider related methods and discuss various aspects of physics-based SBO such as the adaptively adjusted design specification approach, feature-based optimization, response correction methods enhanced by adjoint sensitivities, as well as multi-objective optimization. Finally, we outline applications of response correction techniques for physics-based surrogate modeling. The book contains a large number of practical design cases from various engineering disciplines including electrical engineering, antenna design, aerospace engineering, and hydrodynamics, as well as recommendations for practical use of the presented methods.

Reykjavik, Iceland Slawomir Koziel
Ames, IA, USA Leifur Leifsson

Contents

Chapter 1
Introduction

Computer simulations have become the principal design tool in many engineering disciplines. They are commonly used for verification purposes, but also, more and more often, directly utilized within the design process, e.g., to adjust geometry and/or material parameters of the system of interest so that given performance requirements are satisfied. As a matter of fact, simulation-driven design has become a necessity for a growing number of devices and systems, where traditional approaches (e.g., based on design-ready theoretical models) are no longer adequate. One of the reasons is the increasing level of complexity of engineering systems, as well as various system- and component-level interactions, which have to be taken into account in the design process (Koziel and Ogurtsov 2014a). A reliable evaluation of the system performance can only be obtained (apart from physical measurements of the fabricated prototype) through high-fidelity computer simulations (typically, these simulations are computationally expensive).

Perhaps the most common design task involving multiple evaluations of the computational model is parametric optimization, in which a sequence of candidate designs are produced so as to converge to a solution that is optimal with respect to a predefined objective function (Sobester and Forrester 2015; Koziel et al. 2013a). Other tasks involving numerous simulations include statistical design (e.g., Monte Carlo analysis; Styblinski and Oplaski 1986), construction of behavioral models (Couckuyt et al. 2012), or uncertainty quantification (Hosder 2012; Allaire and Willcox 2014). A wide range of specialized algorithms have been developed to perform each of these tasks (Nocedal and Wright 2000; Yang 2010; Conn et al. 2009; Gorissen et al. 2010). Some of these methods are generic, i.e., applicable to a number of problems in different areas (Nocedal and Wright 2000), and others are more problem specific (e.g., Koziel et al. 2013a).

Unfortunately, in many practical cases, simulation-driven design is a very challenging problem. A fundamental bottleneck is the high computational cost of the simulation models utilized in many areas of engineering and science. Depending on the complexity of the computational model, simulation times may vary from seconds (e.g., for simple electromagnetic (EM) simulations of

S. Koziel, L. Leifsson, *Simulation-Driven Design by Knowledge-Based Response Correction Techniques*, DOI 10.1007/978-3-319-30115-0_1

two-dimensional models, e.g., planar microwave filters), through minutes (e.g., computational fluid dynamics (CFD) simulations of two-dimensional airfoil profiles), to hours and days (CFD simulations of three-dimensional structures such as aircraft wings, or EM simulations of integrated photonic components), or even weeks (simulations of complex 3D structures such as a full aircraft or a ship, evaluation of climate models). This often makes the use of conventional optimization algorithms prohibitive as these techniques require—especially for higher dimensional parameter spaces—a large number of objective function evaluations. Another issue is the numerical noise that is inherent to simulation models. The noise might be, among others, a result of terminating the simulation process before it fully converges or it may be related to adaptive meshing utilized by certain solvers (in which case, very small changes of the structure geometry lead to considerable changes of the mesh topology, and, consequently, to noticeable changes of the simulation results). The noise is particularly an issue for gradient-based methods that normally require smoothness of the objective function. The problem of the high computational cost can be alleviated to some extent by means of techniques such as adjoint sensitivities (Director and Rohrer 1969; Pironneau 1984; Jameson 1988; Papadimitriou and Giannakoglou 2008; El Sabbagh et al. 2006), or automated differentiation (Griewank 2000; Bischof et al. 2008). For many problems, these methods allow for a fast evaluation of gradients of the figures of interest at small additional computational effort (often only one additional simulation) regardless of the number of designable parameters. Thus, the benefits of adjoints are particularly evident for higher dimensional problems. Adjoint sensitivities are currently available in some commercial simulation packages in various areas (electromagnetic solvers: CST 2011; HFSS 2010; computational fluid dynamics solvers: ANSYS Fluent 2015; Star-CCM+ 2015), as well as some noncommercial codes (e.g., Stanford University Unstructured; Palacios et al. 2013).

In general, robust, automated, and computationally efficient optimization of expensive computational models remains an open problem. It should be mentioned here that—given the challenges of simulation-driven design mentioned in the previous paragraph—the most common way of utilizing computer simulations in design work is an interactive one. Although the details may vary in various fields, the basic flow is that an engineer is making predictions about possibly better designs (e.g., parameter setups) using his or her experience and insight, and utilizing simulation tools to verify these predictions. A typical realization would be to select (based on insight) a few parameters and to sweep them within certain ranges (also set up by experience), looking for an optimum. Parameter sweeps are normally executed one parameter at a time, and iterated until a satisfactory design is identified. For an experienced user, especially the one who solved a number of similar problems before, such a procedure usually leads to reasonable results; however, it might be very laborious and time consuming. Obviously, it does not guarantee optimum results, and it often fails when the user does not have a previous experience with similar class of problems, the number of relevant parameters is large, or optimum parameter setups are counterintuitive (which occurs for many contemporary structures and systems, such as miniaturized microwave and antenna components, Koziel et al. 2013b).

Despite the aforementioned difficulties, automated simulation-driven optimization is highly desirable. Simplified models and engineering insight can, in most cases, only lead to initial designs that have to be further tuned. Finding optimum designs normally require simultaneous adjustments of multiple parameters. Thus, numerical optimization techniques are indispensable. A large variety of conventional optimization techniques are available, including gradient-based methods (Nocedal and Wright 2000), e.g., conjugate-gradient, quasi-Newton, sequential quadratic programming, and derivative free (Kolda et al. 2003), e.g., Nelder-Mead, pattern-search techniques, as well as global optimization methods, including a popular class of population-based metaheuristics, e.g., genetic algorithms (Goldberg 1989), evolutionary algorithms (Back et al. 2000), ant systems (Dorigo and Gambardella 1997), particle swarm optimizers (Kennedy 1997), and many others (cf. Yang 2010). A common issue, from the simulation-driven design standpoint, is a large number of objective function evaluations required by majority of these methods, from dozens (for low-dimensional problems solved with gradient-based or pattern-search methods), through hundreds (for medium-size problems), to thousands, and tens of thousands (when using metaheuristics). Certain technologies, such as the aforementioned adjoint sensitivity and its commercial availability, alleviate this difficulty to some extent. Another factor is the continuous increase of available computational power (such as faster processors, parallel and distributed computing, and hardware acceleration). However, its impact is reduced by ever-growing need for more accurate (and therefore more expensive) simulations, including the necessity of simulating more complex systems.

Perhaps the most promising approach to realize computationally efficient optimization of expensive computational models is surrogate-based optimization (SBO) (Forrester and Keane 2009; Queipo et al. 2005; Booker et al. 1999; Koziel et al. 2011; Bandler et al. 2004a). The fundamental concept of SBO is to replace direct optimization of the expensive (high-fidelity) simulation model by means of an iterative construction, enhancement, and re-optimization of its cheap representation, referred to as a surrogate model. The surrogate is utilized as a prediction tool that enables the identification of promising locations of the search space (Forrester and Keane 2009; Koziel and Leifsson 2013a). New candidate solutions obtained through such a prediction are then verified by referring to the high-fidelity model. This additional data is also used to update the surrogate model and, consequently, improve its local (or global, depending on the SBO algorithm version) accuracy. The most important prerequisite of the SBO paradigm is that the surrogate is significantly faster than the high-fidelity model. It is critical because most operations in the SBO process are executed at the surrogate model level (in particular, its optimization is usually performed by means of conventional tools), and its numerous evaluations could become a major contributor to the overall optimization cost if the model is not sufficiently fast. At the same time, in many SBO algorithms, the high-fidelity model is only evaluated once per iteration, and—for a well-working

SBO process—the number of iteration is relatively small. As a result, the computational cost of the SBO-assisted design process may be significantly lower than for most of the conventional optimization methods.

Despite conceptual similarity, a number of rather distinct SBO techniques have evolved over the recent years. The most important difference between various SBO methods lies in the construction of the surrogate model. There are two major types of surrogate models. The first one comprises response surface approximation models constructed from sampled high-fidelity simulation data (Simpson et al. 2001a, b). A large variety of approximation (and interpolation) techniques are available, including artificial neural networks (Haykin 1998), radial basis functions (Wild et al. 2008), Kriging (Forrester and Keane 2009), support vector regression (Smola and Schölkopf 2004), Gaussian process regression (Angiulli et al. 2007; Jacobs 2012), or multidimensional rational approximation (Shaker et al. 2009), to name just the most popular ones. These data-driven surrogates are versatile, and if the design space is sampled with sufficient density the resulting model becomes a reliable representation of the system or device of interest so that it can be used for design purposes. In some cases, given a sufficiently accurate approximation surrogate, optimization process can be completely detached from the high-fidelity model (i.e., the optimum design can be found just by optimizing a surrogate without further reference to the simulation model). Unfortunately, computational overhead related to data acquisition and construction of such accurate approximations may be significant. Depending on the number of designable parameters, the number of training samples necessary to ensure usable (not to mention very high) accuracy might be hundreds, thousands, or even tens of thousands. Moreover, the number of samples quickly grows with the dimensionality of the problem (the so-called curse of dimensionality). As a consequence, globally accurate approximation modeling can only be justified in case of multiple-use library models of specific components described by a limited number of parameters. It is not suitable for ad hoc (one-time) optimization of systems evaluated through expensive simulations.

Iteratively improved approximation surrogates are becoming prevalent for global optimization (Gorissen et al. 2010). Various ways of incorporating new training points into the model (so-called infill criteria) exist, including exploitative models (i.e., models oriented towards improving the design in the vicinity of the current one), explorative models (i.e., models aiming at improving global accuracy), as well as model with balanced exploration and exploitation (Forrester and Keane 2009). Generally, these classes of techniques are often referred to as efficient global optimization (EGO) methods (Jones et al. 1998) or surrogate-assisted evolutionary algorithms (SAEAs) (Jin 2011; Lim et al. 2010).

Physics-based models constitute another class of surrogates whose adaptation has been growing recently due to the fact that SBO methods exploiting such models tend to be computationally more efficient than those exploiting approximation models. At the same time, surrogate construction is rather problem specific, and the related optimization algorithms generally require more complex implementation (Bandler et al. 2004a; Koziel and Leifsson 2013a). Physics-based surrogates are constructed from underlying low-fidelity (or coarse) models of the structures or systems being studied. Perhaps their biggest advantage is the fact that because the low-fidelity models embed system-specific knowledge, usually a small number of high-fidelity simulations are sufficient to configure a reliable surrogate. Low-fidelity models can be obtained in various ways: (1) as analytical models (in practice, a set of design-ready equations offering a considerably simplified description of the system); (2) by simulating the system at a different level (e.g., in microwave engineering: equivalent circuit representation evaluated using circuit theory rules versus full-wave electromagnetic simulation, Bandler et al. 2004a); and (3) lower fidelity or lower resolution simulation (e.g., simulation with coarser discretization of the structure and/or relaxed convergence criteria, Koziel and Ogurtsov 2014a, b). It should also be mentioned that because low-fidelity models are typically simulation ones, their evaluation time cannot—in many cases—be neglected and the aggregated computational cost of the low-fidelity model simulation may be significant (sometimes, a major) contribution to the overall cost of the SBO process (Leifsson et al. 2014a, b, c). Also, one of the important considerations is finding a proper balance between the accuracy and the speed of the low-fidelity model (Koziel and Ogurtsov 2012a), which may affect the SBO algorithm performance.

One of the most popular (although not the simplest in terms of formulation) SBO approaches using physics-based surrogates is space mapping (SM) (Bandler et al. 2004a, b, c; Koziel et al. 2008a). On the other hand, the most straightforward way of constructing the surrogate model from the underlying low-fidelity model is correcting its response. Various response correction techniques and related SBO algorithms can be found in the literature, including AMMO (Alexandrov and Lewis 2001), multi-point correction (Toropov 1989), manifold mapping (Echeverria and Hemker 2005), adaptive response correction (Koziel et al. 2009), or shape-preserving response prediction (Koziel 2010a). Some of the above methods are based on simple analytical formulas, and others (mostly non-parametric ones) aim at tracking the response changes of the low-fidelity model, thus ensuring better generalization capability, although at the cost of increased formulation complexity and more restrictive assumptions regarding applicability of a given method.

This book focuses on surrogate-based optimization techniques exploiting response-corrected physics-based surrogates. At the same time, for the sake of keeping the text self-contained, related background material is also covered to some extent, including the basics of classical and surrogate-based optimization. We begin, in Chap. 2, by formulating the simulation-driven optimization problem and discussing its major challenges. In Chap. 3, we briefly discuss conventional numerical optimization techniques, both gradient-based and derivative-free methods, including population-based metaheuristics. Chapter 4 is an introduction to surrogate-based optimization, where we discuss—on a generic level—the SBO design workflow as well as various aspects of surrogate-based optimization. In the same chapter, an outline of surrogate modeling is given, both regarding function approximation and physics-based models. In Chap. 5, we introduce surrogate modeling through a response correction, outline parametric and non-parametric correction techniques, as well as discuss some practical issues on the construction and trade-offs for low-fidelity models. Chapters 6 and 7 provide formulation and application examples of several specific response correction techniques, both parametric (Chap. 6) and non-parametric (Chap. 7) ones. The remaining chapters discuss various methods and aspects of physics-based SBO such as the adaptively adjusted design specification approach (Chap. 8), optimization using response features (Chap. 9), response correction methods enhanced by adjoint sensitivities (Chap. 10), multi-objective optimization exploiting response correction (Chap. 11), and response correction techniques for physics-based surrogate modeling (Chap. 12). The book is concluded in Chap. 13 where we formulate recommendations for the readers interested in applying the presented algorithms and techniques in their design work and discuss possible future developments concerning automation of simulation-driven optimization. The book is illustrated with a large number of practical design cases from various engineering disciplines including electrical engineering, antenna design, aerospace engineering, hydrodynamics, and others.

Chapter 2
Simulation-Driven Design

In this chapter, the simulation-driven design task is formulated as a nonlinear minimization problem. Among others, we introduce the notation used throughout the book, discuss typical design objectives and constraints, as well as describe common challenges, mostly related to the high computational cost of evaluating simulation models of the devices and systems under consideration. A brief outline of conventional numerical optimization methods is provided in Chap. 3. The concept of surrogate-based optimization is discussed, including various types of surrogate models, in Chap. 4.

2.1 Formulation of the Optimization Problem

We will denote the response of a high-fidelity (or fine) simulation model of a device or system under design by $\boldsymbol{f}(\boldsymbol{x})$. Typically, $\boldsymbol{f}$ will represent an evaluation of the performance characteristics of interest (see Sect. 2.2 for specific examples). The vector $\boldsymbol{x} = [x_1\ x_2\ \ldots\ x_n]^T$ represents the designable parameters to be adjusted (e.g., related to the geometry and/or the material).

In many situations, individual components of the vector $\boldsymbol{f}(\boldsymbol{x})$ will be considered and we will use the notation $\boldsymbol{f}(\boldsymbol{x}) = [f_1(\boldsymbol{x})\ f_2(\boldsymbol{x})\ \ldots\ f_m(\boldsymbol{x})]^T$, where $f_k(\boldsymbol{x})$ is the kth component of $\boldsymbol{f}(\boldsymbol{x})$. In particular, we might have $\boldsymbol{f}(\boldsymbol{x}) = [f(\boldsymbol{x},t_1)\ f(\boldsymbol{x},t_2)\ \ldots\ f(\boldsymbol{x},t_m)]^T$, where t_k, $k = 1, \ldots, m$, parameterizes the components of $\boldsymbol{f}(\boldsymbol{x})$; for example, t might represent time, frequency, or a relevant geometry variable (such as the airfoil thickness). In some cases, the response $\boldsymbol{f}(\boldsymbol{x})$ may actually consist of several scalars and/or vectors representing the performance characteristics of interest.

The simulation-driven design task is usually formulated as a nonlinear minimization problem of the following form (Koziel and Leifsson 2013a):

S. Koziel, L. Leifsson, *Simulation-Driven Design by Knowledge-Based Response Correction Techniques*, DOI 10.1007/978-3-319-30115-0_2

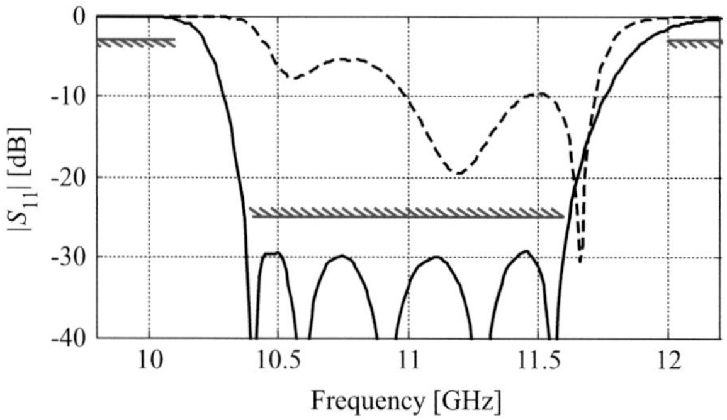

Fig. 2.1 Illustration of minimax design specifications, here, $|S_{11}| \leq -25$ dB for frequencies from 10.4 to 11.6 GHz, and $|S_{11}| \geq -3$ dB for frequencies lower than 10.1 GHz and higher than 12.0 GHz, marked with *thick horizontal lines*. An example waveguide filter response that does not satisfy our specifications (*dashed line*) (specification error, i.e., maximum violation of the requirements is about + 22 dB at 10.4 GHz), and another response that does satisfy the specifications (*solid line*) (specification error, i.e., the minimum margin of satisfying the requirements is about 1.5 dB at 12 GHz)

$$\boldsymbol{x}^* = \arg\min_{\boldsymbol{x}} U(\boldsymbol{f}(\boldsymbol{x})) \tag{2.1}$$

Here, U is the scalar merit function encoding the design specifications, whereas $\boldsymbol{x}^*$ is the optimum design to be found. The composition $U(\boldsymbol{f}(\boldsymbol{x}))$ will be referred to as the objective function. The function U is implemented so that a better design $\boldsymbol{x}$ corresponds to a smaller value of $U(\boldsymbol{f}(\boldsymbol{x}))$.

In some cases, definition of the objective function is pretty straightforward (e.g., minimization of the drag coefficient of an airfoil, and the maximization of the heat transfer of a heat exchanger). Thus, in many cases, the objective function can be defined in a norm sense as $U(\boldsymbol{f}(\boldsymbol{x})) = ||\boldsymbol{f}(\boldsymbol{x}) - \boldsymbol{f}^*||$, where $\boldsymbol{f}^*$ is a target response.

In several areas, optimization problems are formulated in a minimax sense with upper (and/or lower) specifications. Figure 2.1 shows an example of minimax specifications corresponding to typical requirements for microwave filters, here, $|S_{11}| \leq -25$ dB for 10.4–11.6 GHz, and, at the same time, $|S_{11}| \geq -3$ dB for frequencies lower than 10.1 GHz and higher than 12.0 GHz. The value of $U(\boldsymbol{f}(\boldsymbol{x}))$ (also referred to as the minimax specification error) corresponds to the maximum violation of the design specifications within the frequency bands of interest.

To simplify notation, we will occasionally use the symbol $f(\boldsymbol{x})$ as an abbreviation for $U(\boldsymbol{f}(\boldsymbol{x}))$.

In practical situations, the problem (2.1) is almost always constrained. The following types of constraints can be considered:

- Lower and upper bounds for the design variables, i.e., $l_k \leq x_k \leq u_k$, $k = 1, \ldots, n$.
- Inequality constraints, i.e., $c_{ineq.l}(\boldsymbol{x}) \leq 0$, $l = 1, \ldots, N_{ineq}$, where N_{ineq} is the number of constraints.
- Equality constraints, i.e., $c_{eq.l}(\boldsymbol{x}) = 0$, $l = 1, \ldots, N_{eq}$, where N_{eq} is the number of constraints.

Design constraints may be introduced to make sure that the device or system that is to be evaluated by the simulation software is physically valid (e.g., linear dimensions assuming nonnegative values). Also, constraints can be introduced to ensure that the physical dimensions (length, width, area, and volume) or other characteristics (e.g., cost) of the structure do not exceed certain assumed values. In some cases, evaluation of the constraints may be just as expensive as evaluating the objective function itself.

Another important aspect of simulation-driven design is that the majority of practical problems are multi-objective ones; that is, there is more than one performance characteristic that should be considered at a time. Typically, objectives conflict with each other so that the improvement of one is only possible at the expense of degrading the others. In many cases, a priori preference articulation is possible, so that only one specific objective is to be explicitly optimized, whereas the others can be handled through design constraints. This allows for solving the problem through single-objective formulation. Sometimes, however, this is neither possible nor convenient, e.g., when acquiring knowledge about possible trade-offs between competing objectives is important. In those situations a genuine *multi-objective* optimization is necessary. Typically, a solution to multi-objective design is represented as the so-called Pareto front (Tan et al. 2005), which represents the set of the best possible designs which are non-commensurable in the conventional (single-objective) sense.

A popular approach to handle multiple objectives is aggregation, using, for example, the weighted sum method (Tan et al. 2005), where

$$U(\boldsymbol{f}(\boldsymbol{x})) = \sum_{k=1}^{N_{obj}} w_k U_k(\boldsymbol{f}(\boldsymbol{x})) \tag{2.2}$$

Here, $U_k(\boldsymbol{f}(\boldsymbol{x}))$ represents the kth objective, whereas w_k, $k = 1, \ldots, N_{obj}$, are weighting factors (usually, convex combinations are considered, with $0 \leq w_k \leq 1$ and $\sum_k w_k = 1$). It should be emphasized that combining objectives into a single scalar function only allows for finding a single solution, i.e., one point on the Pareto front. Other solutions can be obtained (with some restrictions) by optimizing an aggregated objective function such as (2.2) with different setups of the weighting factors.

A discussion of genuine multi-objective optimization that aims at identifying the entire set of solutions representing the Pareto front exceeds the scope of this book. See for example Deb (2001). Some information, in the context of specific application examples, can be found in Chap. 11.

2.2 Design Challenges

The main focus of this book is on the optimization of simulation models. Perhaps the major challenge is the fact that realistic (in particular, accurate) simulations of real-world devices and systems tend to be computationally expensive. When

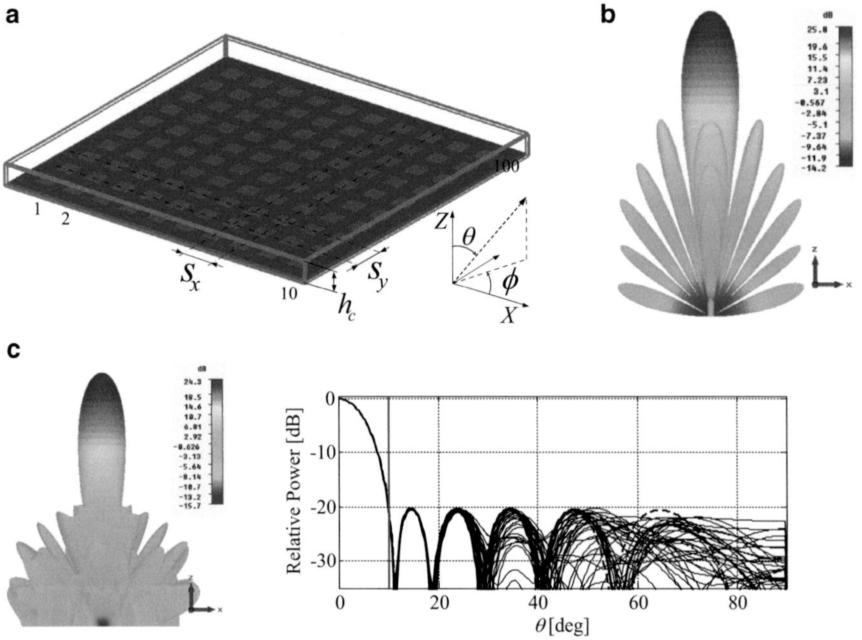

Fig. 2.2 Microstrip array antenna (Koziel and Ogurtsov 2015a): (**a**) Geometry: the structure contains 100 independent element (radiators); radiator geometry, element spacings, as well as excitation amplitudes and phases are designable parameters (over 200 in total); electromagnetic simulation of the structure takes about 1 h; (**b**) radiation pattern: the objective is to control the main beam width while minimizing radiation in other directions (or reduce the so-called side lobes); (**c**) radiation response of the optimized structure (note lower side lobe levels)

combined with other issues such as a large number of designable parameters, complex objective function landscapes, numerical noise, as well as multiple objectives, the simulation-driven optimization becomes a very challenging problem which—in many cases—is essentially unsolvable with conventional techniques unless massively parallel computations are utilized. In this section, we give a few examples of typical simulation-based optimization problems in various engineering areas. The presented problems are concerned with single devices rather than systems (the latter being even more challenging from computational point of view).

Figure 2.2 shows an example of a typical array antenna, where one of the major tasks is the synthesis of the radiation pattern, in particular the reduction of the side lobes (cf. Fig. 2.2b–c). Evaluating a high-fidelity electromagnetic (EM) simulation model of the array antenna can take about 1 h. At the same time, the number of designable parameters is large (in this case the parameters are related to the element geometries, element spacings, locations of feeds, and excitation amplitudes and phases, and are over 200 in total). Moreover, global optimization is necessary due to a rugged objective landscape. Optimization of the structure using conventional methods, such as gradient-based search (not to mention population-based

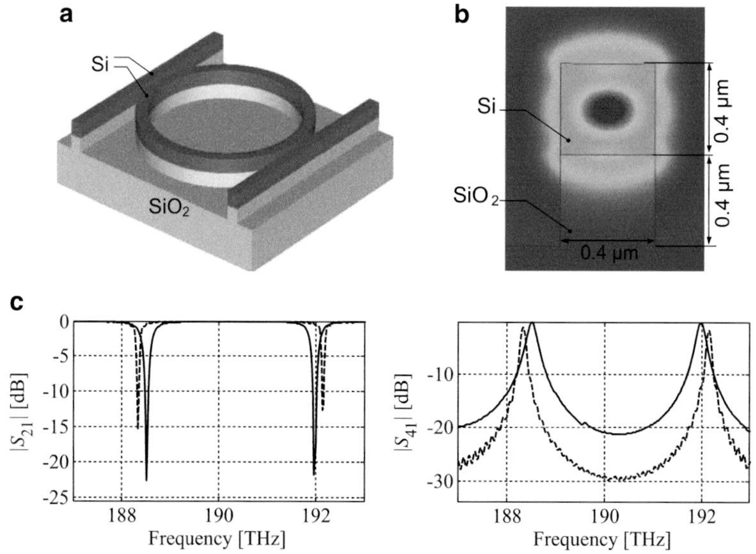

Fig. 2.3 Microring resonator (Koziel and Ogurtsov 2014b): (**a**) 3D view; (**b**) E-field amplitude distribution; (**c**) initial (*dashed line*) and optimized (*solid line*) frequency characteristics. The problem has only a few design variables but the high-fidelity simulation model is very expensive (~20 h with GPU acceleration)

metaheuristic algorithms), is virtually impossible in a reasonable timeframe. Figure 2.2c shows the optimized pattern, obtained using a surrogate-based methodology with an underlying analytical array factor model and a suitably defined response correction (Koziel et al. 2015).

Figure 2.3 shows an example of an integrated photonic component, a so-called add-drop microring resonator (Bogaerts et al. 2012). Although single components like this one are described by a rather small number of parameters (here, just three), the high-fidelity EM simulation is very expensive (~20 h of CPU time with GPU acceleration when using fine discretization of the structure). Simulation times of more involved structures (such as coupled microrings) are significantly longer. Figure 2.3c shows initial and optimized characteristics of the microring for an example optimization problem, here, obtained at a coarse-mesh simulation setup and exploiting the so-called feature-based optimization (less than 20 EM simulations in total; Koziel and Ogurtsov 2014b).

Another example of an expensive engineering problem is the design of a transport aircraft (Fig. 2.4a), which is a highly coupled multidisciplinary system—involving disciplines such as aerodynamics, structures, controls, and weights—with thousands of design variables. One of the major concerns in aircraft design is the aerodynamic performance where the wing shape plays a major role (Holst and Pulliam 2003; Leoviriakit and Jameson 2005; Epstein and Peigin 2006; Mavriplis 2007; Leung 2010). The evaluation time of a single high-fidelity model of the external aerodynamics at transonic speeds can be more than 24 h (depending on the available computational power) (Leifsson and Koziel 2015a).

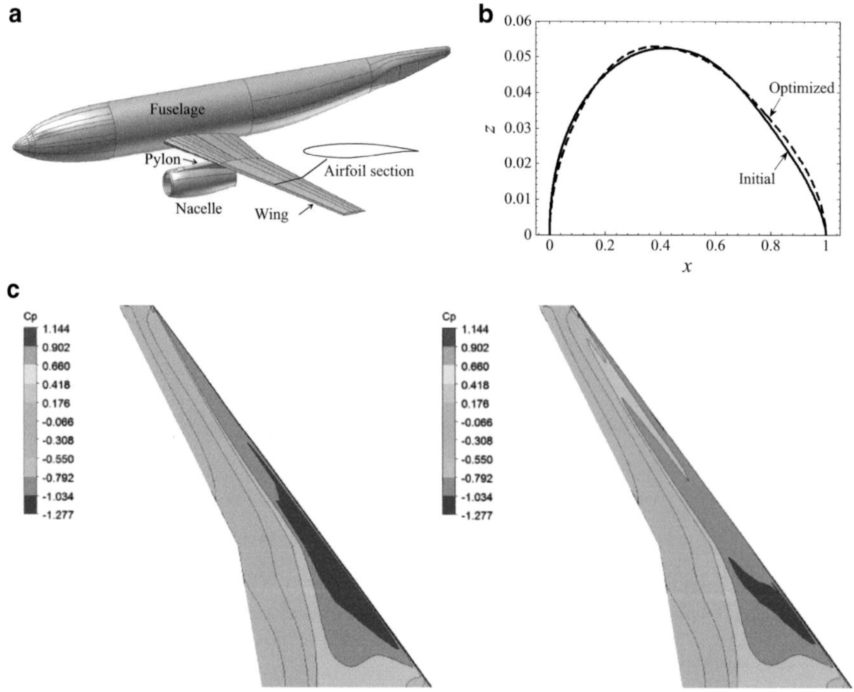

Fig. 2.4 Wing of a typical transport aircraft (Leifsson and Koziel 2015a): (**a**) wing-body-nacelle configuration, (**b**) the upper surface shapes of the initial (*solid*) and optimized (*dashed*) airfoil, (**c**) contours of pressure coefficient of the initial (*left*) and optimized (*right*) wings

Direct aerodynamic shape optimization of the wing (only one of the disciplines) is impractical without adjoint sensitivity information (Jameson 1988). Even with this information, the aerodynamic shape design task is challenging, due to numerical noise in the computational fluid dynamics (CFD) models, as well as the multimodality of the design space. Figure 2.4b shows the shape of the upper surface of one of the wing airfoil sections, before and after optimization with a surrogate-based optimization technique and a shape parameterization with five variables (Jonsson et al. 2013a). The total number of equivalent high-fidelity model evaluations to obtain the optimized design is 15 (the evaluation time of the high-fidelity model is 27.5 h, the surrogate model is 100 times faster), or 17 days. Figure 2.4c shows the pressure distribution of the initial and optimized designs, indicating a significant reduction of upper surface pressure shock strength, leading to a reduction in wing drag by 4.6 %.

The design of trawl-doors—which are an important part of a typical fishing gear (Fig. 2.5a)—is another example involving expensive aerodynamic (or hydrodynamic) CFD simulations, but in a different medium and at much lower speeds (Haraldsson 1996; Jonsson et al. 2013a). Trawl-doors function much like aircraft wings (an example configuration is shown in Fig. 2.5b); that is, they maintain certain lift for a given operating condition, but are always operated at

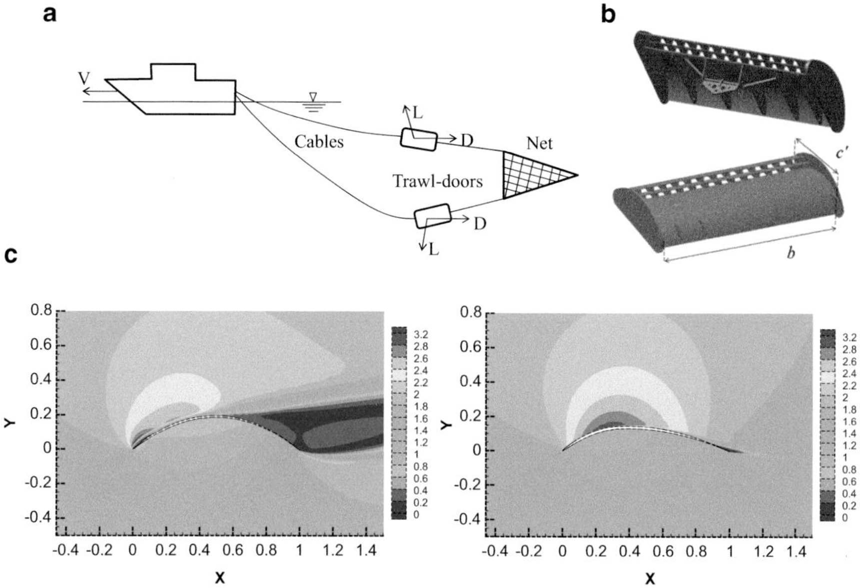

Fig. 2.5 Trawl-doors on a fishing vessel (Jonsson et al. 2013a): (**a**) main parts of the fishing gear (not drawn to scale), and (**b**) a typical trawl-door with two slots at the leading edge, and (**c**) velocity contours (in m/s) of initial (*left*) and optimized designs (*right*)

high lift. The flow physics of these devices are highly complex and become increasingly difficult to simulate with added number of airfoil elements. A two-dimensional CFD simulation of a trawl-door takes around 4 h on a typical desktop computer. Designable parameters are on the order of 10–15 in the two-dimensional single-element case, but can go up to 50 for a full three-dimensional multi-element case. Figure 2.5c shows the velocity contours around an initial and optimized single-element trawl-door (Leifsson et al. 2014a). The initial and optimized designs yield the same lift forces. However, the initial design operates at a high angle of attack (around 25°) with massive flow separation on the upper surface, whereas the optimized design operates at the same lift but at a much lower angle of attack (around 7°) and with almost no flow separation due to an improved shape.

The simulation-driven design examples discussed here illustrate problems that are common for many engineering disciplines. Among these, high computational cost of evaluating simulation models is usually a major bottleneck when it comes to numerical optimization. As also indicated, surrogate-based methods—particularly if properly tailored to a given class of problems—may allow for solving such tasks in a reasonable timeframe.

Chapter 3
Fundamentals of Numerical Optimization

Although the main focus of the book is on surrogate-assisted optimization using physics-based low-fidelity models and response correction techniques, we provide—for the sake of making the material self-contained—some basic information about conventional optimization algorithms. In this book, we refer to conventional (or direct) methods as those that handle the expensive simulation model directly in the optimization scheme (as opposed to surrogate-based approaches where most of the operations are carried out using a fast surrogate). In particular, each candidate solution suggested by the optimizer is evaluated using a computationally expensive high-fidelity simulation. Figure 3.1 shows a generic flow of the direct simulation-driven optimization process.

In this chapter, we provide an outline and a brief overview of conventional optimization techniques, including gradient-based and derivative-free methods, as well as metaheuristics. The readers interested in further information on these techniques are referred to the literature (e.g., Nocedal and Wright 2000; Yang 2010; Conn et al. 2009).

3.1 Optimization Problem

In this chapter, we consider the following formulation of the optimization problem:

$$\boldsymbol{x}^* = \arg\min_{\boldsymbol{x}\in X} f(\boldsymbol{x}) \tag{3.1}$$

where $f(\boldsymbol{x})$ is a scalar objective function, whereas $\boldsymbol{x} \in X \subseteq R^n$ is the search space (the objective function domain). We recall that in practice, the objective function is a composition $U(\boldsymbol{f}(\boldsymbol{x}))$ of the merit function U and (usually vector-valued) system response $\boldsymbol{f}(\boldsymbol{x})$, cf. Chap. 2. The objective function represents the performance of the system of interest. An alternative formulation of the problem with explicit

S. Koziel, L. Leifsson, *Simulation-Driven Design by Knowledge-Based Response Correction Techniques*, DOI 10.1007/978-3-319-30115-0_3

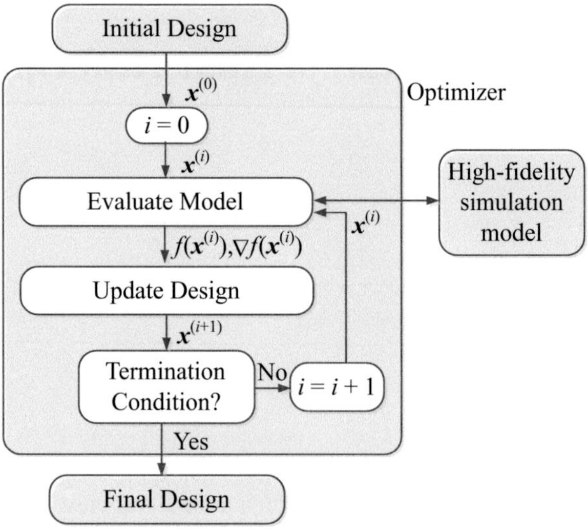

Fig. 3.1 Direct simulation-driven optimization flow. The candidate designs generated by the algorithm are evaluated through a high-fidelity simulation for verification purposes and to provide the optimizer with information to search for better designs. The search process may be guided by the model response only, or by (if available) the response as well as its derivatives (gradient)

constraints (lower/upper bounds, linear and nonlinear inequality or equality constraints) has been presented in Chap. 2.

In the remaining section of this chapter we outline the conventional optimization methods, both gradient based and derivative free (including population-based metaheuristics).

3.2 Gradient-Based Optimization Methods

Gradient-based algorithms belong to the most popular and widely used optimization techniques (Nocedal and Wright 2000). The fundamental piece of information directing the search for a better solution is the gradient $\nabla f(\boldsymbol{x})$ of the objective function $f(\boldsymbol{x})$ defined as

$$\nabla f(\boldsymbol{x}) = \left[\frac{\partial f}{\partial x_1}(\boldsymbol{x}) \quad \frac{\partial f}{\partial x_2}(\boldsymbol{x}) \quad \ldots \quad \frac{\partial f}{\partial x_n}(\boldsymbol{x}) \right]^T \tag{3.2}$$

Assuming that f is sufficiently smooth (i.e., at least continuously differentiable), the gradient provides information about the local behavior of f in the neighborhood of $\boldsymbol{x}$. In particular, we have

$$f(\boldsymbol{x} + \boldsymbol{h}) \cong f(\boldsymbol{x}) + \nabla f(\boldsymbol{x})^T \cdot \boldsymbol{h} < f(\boldsymbol{x}) \tag{3.3}$$

for a sufficiently small vector $\boldsymbol{h}$, provided that $\nabla f(\boldsymbol{x})^T \cdot \boldsymbol{h} < 0$. Furthermore, $\boldsymbol{h} = -\nabla f(\boldsymbol{x})$ determines the direction of the (locally) steepest descent. There are two basic ways to utilize the gradient in the search process: (1) by moving along a descent direction, or (2) by exploiting the local approximation model of the objective function constructed with ∇f. In some algorithms, the second-order derivatives are also used, in particular, the Hessian $\boldsymbol{H}(\boldsymbol{x})$, which is defined as

$$\boldsymbol{H}(\boldsymbol{x}) = \begin{bmatrix} \dfrac{\partial^2 f}{\partial x_1 \partial x_1}(\boldsymbol{x}) & \cdots & \dfrac{\partial^2 f}{\partial x_1 \partial x_n}(\boldsymbol{x}) \\ \vdots & \ddots & \vdots \\ \dfrac{\partial^2 f}{\partial x_n \partial x_1}(\boldsymbol{x}) & \cdots & \dfrac{\partial^2 f}{\partial x_n \partial x_n}(\boldsymbol{x}) \end{bmatrix} \tag{3.4}$$

3.2.1 Descent Methods

In this section, we outline iterative algorithms that generate a sequence $\boldsymbol{x}^{(i)}$, $i = 0, 1, \ldots,$ of approximation solutions to (3.1). The basic requirement is that $f(\boldsymbol{x}^{(i+1)}) < f(\boldsymbol{x}^{(i)})$. Given $\boldsymbol{x}^{(0)}$ as the initial approximation, the generic structure of a descent method can be summarized as follows (Nocedal and Wright 2000):

1. Set $i = 0$.
2. Find a search direction $\boldsymbol{h}_d$.
3. Find a step length α along $\boldsymbol{h}_d$.
4. Set $\boldsymbol{x}^{(i)} = \boldsymbol{x}^{(i)} + \alpha \boldsymbol{h}_d$.
5. Set $i = i + 1$.
6. If the termination condition is not satisfied, go to 2.
7. END.

The search direction of Step 2 is supposed to be a descent direction for which $\nabla f(\boldsymbol{x})^T \cdot \boldsymbol{h}_d < 0$ (cf. (3.3)). A separate algorithm is utilized to determine the step size $\alpha \boldsymbol{h}_d$ so that $f(\boldsymbol{x}^{(i+1)}) < f(\boldsymbol{x}^{(i)})$. A variety of termination conditions are used in practical algorithms, including $||\boldsymbol{x}^{(i+1)} - \boldsymbol{x}^{(i)}|| \le \varepsilon_1$ (convergence in the argument), $||\nabla f(\boldsymbol{x}^{(i)})|| \le \varepsilon_2$ (vanishing of the gradient), $f(\boldsymbol{x}^{(i)}) - f(\boldsymbol{x}^{(i+1)}) \le \varepsilon_3$ (convergence in the function value), or a combination of the above.

One way of finding the step length is by performing a line search (Nocedal and Wright 2000), in which one considers a function $\varphi(\alpha) = f(\boldsymbol{x} + \alpha \boldsymbol{h}_d)$ defined for $\alpha > 0$. In *exact line search*, the step length α_e is found as an exact minimum of $\varphi(\alpha)$, i.e., $\alpha_e = \text{argmin}\{\alpha > 0: f(\boldsymbol{x} + \alpha \boldsymbol{h}_d)\}$. *Soft line search* only requires satisfaction of certain (usually rather loose) conditions such as the Wolfe conditions: (1) $\varphi(\alpha_s) \le \lambda(\alpha_s)$, where $\lambda(\alpha) = \varphi(0) + \rho\varphi'(0) \cdot \alpha$, with $0 < \rho < 0.5$, and (2) $\varphi'(\alpha) \ge \beta \cdot \varphi'(0)$, with $\rho < \beta < 1$. Satisfaction of these conditions ensures a sufficient decrease in the objective function f without making the step too short. Descent methods with a line search governed by the above conditions are normally convergent to at least a first-order stationary point of f. Soft line search is generally more advantageous than the exact line search in terms of the overall computational cost of the optimization process (the use of the exact line search results in a smaller number but much more expensive iterations).

At this point, it is necessary to mention that the steepest descent algorithm, in which the steepest descent direction $\boldsymbol{h} = -\nabla f(\boldsymbol{x})$ is used in each iteration, has a poor practical performance when in the vicinity of the optimum. However, it might be useful at the initial stages of the search process.

Conjugate-gradient methods (Yang 2010) ensure a better selection of the search direction by combining the previous direction $\boldsymbol{h}_{prev}$ and the current gradient, i.e.,

$$\boldsymbol{h} = -\nabla f\left(\boldsymbol{x}^{(i)}\right) + \gamma \boldsymbol{h}_{prev} \tag{3.5}$$

The two most popular schemes for calculating the coefficient γ are a Fletcher-Reeves method with

$$\gamma = \frac{\nabla f(\boldsymbol{x})^T \nabla f(\boldsymbol{x})}{\nabla f\left(\boldsymbol{x}_{prev}\right)^T \nabla f\left(\boldsymbol{x}_{prev}\right)} \tag{3.6}$$

and a Polak-Ribiére method with

$$\gamma = \frac{\left(\nabla f(\boldsymbol{x}) - \nabla f\left(\boldsymbol{x}_{prev}\right)\right)^T \nabla f(\boldsymbol{x})}{\nabla f\left(\boldsymbol{x}_{prev}\right)^T \nabla f\left(\boldsymbol{x}_{prev}\right)} \tag{3.7}$$

An alternative way of realizing descent methods is through trust regions (Conn et al. 2000). More specifically, the iteration step is determined using the local approximation of the objective function f, such as a first- or second-order Taylor expansion, i.e., $f(\boldsymbol{x}+\boldsymbol{h}) \approx q(\boldsymbol{h})$ with $q(\boldsymbol{h}) = f(\boldsymbol{x}) + \nabla f(\boldsymbol{x}) \cdot \boldsymbol{h}^T$ (first-order) or $q(\boldsymbol{h}) = f(\boldsymbol{x}) + \nabla f(\boldsymbol{x}) \cdot \boldsymbol{h}^T + 0.5 \cdot \boldsymbol{h}^T \boldsymbol{H}(\boldsymbol{x}) \boldsymbol{h}$ (second order). In both cases, $q(\boldsymbol{h})$ is a good approximation of $f(\boldsymbol{x}+\boldsymbol{h})$ for sufficiently small $\boldsymbol{h}$. In the trust region (TR) framework, the iteration step size $\boldsymbol{h}_{tr}$ is determined as ($\delta > 0$ is the TR radius)

$$\boldsymbol{h}_{tr} = \arg \min_{\boldsymbol{h},\ \|\boldsymbol{h}\| \le \delta} q(\boldsymbol{h}) \tag{3.8}$$

$\boldsymbol{h}_{tr}$ is accepted if $f(\boldsymbol{x}+\boldsymbol{h}_{tr}) < f(\boldsymbol{x})$; otherwise it is rejected. The size of the trust region radius is adjusted based on the $r = [f(\boldsymbol{x}) - f(\boldsymbol{x}+\boldsymbol{h}_{tr})]/[q(0) - q(\boldsymbol{h}_{tr})]$. A typical adjustment scheme is $\delta \leftarrow 2\delta$ if $r > 0.75$, and $\delta \leftarrow \delta/3$ if $r < 0.25$. Note that the condition $f(\boldsymbol{x}+\boldsymbol{h}_{tr}) < f(\boldsymbol{x})$ would be satisfied for sufficiently small δ (assuming that f is sufficiently smooth given the order of the local model q).

3.2.2 Newton and Quasi-Newton Methods

If f is at least twice continuously differentiable, f can be locally represented by its second-order Taylor expansion $q(\boldsymbol{h}) = f(\boldsymbol{x}) + \nabla f(\boldsymbol{x}) \cdot \boldsymbol{h}^T + 0.5 \cdot \boldsymbol{h}^T \boldsymbol{H}(\boldsymbol{x})\boldsymbol{h}$. If $\boldsymbol{H}(\boldsymbol{x})$ is positive definite, then the model $q(\boldsymbol{h})$ has a unique minimizer with respect to $\boldsymbol{h}$ such that $\nabla q(\boldsymbol{h}) = 0$, i.e., where

$$\nabla f(\boldsymbol{x}) + \boldsymbol{H}(\boldsymbol{x})\boldsymbol{h} = 0 \tag{3.9}$$

Solution to (3.9) gives rise to Newton's method which is formulated as the following iterative process:

$$\boldsymbol{x}^{(i+1)} = \boldsymbol{x}^{(i)} - \left[\boldsymbol{H}\left(\boldsymbol{x}^{(i)}\right)\right]^{-1} \nabla f\left(\boldsymbol{x}^{(i)}\right) \tag{3.10}$$

The algorithm (3.10) is well defined assuming that $\boldsymbol{H}(\boldsymbol{x})$ is non-singular. It is characterized by a fast (quadratic) convergence if the starting point is sufficiently close to the optimum. Despite its apparent simplicity, the formulation (3.10) is impractical: the algorithm is not globally convergent; it may converge to a maximum or a saddle point, and (3.10) may be ill conditioned. Also, the method requires analytical second-order derivatives.

A simple way of alleviating the difficulties related to positive definiteness of the Hessian is a damped Newton method, in which the iteration (3.10) is modified to

$$\boldsymbol{x}^{(i+1)} = \boldsymbol{x}^{(i)} - \left[\boldsymbol{H}\left(\boldsymbol{x}^{(i)}\right) + \mu \boldsymbol{I}\right]^{-1} \nabla f\left(\boldsymbol{x}^{(i)}\right) \tag{3.11}$$

with $\boldsymbol{I}$ being the identity matrix. For sufficiently large $\mu > 0$, the matrix $\boldsymbol{H}(\boldsymbol{x}) + \mu \boldsymbol{I}$ is positive definite. Consequently, the step $\boldsymbol{h}_\mu$ found as a solution to the problem $[\boldsymbol{H}(\boldsymbol{x}) + \mu \boldsymbol{I}]\boldsymbol{h}_\mu = -\nabla f(\boldsymbol{x})$ is a minimizer of the model

$$q_\mu(\boldsymbol{h}) = q(\boldsymbol{h}) + \frac{1}{2}\mu \boldsymbol{h}^T \boldsymbol{h} = f(\boldsymbol{x}) + \nabla f(\boldsymbol{x}) \cdot \boldsymbol{h}^T + \frac{1}{2}\boldsymbol{h}^T [\boldsymbol{H}(\boldsymbol{x}) + \mu \boldsymbol{I}]\boldsymbol{h} \tag{3.12}$$

It should be noted that $\boldsymbol{h}_\mu$ is a descent direction for f at $\boldsymbol{x}$. Moreover, the term $\mu \boldsymbol{h}^T \boldsymbol{h}/2$ penalizes large steps. In particular, for very large μ, the algorithm defaults to the steepest descent method as $\boldsymbol{h}_\mu \approx -\nabla f(\boldsymbol{x})/\mu$.

In practical implementations, large values of μ are used at the initial iterations of (3.11), whereas μ is reduced when close to convergence, which allows for finding the minimizer of f. A Levenberg-Marquardt (LM) algorithm is a popular realization of an adaptive adjustment of the value of μ, using the following formula: $\mu \leftarrow \mu \cdot \max\{1/3,\ 1-(2r-1)^3\}$ if $r > 0$, and $\mu \leftarrow 2\mu$ otherwise (Nielsen 1999). Thus, μ is increased at the beginning of each iteration until $\boldsymbol{H}(\boldsymbol{x}) + \mu \boldsymbol{I}$ becomes positive definite, and then updated based on the value of the gain ratio r (cf. Sect. 3.2.1).

Although damped Newton methods overcome most of the disadvantages of the Newton's algorithm, a potentially high cost of obtaining second-order derivatives of f is still an issue. In quasi-Newton methods, exact Hessian (or its inverse) is replaced by approximations obtained using suitable updating formulas. One of the best and the most popular updating formulas, preserving both positive symmetry and definiteness of the approximation, is a BFGS one (Nocedal and Wright 2000). It is defined as follows for Hessian approximation $\boldsymbol{B}$:

$$\begin{gathered}\boldsymbol{B}_{new} = \boldsymbol{B} + \frac{1}{\boldsymbol{h}^T\boldsymbol{y}}\boldsymbol{y}\boldsymbol{y}^T - \frac{1}{\boldsymbol{h}^T\boldsymbol{u}}\boldsymbol{u}\boldsymbol{u}^T, \\ \text{where} \quad \boldsymbol{h} = \boldsymbol{x}_{new} - \boldsymbol{x}, \ \ \boldsymbol{y} = \nabla f(\boldsymbol{x}_{new}) - \nabla f(\boldsymbol{x}), \quad \boldsymbol{u} = \boldsymbol{B}\boldsymbol{h}\end{gathered} \tag{3.13}$$

Approximation of the inverse of the Hessian $\boldsymbol{D}$ is defined as

$$\begin{gathered}\boldsymbol{D}_{new} = \boldsymbol{D} + \kappa_1 \boldsymbol{h}\boldsymbol{h}^T - \kappa_2\left(\boldsymbol{h}\boldsymbol{v}^T + \boldsymbol{v}\boldsymbol{h}^T\right), \\ \text{where} \quad \boldsymbol{h} = \boldsymbol{x}_{new} - \boldsymbol{x}, \qquad \boldsymbol{y} = \nabla f(\boldsymbol{x}_{new}) - \nabla f(\boldsymbol{x}), \qquad \mathbf{v} = \boldsymbol{B}\boldsymbol{y} \\ \kappa_2 = \frac{1}{\boldsymbol{h}^T\boldsymbol{y}}, \qquad \kappa_1 = \kappa_2\left(1 + \kappa_2\left(\boldsymbol{y}^T\boldsymbol{v}\right)\right)\end{gathered} \tag{3.14}$$

Quasi-Newton algorithms belong to the most efficient methods for unconstrained optimization. On the other hand, conjugate gradient methods may be better for a large number of design variables (matrix operations required by the Newton-like methods are computationally more involved).

3.2.3 Remarks on Constrained Optimization

The methods discussed so far allow unconstrained optimization. Most practical problems are constrained, with typical constraints being lower/upper bounds for design variables, as well as linear/nonlinear equality and inequality constraints. Here, we make a few comments on constrained optimization methods. More details can be found in numerical optimization textbooks (e.g., Nocedal and Wright 2000).

Formulation of a constrained optimization problem was presented in Sect. 2.1. Here, we consider the problem (3.1) assuming N inequality and M equality constraints:

$$g_k(\boldsymbol{x}) \le 0, k = 1, \ldots, N \tag{3.15}$$

$$h_k(\boldsymbol{x}) = 0, k = 1, \ldots, M \tag{3.16}$$

The set of points that satisfy the above constraints is referred to as a feasible region.

The first-order necessary conditions (Karush-Kuhn-Tucker or KKT conditions; Kuhn and Tucker 1951) for an optimum of a constrained problem (3.1), (3.15), and (3.16) state that—at the local optimum $\boldsymbol{x}^*$ of f—there exist constants $\mu_1, \ldots, \mu_N$, and $\lambda_1, \lambda_2, \ldots, \lambda_M$, such that

$$\nabla f\left(\boldsymbol{x}^*\right) + \sum_{k=1}^{N} \mu_k \nabla g_k\left(\boldsymbol{x}^*\right) + \sum_{k=1}^{M} \lambda_k \nabla h_k\left(\boldsymbol{x}^*\right) = 0 \tag{3.17}$$

and $g_k(\boldsymbol{x}^*) \le 0$, $\mu_k g_k(\boldsymbol{x}^*) = 0$, where $\mu_k \ge 0$ for $k = 1, \ldots, N$, and provided that all the functions (both objective and constraints) are continuously differentiable. The function at the left-hand side of (3.17) is called the Lagrangian function, whereas coefficients μ and λ are the Lagrange multipliers.

The methods of handling constrained optimization generally fall into two categories. The so-called active set methods search for the optimum along the boundary of the feasible region (assuming that the unconstrained minimum of f is not in the interior of the feasible region, it has to be on its boundary). Inequality constraints are handled by keeping track of the set of active constraints while moving downhill along the boundary. Another class of methods attempt to approach the optimum in an iterative way, either from within the feasible region (so-called interior point methods) or by a possible use of infeasible points (however, not by moving along the feasible region boundary). In the latter approaches, the objective function is modified and the corresponding unconstrained optimization problems are solved in each iteration.

The simplest way of handling constraints is through penalty functions, where the original problem (3.1) with constraints (3.15), (3.16) is replaced by

$$\arg\min_{\boldsymbol{x}} \phi(\boldsymbol{x}, \boldsymbol{\alpha}, \boldsymbol{\beta}) = \arg\min_{\boldsymbol{x}} \left\{ f(\boldsymbol{x}) + \sum_{k=1}^{N} \alpha_k \overline{g}_k^2(\boldsymbol{x}) + \sum_{k=1}^{M} \beta_k h_k^2(\boldsymbol{x}) \right\} \tag{3.18}$$

where $\boldsymbol{\alpha} = [\alpha_1 \ \ldots \ \alpha_N]$, $\boldsymbol{\beta} = [\beta_1 \ \ldots \ \beta_M]$, and α_k, $\beta_k \gg 1$ are penalty factors, and $\overline{g}_k(\boldsymbol{x}) = \max\{0, g_k(\boldsymbol{x})\}$. In practice, a sequence of problems (3.18) is solved for increasing values of the penalty factors using the solution of the previous iteration as the starting point for the next one.

The barrier method used the following formulation (here, only inequality-type constraints are assumed for simplicity):

$$\arg\min_{\boldsymbol{x}} \phi(\boldsymbol{x}, \mu) = \arg\min_{\boldsymbol{x}} \left\{ f(\boldsymbol{x}) - \mu \sum_{k=1}^{N} \ln(-g_k(\boldsymbol{x})) \right\} \tag{3.19}$$

where $\mu > 0$ is a barrier parameter. The search starts from a feasible point. A sequence of unconstrained problems is solved with μ decreasing to zero so that the minimum of $\phi(\boldsymbol{x},\mu)$ converges to a solution of f.

Yet another set of approaches are the augmented Lagrangian methods which are similar to the penalty methods in the sense that the original constrained problem is solved as a sequence of suitably formulated unconstrained ones. The unconstrained objective is the Lagrangian of the constrained problem with an additional penalty term (the so-called augmentation). Assuming for simplicity equality-only constraints $h_k(\boldsymbol{x})$, $k = 1, \ldots, M$, the augmented Lagrangian algorithm generates a new approximation of the constrained solution to (3.1) as

$$\boldsymbol{x}^{(i+1)} = \arg\min_{\boldsymbol{x}} \phi^{(i)}(\boldsymbol{x}) = \arg\min_{\boldsymbol{x}} \left\{ f(\boldsymbol{x}) + \frac{\mu^{(i)}}{2} \sum_{k=1}^{M} h_k^2(\boldsymbol{x}) - \sum_{k=1}^{M} \lambda_k h_k(\boldsymbol{x}) \right\} \tag{3.20}$$

The starting point is the previous approximation $\boldsymbol{x}^{(i)}$. In each iteration, the coefficients μ_k are updates as follows: $\lambda_k \leftarrow \lambda_k - \mu^{(i)} h_k(\boldsymbol{x}^{(i+1)})$. The coefficients λ_k are estimates of the Lagrange multipliers (cf. (3.17)), and their accuracy increases as the optimization process progresses. The value of μ is increased in each iteration; however, unlike in the penalty method, it is not necessary to take $\mu \rightarrow \infty$.

One of the most popular methods for solving constrained optimization problems is sequential quadratic programming (SQP; Han 1977; Nocedal and Wright 2000). At each iteration, the following quadratic programming subproblem is used to compute a search direction $\boldsymbol{h}^{(i)}$:

$$\boldsymbol{h}^{(i)} = \arg\min_{\boldsymbol{h}} \left\{ f\left(\boldsymbol{x}^{(i)}\right) - \nabla f\left(\boldsymbol{x}^{(i)}\right)^T \boldsymbol{h} + \frac{1}{2} \boldsymbol{h}^T \boldsymbol{H}\left(\boldsymbol{x}^{(i)}\right) \boldsymbol{h} \right\} \tag{3.21}$$

so that $g_k(\boldsymbol{x}^{(i)}) + \nabla g_k(\boldsymbol{x}^{(i)})^T \boldsymbol{h} \leq 0$, $k = 1, \ldots, N$, and $h_k(\boldsymbol{x}^{(i)}) + \nabla h_k(\boldsymbol{x}^{(i)})^T \boldsymbol{h} = 0$, $k = 1, \ldots, M$. Here, $\boldsymbol{x}^{(i)}$ is a current iteration point, whereas $\boldsymbol{H}$ is a symmetric, positive definite matrix (preferably an approximation of the Hessian of the Lagrangian of f). The new approximation $\boldsymbol{x}^{(i+1)}$ is obtained using a line search.

3.3 Derivative-Free Optimization Methods

Derivative-free techniques (Conn et al. 2009) are optimization methods that do not rely on derivative information in the search process. In many situations, derivatives of the objective functions are not available or too expensive to compute. Also, gradient-based search may not be a good option for noisy functions (numerical noise is inherent to a majority of simulation models).

Derivative-free methods include various types of pattern search algorithms (Kolda et al. 2003), global optimization methods, such as metaheuristics (Yang 2010), as well as surrogate-based optimization (SBO) algorithms (Queipo et al. 2005; Koziel and Leifsson 2013b, c). As SBO is the main topic of this book; its more detailed exposition is provided in Chap. 4 (fundamentals) and Chaps. 5–12 (specific methods and applications). In this section, we briefly outline selected methods and concepts such as the pattern search (Sect. 3.3.1), Nelder-Mead algorithm (Sect. 3.3.2), global optimization methods exploiting population-based metaheuristics, including genetic algorithms, particle swarm optimizers, and differential evolution (Sects. 3.3.3 through 3.3.5).

3.3.1 Pattern Search

Pattern search methods are popular optimization algorithms in which the search is restricted to a predefined grid. The objective function is evaluated on a stencil determined by a set of directions that are suitable from geometric or algebraic point of view. The initial grid is subsequently modified (in particular, refined) during the optimization run.

The overall concept of the pattern search approach (Kolda et al. 2003) is shown in Fig. 3.2, where—for the sake of illustration—the optimization process is restricted to the rectangular grid and thus explores a grid-restricted vicinity of the current design. If the operations performed on the current grid fail to lead to improvement of the objective function, the grid is refined to allow smaller steps. A typical termination criterion involves reaching a required resolution (e.g., a user-defined minimum stencil size). There are numerous variations of the pattern search algorithms (see, e.g., Kolda et al. 2003, for a review).

While being relatively robust, pattern search and similar methods usually converge slowly when compared to gradient-based algorithms. On the other hand, they do not need derivative information and, even more importantly, they are relatively immune to numerical noise. Interest in these types of methods has recently been revived, not only due to the development of rigorous convergence theory (e.g., Conn et al. 2009), but also because pattern search may easily benefit from availability of parallel computing resources.

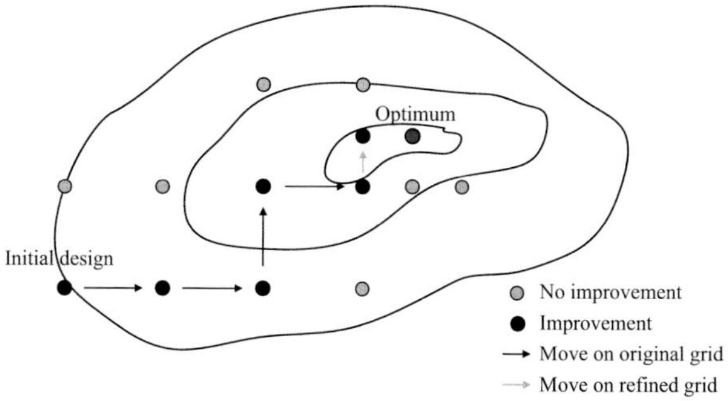

Fig. 3.2 Conceptual illustration of the pattern search process, here using a rectangular grid. The exploratory movements are restricted to the grid around the initial design. In the case of failure of making a successful move, the grid is refined to allow smaller steps. Practical implementations of pattern search routines typically use more sophisticated strategies (e.g., grid-restricted line search)

3.3.2 Nelder-Mead Algorithm

One of the most popular derivative-free methods is the Nelder-Mead algorithm which is a downhill simplex procedure for unconstrained optimization (Nelder and Mead 1965). The key concept of the method is the simplex (or an n-simplex), i.e., a convex hull of a set of $n+1$ affinely independent points $\boldsymbol{x}^{(i)}$ in a vector space (here, we consider an n-simplex in an n-dimensional space). The set $X = \{\boldsymbol{x}^{(i)}\}_{i\ =\ 1,...,k}$ is affinely independent if the vectors $\boldsymbol{v}^{(i)} = \boldsymbol{x}^{(i+1)} - \boldsymbol{x}^{(1)}$, $i = 1,..., k-1$ are linearly independent. A convex hull $H(X)$ of X is defined as a set of all convex combinations of the points from X, i.e., $H(X) = \{\sum_{i\ =\ 1,...,k} a_i\, \boldsymbol{x}^{(i)} : \boldsymbol{x}^{(i)} \in X,\ a_i \geq 0,\ \sum_{i\ =\ 1,...,k} a_i = 1\}$. A triangle and a tetrahedron are examples of a 2-simplex and a 3-simplex, respectively.

An iteration of the Nelder-Mead algorithm is a modification of the current simplex, which aims at finding a better solution (i.e., corresponding to a lower value of the objective function). The steps carried out in the process are the following (see also Fig. 3.3):

- Order the vertices so that $f(\boldsymbol{x}^{(1)}) \leq f(\boldsymbol{x}^{(2)}) \leq \ldots f(\boldsymbol{x}^{(n+1)})$. An auxiliary point $\boldsymbol{x}^{(0)}$ is then defined to be the center of gravity of all the vertices but $\boldsymbol{x}^{(n+1)}$.
- Compute a *reflection* $\boldsymbol{x}^r = (1+\alpha)\boldsymbol{x}^{(0)}\text{-}\alpha\boldsymbol{x}^{(n+1)}$ (a typical value of α is 1). If $f(\boldsymbol{x}^{(1)}) \leq f(\boldsymbol{x}^r) < f(\boldsymbol{x}^{(n)})$ reject $\boldsymbol{x}^{(n+1)}$ and update the simplex using $\boldsymbol{x}^r$.
- If $f(\boldsymbol{x}^r) < f(\boldsymbol{x}^{(1)})$ compute an *expansion* $\boldsymbol{x}^e = \rho\boldsymbol{x}^r + (1\text{-}\rho)\boldsymbol{x}^{(0)}$ (a typical value of ρ is 2). If $f(\boldsymbol{x}^e) < f(\boldsymbol{x}^r)$ reject $\boldsymbol{x}^{(n+1)}$ and update the simplex using $\boldsymbol{x}^e$.
- If $f(\boldsymbol{x}^{(n)}) \leq f(\boldsymbol{x}^r) < f(\boldsymbol{x}^{(n+1)})$ compute a *contraction* $\boldsymbol{x}^c = (1+\gamma)\boldsymbol{x}^{(0)}\text{-}\gamma\boldsymbol{x}^{(n+1)}$ (a typical value of γ is 0.5). If $f(\boldsymbol{x}^c) \leq f(\boldsymbol{x}^r)$ reject $\boldsymbol{x}^{(n+1)}$ and update the simplex using $\boldsymbol{x}^c$.
- If $f(\boldsymbol{x}^c) > f(\boldsymbol{x}^r)$ or $f(\boldsymbol{x}^r) \geq f(\boldsymbol{x}^{(n+1)})$ shrink the simplex: $\boldsymbol{x}^{(i)} = \boldsymbol{x}^{(1)} + \sigma(\boldsymbol{x}^{(i)}\text{-}\boldsymbol{x}^{(1)})$, $i = 1, \ldots, n+1$ (a typical value of σ is 0.5).

Iterations such as described above are continued until the termination condition is met (typically, reduction of the simplex size below a user-defined threshold). The algorithm is computationally cheap (up to a few objective function evaluations per iteration); however, its convergence is rather slow.

3.3.3 Metaheuristics and Global Optimization

Metaheuristics are derivative-free methods that exhibit features attractive from the point of view of handling many practical optimization problems: (1) global search capability; (2) ability to handle non-differentiable, discontinuous, or noisy objective functions; and (3) ability to handle the presence of multiple local optima and multiple objectives. Metaheuristics are developed based on observation of natural processes (in particular, biological or social systems), and typically process sets (populations) of potential solutions to the optimization problem at hand (referred to as individuals or agents). In the course of optimization, individuals are interacting

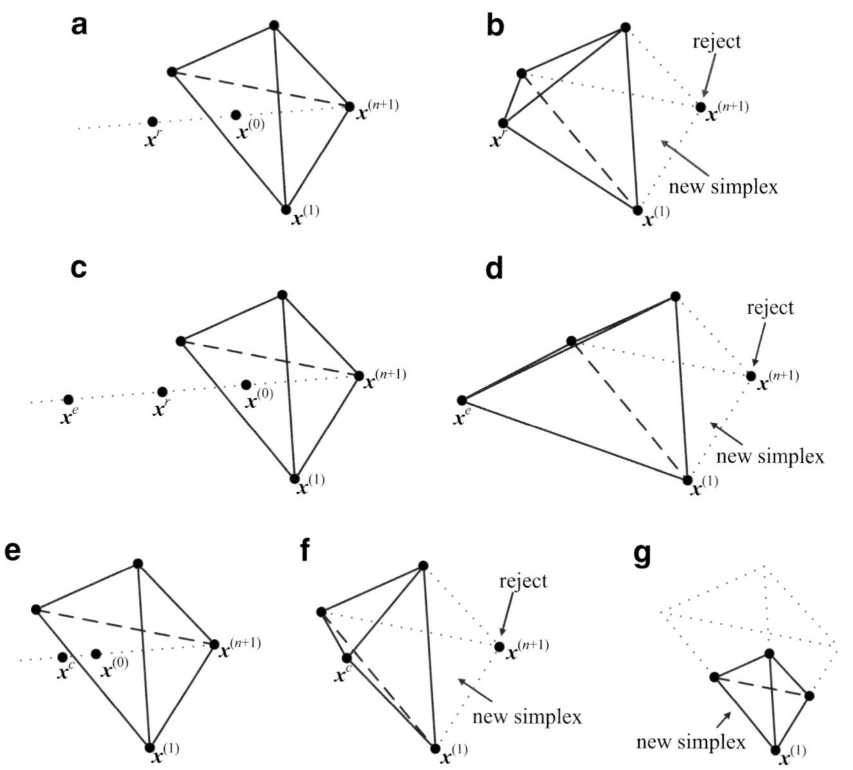

Fig. 3.3 Operations of the Nelder-Mead algorithm: (**a**) reflection: $x^r = (1+\alpha)x^{(0)} - \alpha x^{(n+1)}$, (**b**) acceptance of the reflection point x^r if $f(x^{(1)}) \le f(x^r) < f(x^{(n)})$, (**c**) expansion: $x^e = \rho x^r + (1-\rho)x^{(0)}$, (**d**) acceptance of the expansion point x^e if $f(x^e) < f(x^r)$, (**e**) contraction: $x^c = (1+\gamma)x^{(0)} - \gamma x^{(n+1)}$, (**f**) acceptance of the contraction point x^c if $f(x^c) \le f(x^r)$, (**g**) shrinking of the simplex if neither of the above operations led to the objective function improvement

with each other by means of method-specific operators, both explorative (e.g., crossover) and exploitative (e.g., mutation in genetic algorithms). Some algorithms (e.g., simulated annealing; Kirkpatrick et al. 1983) process a single solution rather than a population.

The most popular metaheuristics include genetic algorithms (GAs) (Goldberg 1989), evolutionary algorithms (EAs) (Back et al. 2000), evolution strategies (ES) (Back et al. 2000), particle swarm optimizers (PSO) (Kennedy et al. 2001), differential evolution (DE) (Storn and Price 1997), and, more recently, firefly algorithm (Yang 2010).

A typical flow of the population-based metaheuristic algorithm can be described as follows (here, P is used to denote the population, i.e., the set of solutions processed by the algorithm):

1. Initialize the population P.
2. Evaluate individuals in the population P.

3. Select parent individuals S from P.
4. Apply recombination operators to create a new population P from parent individuals S.
5. Apply mutation operators to introduce local perturbations in individuals of P.
6. If termination condition is not satisfied go to 2.
7. END.

In most cases, the population is initialized randomly. The "fitness" value assigned to an individual in Step 2 is usually strictly related to the value of the objective function to be minimized. Step 3 may be deterministic (i.e., only the best individuals are selected, ES) or partially random (with higher probability of being selected assigned to better individuals but nonzero even for the poor ones; GAs, EAs). In some of the modern algorithms (e.g., PSO or DE) there is no selection (i.e., the individual life-span is the entire optimization run).

The two major types of operators utilized in metaheuristic algorithms include exploratory ones (such as crossover in EAs or ES) and exploitative ones (such as mutation in GAs). The first group combine (or mix) information embedded in the parent individuals to create a new individual (child, offspring). A typical example would be an arithmetical crossover used in EAs with floating point representation, where a new individual $\boldsymbol{c}$ can be created as a convex combination of the parents $\boldsymbol{p}_1$ and $\boldsymbol{p}_2$, i.e., $\boldsymbol{c} = \alpha\boldsymbol{p}_1 + (1-\alpha)\boldsymbol{p}_2$, where $0 < \alpha < 1$ is a random number. The aim of exploratory operators is to make large "steps" in the design space in order to identify its new and potentially promising regions.

The purpose of exploitative operators is to introduce small perturbations into an individual (e.g., $\boldsymbol{p} \leftarrow \boldsymbol{p} + \boldsymbol{r}$, where $\boldsymbol{r}$ is a random vector selected according to a normal probability distribution with zero mean and problem-dependent variance) so as to allow exploitation of a given region of the design space, thus improving local search properties of the algorithm. In certain algorithms such as PSO, the difference between both types of the operators is not that clear (modification of the individual may be based on the best solution found so far by that given individual as well as the best solution found by the entire population). On the other hand, the operators used by the algorithms such as PSO should be interpreted as an exchange of information between individuals that affect their behavior rather than the individuals themselves.

In Sects. 3.3.4 and 3.3.5 below, we briefly describe representative examples of modern metaheuristic algorithms, i.e., PSO and DE. A compact summary of selected modern metaheuristics can be found in the literature (e.g., Yang 2010).

3.3.4 Particle Swarm Optimization

Particle swarm optimization (PSO) belongs to the most popular metaheuristic algorithms these days. PSO is based on mimicking the swarm behavior observed in nature, such as fish or bird schooling (Kennedy et al. 2001). The PSO algorithm simulates a set of individuals that interact with each other using social influence and social learning. A typical model of the swarm is particles represented by their positions and velocities (vectors). The particles exchange information about good (in terms of the corresponding objective function value) positions and adjust their own velocity and position accordingly. The particle movement has two components: a stochastic one and a deterministic one. Each particle is attracted toward the position of the current global best (position) $\boldsymbol{g}$ and its own best location $\boldsymbol{x}_i^*$ found during the optimization run. At the same time, it has a tendency to move randomly.

Let $\boldsymbol{x}_i$ and $\boldsymbol{v}_i$ denote the position and the velocity vectors of the ith particle. Both vectors are updated according to the following rules:

$$\boldsymbol{v}_i \leftarrow \chi\left[\boldsymbol{v}_i + c_1\boldsymbol{r}_1\bullet\left(\boldsymbol{x}_i^* - \boldsymbol{x}_i\right) + c_2\boldsymbol{r}_2\bullet(\boldsymbol{g} - \boldsymbol{x}_i)\right] \tag{3.22}$$

$$\boldsymbol{x}_i \leftarrow \boldsymbol{x}_i + \boldsymbol{v}_i \tag{3.23}$$

where $\boldsymbol{r}_1$ and $\boldsymbol{r}_2$ are vectors with components being uniformly distributed random numbers between 0 and 1; • denotes component-wise multiplication. The parameter χ is typically equal to 0.7298 (Kennedy et al. 2001); c_1 and c_2 are acceleration constants determining how much the particle is directed toward good positions (they represent a "cognitive" and a "social" component, respectively); usually, c_1, $c_2 \approx 2$. The sum of $c_1 + c_2$ should not be smaller than 4.0; as the values of these parameters increase, χ gets smaller as does the damping effect. The initial locations of all particles should distribute relatively uniformly so that they can sample over most regions. The initial velocity of a particle can be taken as zero. Different variants and extensions of the standard PSO algorithm can be found in the literature (Clerc and Kennedy 2002; Kennedy et al. 2001).

3.3.5 *Differential Evolution*

Differential evolution (DE) was developed in the late 1990s by R. Storn and K. Price (Storn and Price 1997). It is a derivative-free stochastic search algorithm with self-organizing tendency. DE is considered by some to be one of the best population-based metaheuristics for continuous optimization. The design parameters are represented as n-dimensional vectors (agents). New agents are created by combining the positions of the existing agents. More specifically, an intermediate agent is generated from two agents randomly chosen from the current population. This temporary agent is then mixed with a predetermined target agent. The new agent is accepted for the next generation if and only if it yields reduction in the objective function. Formally, the procedure of creating a new agent $\boldsymbol{y} = [y_1 \ldots y_n]$ from the existing one $\boldsymbol{x} = [x_1 \ldots x_n]$ works as follows:

1. Randomly select three agents $\boldsymbol{a} = [a_1 \ldots a_n]$, $\boldsymbol{b} = [b_1 \ldots b_n]$, and $\boldsymbol{c} = [c_1 \ldots c_n]$ so that $\boldsymbol{a} \neq \boldsymbol{b} \neq \boldsymbol{c} \neq \boldsymbol{x}$.
2. Randomly select a position index $p \in \{1,\ldots,N\}$ (N is the population size).

3. Determine the new agent $\boldsymbol{y}$ as follows (separately for each of its components $i = 1, \ldots, n$):

(a) Select a random number $r_i \in (0,1)$ with a uniform probability distribution.
(b) If $i = p$ or $r_i < CR$ let $y_i = a_i + F(b_i - c_i)$; otherwise let $y_i = x_i$; here, $0 \le F \le 2$ is the differential weight and $0 \le CR \le 1$ is the crossover probability, both defined by the user.
(c) If $f(\boldsymbol{y}) < f(\boldsymbol{x})$ then replace $\boldsymbol{x}$ by $\boldsymbol{y}$; otherwise reject $\boldsymbol{y}$ and keep $\boldsymbol{x}$.

Although DE resembles other stochastic optimization techniques, unlike traditional EAs, DE perturbs the solutions in the current generation vectors with scaled differences of two randomly selected agents. As a consequence, no separate probability distribution has to be used, and thus the scheme presents some degree of self-organization. Furthermore, DE is simple to implement, uses very few control parameters, and has been observed to perform satisfactorily in a number of multimodal optimization problems (Chakraborty 2008).

Some research studies have been focused on the choice of the control parameters F and CR as well as various modifications of the updating scheme (Step 3 above). Interested readers are referred to Price et al. (2005) for details.

3.3.6 *Other Methods*

Population-based metaheuristic algorithms are quite popular primarily because of their global optimization capabilities. In recent years, many modifications of the existing methods, as well as new algorithms, have been developed, such as the ant colony optimization algorithm (Dorigo and Stutzle 2004), bee algorithm (Yang 2005), harmony search (Geem et al. 2001), or firefly algorithm (Yang 2008). Majority of these techniques are based on observations of specific natural or social phenomena; however, the general concept is usually similar. A good exposition of the recent methods can be found in Yang (2010). On the other hand, some research efforts have recently been focused on global optimization of expensive models, where metaheuristics are combined with surrogate modeling techniques, e.g., efficient global optimization (Jones et al. 1998) or surrogate-assisted evolutionary algorithms (Jin 2011).

3.4 Summary and Discussion

Optimization techniques described in this chapter are well established and they constitute attractive methods for solving various engineering design problems. Gradient-based algorithms are suitable for optimizing smooth functions, particularly if it is possible to obtain cheap sensitivity information (e.g., through adjoints (see, e.g., Pironneau 1984; and Jameson 1988)). Derivative-free methods are more suitable for handling analytically intractable (noisy, discontinuous) functions. Among them, metaheuristics allow for global optimization as well as handling problems with multiple objectives.

From the point of view of contemporary engineering design problems, majority of the methods outlined in this chapter suffer from a fundamental drawback, which is the high computational cost of the optimization process. Conventional algorithms such as gradient-based routines with numerical derivatives require—depending on the problem dimensionality—tens, hundreds, or even thousands of objective function evaluations. Numerical simulations can be very time consuming, from minutes to hours or even days and weeks per design. Consequently, straightforward attempts to optimize computational models by embedding the simulator directly in the optimization loop may often be impractical, even when adjoint sensitivity is employed. Availability of massive computing resources is not always translated into a computational speedup due to a growing demand for simulation accuracy, both by including more accurate models (governing equations), multi-physics, and second-order effects and by using finer discretization of the structure under consideration.

Population-based metaheuristics are even more expensive. A typical population size is anything between 10 and 100, whereas the number of iteration may be a few dozen to a few hundred, which results in thousands and tens of thousands of function calls necessary to converge. Also, their performance may be heavily dependent on the values of control parameters, which may not be easy to determine beforehand. At the same time, repeatability of the results is limited due to the stochastic nature of these methods. From an engineering design standpoint, metaheuristics are attractive approaches for problems where the evaluation time of the objective function is not of a primary concern and when multiple local optima are expected. On the other hand, combining metaheuristics with surrogate modeling techniques may result in a considerable reduction of the computational cost of the optimization process (Couckuyt 2013; Jin 2011).

Surrogate-based optimization (SBO) is one of the most promising ways of handling computationally expensive design problems, where direct optimization of the high-fidelity model is replaced by optimization of its cheap and analytically tractable representation referred to as a surrogate model. An introduction to SBO is provided in the next chapter (Chap. 4). The remaining part of the book is devoted to specific type of SBO methods, which are the knowledge-based response correction techniques.

Chapter 4
Introduction to Surrogate Modeling and Surrogate-Based Optimization

Surrogate-based optimization (SBO) is the main focus of this book. We provide a brief introduction to the subject in this chapter. In particular, we recall the SBO concept and the optimization flow, discuss the principles of surrogate modeling and typical approaches to construct surrogate models. We also discuss the distinction between function approximation (or data-driven) surrogates and physics-based surrogates, as well as outline the algorithm of SBO exploiting the two aforementioned classes of models. More detailed information about the selected types of SBO algorithms (especially those involving response correction techniques) as well as illustration and application examples in various fields of engineering are provided in the remaining part of the book.

4.1 Surrogate-Based Optimization Concept

Conventional numerical optimization techniques are—in their majority—well established and robust methods as outlined in Chap. 3. However, their applicability for solving contemporary design problems is limited due to the fundamental challenge of the high computational cost of accurate, high-fidelity simulations. A few typical engineering design problems involving expensive computational models were briefly described in Chap. 2. Difficulties of the conventional optimization techniques, in the context of simulation-driven design, were the main incentives for developing alternative design methods. Perhaps the most promising way to address these issues, in particular, to conduct parametric optimization of expensive simulation models in a reasonable timeframe, is surrogate-based optimization (SBO) (Queipo et al. 2005; Koziel et al. 2011; Koziel and Leifsson 2013a; Forrester and Keane 2009).

The main concept behind SBO is to replace direct optimization of an expensive computational model by an iterative process in which a sequence of designs approximating the solution to the original optimization problem (3.1) is generated

S. Koziel, L. Leifsson, *Simulation-Driven Design by Knowledge-Based Response Correction Techniques*, DOI 10.1007/978-3-319-30115-0_4

by means of optimizing a fast yet reasonably accurate representation of the high-fidelity model, referred to as a surrogate. In each iteration, the surrogate model is updated using the high-fidelity model evaluation at the most recent design (and, sometimes, some other suitably selected designs). Formally speaking, the SBO process can be written as (Koziel et al. 2011)

$$\boldsymbol{x}^{(i+1)} = \arg\min_{\boldsymbol{x}} U\left(\boldsymbol{s}^{(i)}(\boldsymbol{x})\right), \tag{4.1}$$

where $\boldsymbol{x}^{(i)}$, $i = 0, 1, \ldots$, is a sequence of approximate solutions to the original problem (3.1), whereas $\boldsymbol{s}^{(i)}$ is the surrogate model at the ith iteration. Here, $\boldsymbol{x}^{(0)}$ is an initial design, typically obtained using engineering insight or by an optimization of any available lower fidelity model.

The main prerequisite for the process (4.1) to be computationally efficient is the surrogate model being significantly faster than the high-fidelity model $\boldsymbol{f}$. At the same time, the surrogate has to be sufficiently accurate (in terms of representing the high-fidelity model). In case of local search methods, reasonable accuracy is only requested in the vicinity of the current design $\boldsymbol{x}^{(i)}$. If both conditions are satisfied, the algorithm (4.1) is likely to quickly converge to the high-fidelity model optimum $\boldsymbol{x}^*$. In many implementations, the high-fidelity model is evaluated only once per iteration (specifically, at every new design $\boldsymbol{x}^{(i+1)}$). As mentioned before, the data set $\{\boldsymbol{x}^{(i+1)},\boldsymbol{f}(\boldsymbol{x}^{(i+1)})\}$ is used for design verification, but it is also to update the surrogate model (Fig. 4.1).

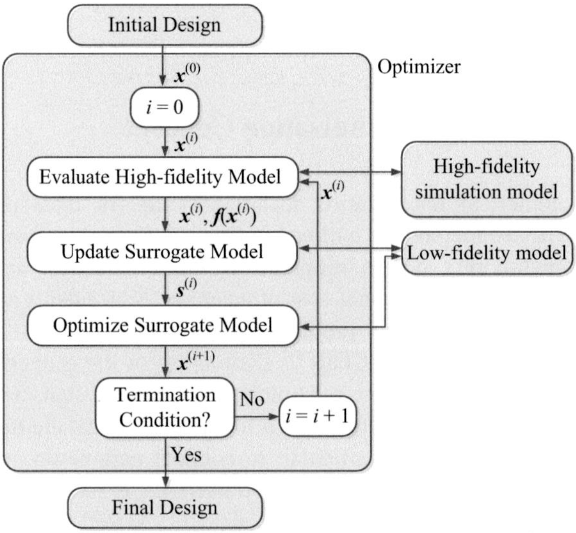

Fig. 4.1 The concept and the flow of the surrogate-based optimization process: the surrogate model $\boldsymbol{s}$ is iteratively refined and optimized to yield an approximate high-fidelity model optimum. The high-fidelity model $\boldsymbol{f}$ is evaluated for verification purposes (in most cases, only once per iteration, at the every new design). The number of iterations needed in SBO is often substantially smaller than for conventional optimization techniques

Ideally, that is, when the surrogate model is very fast, the computational overhead related to solving (4.1) can be neglected, in which case the overall optimization cost is merely determined by the high-fidelity model evaluations throughout the process. Because the number of iterations required by the SBO algorithm is substantially smaller than for the majority of conventional methods that optimize the high-fidelity model directly (such as the gradient-based schemes with numerical derivatives) (Koziel et al. 2006a), the total optimization cost of the SBO procedure is also lower accordingly. In practice, very fast surrogates can be derived either from analytical models, simplified physics models (e.g., equivalent circuit representations in microwave engineering, Bandler et al. 2004a), or response surface approximation (RSA) models (e.g., Liu et al. 2014). In the last case, the RSA models may be constructed beforehand (by extensive sampling of the design space, Koziel and Ogurtsov 2011b) or iteratively (by adaptive sampling techniques, e.g., Forrester and Keane 2009). In most of these cases, the evaluation time of the surrogate is indeed negligible.

Unfortunately, for many real-world design problems in various fields of engineering, very fast (compared to the high-fidelity model) surrogate models are hardly available. This is usually due to the complexity of the system at hand so that simplified analytical models either do not exist or are too inaccurate to be used in the SBO process. Often, the only way of obtaining cheaper representations of the high-fidelity model is also through numerical simulation but with a coarser discretization of the structure and/or relaxed convergence criteria (see Sect. 5.3 for more in-depth treatment of this subject). One of the important issues when setting up such lower fidelity simulation models is a proper balance between the model accuracy and speed (Koziel and Ogurtsov 2012a; Leifsson and Koziel 2014). For practical problems, the time evaluation ratio between the high- and low-fidelity models can be anything between about ten to over a hundred; however, for some problems, it is hard to get the ratio higher than 10 while still maintaining acceptable accuracy. In any case, the computational cost of the surrogate model can no longer be neglected when the surrogate is based on the low-fidelity simulation model. In particular, the cost of solving the iteration (4.1) may become a significant (sometimes even a major) contributor to the overall optimization cost. This results in an additional challenge for the development and implementation of the SBO algorithms where not only the number of high-fidelity but also the number of surrogate model evaluations becomes of concern. Possible solutions to this problem include the use of adjoint sensitivities (e.g., Koziel et al. 2013c), multilevel algorithms (Koziel and Ogurtsov 2013a; Koziel and Leifsson 2013a; Tesfahunegn et al. 2015a), as well as simultaneous utilization of variable-fidelity simulations and RSA models (Koziel and Ogurtsov 2013b).

Let us assume that the surrogate model satisfies zero- and first-order consistency conditions with the high-fidelity model, that is, $\boldsymbol{s}^{(i)}(\boldsymbol{x}^{(i)}) = \boldsymbol{f}(\boldsymbol{x}^{(i)})$ and $\boldsymbol{J}[\boldsymbol{s}^{(i)}(\boldsymbol{x}^{(i)})] = \boldsymbol{J}[\boldsymbol{f}(\boldsymbol{x}^{(i)})]$ (Alexandrov and Lewis 2001), where $\boldsymbol{J}[\cdot]$ stands for the model Jacobian. If, additionally, the surrogate-based algorithm is embedded in a trust-region framework (Conn et al. 2000)

$$\boldsymbol{x}^{(i+1)} = \arg \min_{\boldsymbol{x},\, \left\| \boldsymbol{x}-\boldsymbol{x}^{(i)} \right\| \le \delta^{(i)}} U\left(\boldsymbol{s}^{(i)}(\boldsymbol{x})\right), \tag{4.2}$$

where $\delta^{(i)}$ denotes the trust-region radius at iteration i, and the models involved are sufficiently smooth, then the SBO algorithm (4.1) is provably convergent to a local optimum of $\boldsymbol{f}$ (Alexandrov et al. 1998).

The trust region in (4.2) is updated at every iteration depending on the actual and predicted by the surrogate model improvement of the high-fidelity objective function (increased if the improvement is consistent with the prediction and decreased otherwise). The idea of the trust-region approach is that if the vicinity of $\boldsymbol{x}^{(i)}$ is sufficiently small, the first-order consistent surrogate becomes—in this vicinity—a sufficiently good representation of $\boldsymbol{f}$ to produce a design that reduces the high-fidelity objective function value upon completing (4.2).

Clearly, sensitivity information from both the high-fidelity model and the surrogate is required to satisfy the first-order consistency condition. Furthermore, some additional assumptions concerning the smoothness of the functions involved are also necessary for convergence (Echeverria and Hemker 2005). Convergence of various SBO algorithms can also be ensured under various other scenarios, see e.g., Koziel et al. (2006a) and Koziel et al. (2008b) (space mapping (SM)), Echeverria and Hemker (2005) (manifold mapping) or Booker et al. (1999) (surrogate management framework).

The SBO algorithm of the form (4.1) may be implemented either as a local or as a global optimization process. For local search, the algorithm (4.2) is often using the trust-region-like convergence safeguards and at least zero-order consistency conditions are ensured by appropriate construction of the surrogate model (Koziel et al. 2010a). In order to enable global search, the surrogate model is usually constructed in the larger portion of the design space; the surrogate model optimization is executed using global methods such as evolutionary algorithms, and the surrogate itself is updated using appropriate statistical infill criteria based on the expected improvement of the objective function or minimization of the (global) modeling error (Forrester and Keane 2009; Liu et al. 2014).

4.2 Surrogate Modeling: Approximation-Based Surrogates

Approximation-based models constitute the largest and probably the most popular category of surrogates. There is a large variety of techniques described in the literature (see, e.g., Simpson et al. 2001b; Søndergaard 2003; Forrester and Keane 2009; Couckuyt 2013) and many are available through various toolboxes (Lophaven et al. 2002; Gorissen et al. 2010), mostly implemented in Matlab. The popularity of these techniques comes from several appealing features:

- Approximation-based models are data-driven, i.e., can be constructed without a priori knowledge about the physical system of interest,

- They are generic and, therefore, easily transferable between various classes of problems,
- They are usually based on algebraic (often explicitly formulated) models,
- They are usually very cheap to evaluate.

On the other hand, approximation-based models usually require considerable amount of data to ensure reasonable accuracy. Also, the number of training samples grows very quickly with the dimensionality of the design space (so-called curse of dimensionality). This is a serious bottleneck, particularly for one-time ad hoc optimization of a given system or a component because the computational cost of setting up a sufficiently accurate surrogate model may be higher than the cost of a direct optimization of the high-fidelity model.

In the remaining part of this section, we briefly outline the surrogate modeling flow for approximation-based models, discuss various design of experiments (DOEs) schemes, approximation techniques, as well as model validation methods.

4.2.1 *Surrogate Modeling Flow*

As said before, approximation-based surrogates are data-driven models, i.e., they are constructed from the training samples obtained from the high-fidelity simulation model. The overall flowchart of the surrogate modeling process is shown in Fig. 4.2. There are four stages here:

- DOEs, which aims at allocating a given number of training samples in the design space using a selected strategy. Typically, the number of samples is limited due to available computational budget. When the training data comes from computer

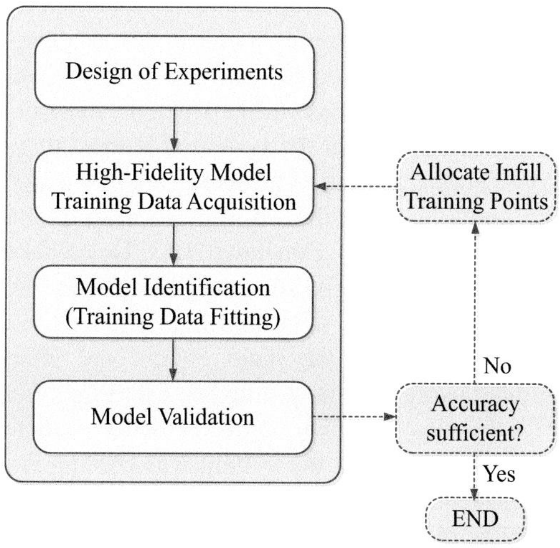

Fig. 4.2 A generic surrogate model construction flowchart. The *shaded box* shows the main flow, whereas the *dashed lines* indicate an optional iterative procedure in which additional training data points are added (using a suitable infill strategy) and the model is updated accordingly

simulations, space-filling DOEs are usually preferred (Simpson et al. 2001b). An outline of popular DOE techniques is provided in Sect. 4.2.2.
- Acquisition of the training data at the points allocated at the first stage.
- Model identification using an approximation technique of choice. In most cases (e.g., kriging (Kleijnen 2009) or neural networks (Rayas-Sanchez 2004)), determination of the surrogate model parameters requires solving a suitably defined minimization problem; however, in some situations the model parameters can be found using explicit formulas by solving an appropriate regression problem (e.g., polynomial approximation, Queipo et al. 2005).
- Model validation: this step is necessary to verify the model accuracy. Usually, the generalization capability of the surrogate is of major concern, i.e., the predictive power at the designs not seen during the identification stage. Consequently, the model testing should involve a separate set of testing samples. Maintaining a proper balance between surrogate approximation and generalization (accuracy at the known and at the unknown data sets) is also of interest.

In practice, the surrogate model construction may be an iterative process so that the steps shown in Fig. 4.2 constitute just a single iteration. Specifically, a new set of samples and the corresponding high-fidelity model evaluations may be added to the training set upon model validation and utilized to reidentify the model. This adaptive sampling process is then continued until the accuracy goals are met. In the optimization context, the surrogate update may be oriented more toward finding better designs rather than toward ensuring global accuracy of the model (Forrester and Keane 2009).

4.2.2 Design of Experiments

DOEs (Giunta et al. 2003; Santner et al. 2003; Koehler and Owen 1996) is a strategy for allocating the training points in the design space. The high-fidelity model data is acquired at the selected locations and utilized to construct the surrogate model. Clearly, there is a trade-off between the number of training samples and the amount of information about the system of interest that can be extracted from these points. As mentioned before, information about the model performance may be fed back to the sampling algorithm to allocate more points in the regions that exhibit more nonlinear behavior (Couckuyt 2013; Devabhaktuni et al. 2001).

A factorial design (allocating samples in the corners, edges, and/or faces of the design space), Fig. 4.3a, is a traditional DOE approach (Giunta et al. 2003) that allows estimating the main effects and interactions between design variables without using too many samples. Spreading the samples minimizes possible errors of estimating the main trends as well as interactions between design variables, in case the data about the system was coming from physical measurements. On the other hand, some of the factorial designs are rather "economical" in terms of the number of samples, which is important if the computational budget for data

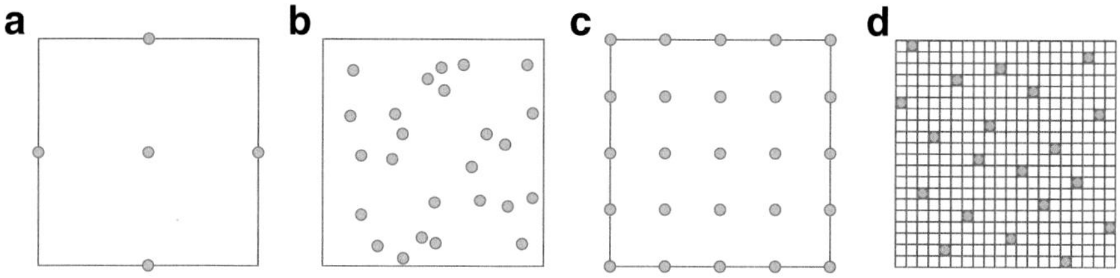

Fig. 4.3 Popular DOE techniques: (**a**) factorial designs (here, star distribution); (**b**) random sampling; (**c**) uniform grid sampling; and (**d**) Latin hypercube sampling (LHS)

acquiring is very limited. One of such designs, the so-called star distribution (Fig. 4.3a) is often used in conjunction with SM (Cheng et al. 2006).

Majority of recent DOE algorithms are space-filling designs attempting to allocate training points uniformly within the design space (Queipo et al. 2005).

This is especially useful for constructing an initial surrogate model when the knowledge about the system is limited. A number of space-filling DOEs are available. The simple ones include pseudorandom sampling (Giunta et al. 2003; Fig. 4.3b) or uniform grid sampling (Fig. 4.3c). Unfortunately, the former does not ensure sufficient uniformity, the latter is only practical for low-dimensional spaces as the number of samples is restricted to $N_1 \cdot N_2 \cdot \ldots \cdot N_n$, where N_j is the number of samples along jth axis of the design space. Probably the most popular DOE for uniform sample distributions is Latin hypercube sampling (LHS) (McKay et al. 1979). In order to allocate p samples with LHS, the range for each parameter is divided into p bins, which for n design variables, yields a total number of p^n bins in the design space. The samples are randomly selected in the design space so that (1) each sample is randomly placed inside a bin, and (2) for all one-dimensional projections of the p samples and bins, there is exactly one sample in each bin. Figure 4.3d shows an example LHS realization of 20 samples in the two-dimensional space. Various improvements of the standard LHS have been developed that provide more uniform sampling distributions (e.g., Beachkofski and Grandhi 2002; Leary et al. 2003; Ye 1998; Palmer and Tsui 2001). Other uniform sampling techniques that are commonly used include orthogonal array sampling (Queipo et al. 2005), quasi-Monte Carlo sampling (Giunta et al. 2003), or Hammersley sampling (Giunta et al. 2003). Space-filling DOE can also be realized as an optimization problem by minimizing a suitably defined nonuniformity measure, e.g., $\sum_{i=1,\ldots,p}\sum_{j=i+1,\ldots,p} d_{ij}^{-2}$ (Leary et al. 2003), where d_{ij} is the Euclidean distance between samples i and j.

4.2.3 Approximation Techniques

There is a large number of approximation methods available, some of which are very popular, including polynomial regression (Queipo et al. 2005), radial-basis functions (Wild et al. 2008), kriging (Forrester and Keane 2009), neural networks

(Haykin 1998), support vector regression (SVR) (Gunn 1998), Gaussian process regression (GPR) (Rasmussen and Williams 2006), and multidimensional rational approximation (Shaker et al. 2009). In this section, we give a brief outline of the selected methods. Interested readers can find more insightful information in the aforementioned literature as well as in the review works (e.g., Simpson et al. 2001b).

Throughout this section we will denote the training data samples as $\{\boldsymbol{x}^{(i)}\}$, $i = 1, \ldots, p$ and the corresponding high-fidelity model evaluations as $f(\boldsymbol{x}^{(i)})$. The surrogate model is constructed by approximating the data pairs $\{\boldsymbol{x}^{(i)}, f(\boldsymbol{x}^{(i)})\}$.

4.2.3.1 Polynomial Regression

Polynomial regression is one of the simplest approximation techniques (Queipo et al. 2005). The surrogate model is defined as

$$\boldsymbol{s}(\boldsymbol{x}) = \sum_{j=1}^{K} \beta_j v_j(\boldsymbol{x}), \tag{4.3}$$

where β_j are unknown coefficients and v_j are the (polynomial) basis functions. The model parameters can be found as a least-square solution to the linear system

$$\boldsymbol{f} = \boldsymbol{X\beta}, \tag{4.4}$$

where $\boldsymbol{f} = [f(\boldsymbol{x}^{(1)})\ f(\boldsymbol{x}^{(2)})\ \ldots\ f(\boldsymbol{x}^{(p)})]^T$, $\boldsymbol{X}$ is a $p \times K$ matrix containing the basis functions evaluated at the sample points, and $\boldsymbol{\beta} = [\beta_1 \beta_2\ \ldots\ \beta_K]^T$. The number of sample points p should be consistent with the number of basis functions considered K (typically $p \geq K$). If the sample points and basis functions are taken arbitrarily, some columns of $\boldsymbol{X}$ can be linearly dependent. If $p \geq K$ and rank($\boldsymbol{X}$) $= K$, a solution to (4.4) in the least-squares sense can be computed through $\boldsymbol{X}^+$, the pseudoinverse of $\boldsymbol{X}$ (Golub and Van Loan 1996), i.e., $\boldsymbol{\beta} = \boldsymbol{X}^+\boldsymbol{f} = (\boldsymbol{X}^T\boldsymbol{X})^{-1}\boldsymbol{X}^T\boldsymbol{f}$.

One of the simplest yet useful examples of a regression model is a second-order polynomial one defined as

$$\boldsymbol{s}(\boldsymbol{x}) = \boldsymbol{s}\left(\begin{bmatrix} x_1 & x_2 & \ldots & x_n \end{bmatrix}^T\right) = \beta_0 + \sum_{j=1}^{n} \beta_j x_j + \sum_{i=1}^{n} \sum_{j \leq i}^{n} \beta_{ij} x_i x_j, \tag{4.5}$$

with the basis functions being monomials: 1, x_j, and $x_i x_j$.

4.2.3.2 Radial-Basis Functions

Radial-basis function model interpolation/approximation (Forrester and Keane 2009; Wild et al. 2008) is a regression model that exploits combinations of K radially symmetric functions ϕ

$$s(\boldsymbol{x}) = \sum_{j=1}^{K} \lambda_j \phi\left(\left|\left|\boldsymbol{x} - \boldsymbol{c}^{(j)}\right|\right|\right), \tag{4.6}$$

where $\boldsymbol{\lambda} = [\lambda_1 \lambda_2 \ldots \lambda_K]^T$ is the vector of model parameters, and $\boldsymbol{c}^{(j)}, j = 1, \ldots, K$, are the (known) basis function centers. The model parameters can be calculated as $\boldsymbol{\lambda} = \boldsymbol{\Phi}^{+}\boldsymbol{f} = (\boldsymbol{\Phi}^T\boldsymbol{\Phi})^{-1}\boldsymbol{\Phi}^T\boldsymbol{f}$, where $\boldsymbol{f} = [f(\boldsymbol{x}^{(1)})\ f(\boldsymbol{x}^{(2)}) \ldots f(\boldsymbol{x}^{(p)})]^T$, and the $p \times K$ matrix $\boldsymbol{\Phi} = [\Phi_{kl}]_{k=1,\ldots,p;\ l=1,\ldots,K}$, with the entries defined as

$$\Phi_{kl} = \phi(||\boldsymbol{c}^{(k)} - \boldsymbol{c}^{(l)}||) \tag{4.7}$$

If the number of basis functions is equal to the number of samples, i.e., $p = K$, the centers of the basis functions coincide with the data points and are all different, $\boldsymbol{\Phi}$ is a regular square matrix. In such a case, we have $\boldsymbol{\lambda} = \boldsymbol{\Phi}^{-1}\boldsymbol{f}$.

A popular choice of the basis function is a Gaussian, $\phi(r) = \exp(-r^2/2\sigma^2)$, where σ is the scaling parameter.

4.2.3.3 Kriging

Kriging is a popular technique for interpolating deterministic noise-free data (Journel and Huijbregts 1981; Simpson et al. 2001b; Kleijnen 2009; O'Hagan 1978). Kriging is a Gaussian process based modeling method, which is compact and cheap to evaluate (Rasmussen and Williams 2006). In its basic formulation, kriging (Journel and Huijbregts 1981; Simpson et al. 2001b) assumes that the function of interest is of the following form:

$$f(\boldsymbol{x}) = \boldsymbol{g}(\boldsymbol{x})^T\boldsymbol{\beta} + Z(\boldsymbol{x}), \tag{4.8}$$

where $\boldsymbol{g}(\boldsymbol{x}) = [g_1(\boldsymbol{x})\ \ g_2(\boldsymbol{x})\ \ \ldots\ \ g_K(\boldsymbol{x})]^T$ are known (e.g., constant) functions, $\boldsymbol{\beta} = [\beta_1\beta_2 \ldots \beta_K]^T$ are the unknown model parameters (hyperparameters), and $Z(\boldsymbol{x})$ is a realization of a normally distributed Gaussian random process with zero mean and variance σ^2. The regression part $\boldsymbol{g}(\boldsymbol{x})^T\boldsymbol{\beta}$ is a trend function for f, and $Z(\boldsymbol{x})$ takes into account localized variations. The covariance matrix of $Z(\boldsymbol{x})$ is given as

$$Cov\left[Z\left(\boldsymbol{x}^{(i)}\right)Z\left(\boldsymbol{x}^{(j)}\right)\right] = \sigma^2\boldsymbol{R}\left(\left[R\left(\boldsymbol{x}^{(i)}, \boldsymbol{x}^{(j)}\right)\right]\right), \tag{4.9}$$

where $\boldsymbol{R}$ is a $p \times p$ correlation matrix with $R_{ij} = R(\boldsymbol{x}^{(i)}, \boldsymbol{x}^{(j)})$. Here, $R(\boldsymbol{x}^{(i)}, \boldsymbol{x}^{(j)})$ is the correlation function between sampled data points $\boldsymbol{x}^{(i)}$ and $\boldsymbol{x}^{(j)}$. The most popular choice is the Gaussian correlation function

$$R(\boldsymbol{x}, \boldsymbol{y}) = \exp\left[-\sum\nolimits_{k=1}^{n} \theta_k |x_k - y_k|^2\right], \tag{4.10}$$

where θ_k are the unknown correlation parameters, and x_k and y_k are the kth components of the vectors $\boldsymbol{x}$ and $\boldsymbol{y}$, respectively. The kriging predictor (Simpson et al. 2001b; Journel and Huijbregts 1981) is defined as

$$s(\boldsymbol{x}) = \boldsymbol{g}(\boldsymbol{x})^T\boldsymbol{\beta} + \boldsymbol{r}^T(\boldsymbol{x})\boldsymbol{R}^{-1}(\boldsymbol{f} - \boldsymbol{G\beta}), \tag{4.11}$$

where $\boldsymbol{r}(\boldsymbol{x}) = [R(\boldsymbol{x}, \boldsymbol{x}^{(1)}) \ldots R(\boldsymbol{x}, \boldsymbol{x}^{(p)})]^T$, $\boldsymbol{f} = [f(\boldsymbol{x}^{(1)})\ f(\boldsymbol{x}^{(2)}) \ldots f(\boldsymbol{x}^{(p)})]^T$, and $\boldsymbol{G}$ is a $p \times K$ matrix with $G_{ij} = g_j(\boldsymbol{x}^{(i)})$. The vector of model parameters $\boldsymbol{\beta}$ can be computed as $\boldsymbol{\beta} = \left(\boldsymbol{G}^T\boldsymbol{R}^{-1}\boldsymbol{G}\right)^{-1}\boldsymbol{G}^T\boldsymbol{R}^{-1}\boldsymbol{f}$. Model fitting is accomplished by maximum likelihood for θ_k (Journel and Huijbregts 1981).

An important property of kriging is that the random process $Z(\boldsymbol{x})$ gives information on the approximation error that can be used for improving the surrogate, e.g., by allocating additional training samples at the locations where the estimated model error is the highest (Forrester and Keane 2009; Journel and Huijbregts 1981). This feature is also utilized in various global optimization methods (see Couckuyt 2013, and references therein).

4.2.3.4 Artificial Neural Networks

Artificial neural networks (ANNs) is a large area of research (Haykin 1998). For the purpose of this book, ANNs can be considered as just another way of approximating sampled high-fidelity model data to create a surrogate model.

The most important component of a neural network (Haykin 1998; Minsky and Papert 1969) is the neuron (or single-unit perceptron). A neuron realizes a nonlinear operation illustrated in Fig. 4.4a, where w_1 through w_n are regression coefficients, β is the bias value of the neuron, and T is a user-defined slope parameter. The most common neural network architecture is the multilayer feed-forward network shown in Fig. 4.4b.

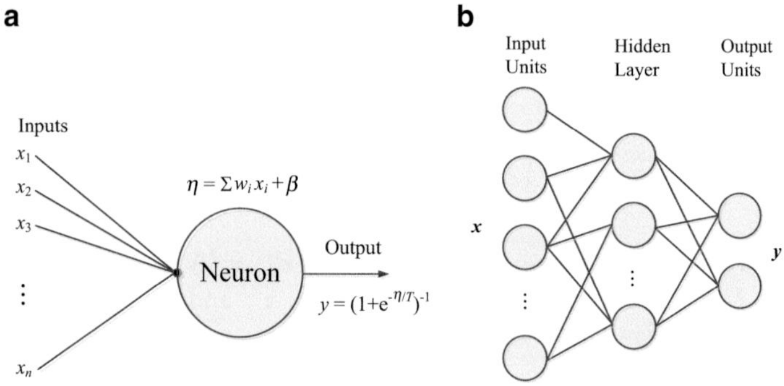

Fig. 4.4 Basic concepts of artificial neural networks: (**a**) structure of a neuron and (**b**) two-layer feed-forward neural network architecture

The construction of a neural network model involves selection of its architecture selection and the training, i.e., the assignment of the values to the regression parameters. The network training can be formulated as a nonlinear least-squares regression problem. A popular technique for solving this regression problem is the error back-propagation algorithm (Simpson et al. 2001b; Haykin 1998). For large networks with complex architecture, the use of global optimization methods might be necessary (Alba and Marti 2006).

4.2.3.5 Support Vector Regression

SVR (Gunn 1998) exhibits good generalization capability (Angiulli et al. 2007) and easy training by means of quadratic programming (Smola and Schölkopf 2004). SVM exploits the Structural Risk Minimization principle, which has been shown to be superior (Gunn 1998) to traditional Empirical Risk Minimization principle, employed by, e.g., neural networks. SVR has been gaining popularity in various areas including electrical engineering and aerodynamic design (Ceperic and Baric 2004; Rojo-Alvarez et al. 2005; Yang et al. 2005; Meng and Xia 2007; Xia et al. 2007; Andrés et al. 2012; Zhang and Han 2013).

Let $r^k = f(\boldsymbol{x}^k)$, $k = 1, 2, \ldots, N$, denote the sampled high-fidelity model responses (we formulate SVR for vector-valued f). We want to use SVR to approximate r^k at all base points $\boldsymbol{x}^k$, $k = 1, 2, \ldots, N$. We shall also use the notation $r^k = [r_1{}^k r_2{}^k \ldots r_m{}^k]^T$ to denote components of the vector r^k. For linear regression, we aim at approximating a training data set, here, the pairs $D_j = \{(\boldsymbol{x}^1, r_j{}^1), \ldots, (\boldsymbol{x}^N, r_j{}^N)\}$, $j = 1, 2, \ldots, m$, by a linear function $f_j(\boldsymbol{x}) = \boldsymbol{w}_j{}^T\boldsymbol{x} + b_j$. The optimal regression function is given by the minimum of the following functional (Smola and Schölkopf 2004)

$$\Phi_j(\boldsymbol{w}, \xi) = \frac{1}{2}||\boldsymbol{w}_j||^2 + C_j\sum_{i=1}^{N}\left(\xi_{j.i}^+ + \xi_{j.i}^-\right). \tag{4.12}$$

Here, C_j is a user-defined value, and $\xi_{j.i}{}^+$ and $\xi_{j.i}{}^-$ are the slack variables representing upper and lower constraints on the output of the system. The typical cost function used in SVR is an ε-insensitive loss function defined as

$$L_\varepsilon(y) = \begin{cases} 0 & \text{for } |f_j(\boldsymbol{x}) - y| < \varepsilon \\ |f_j(\boldsymbol{x}) - y| & \text{otherwise} \end{cases} \tag{4.13}$$

The value of C_j determines the trade-off between the flatness of f_j and the amount up to which deviations larger than ε are tolerated (Gunn 1998).

Here, we describe nonlinear regression employing the kernel approach, in which the linear function $\boldsymbol{w}_j{}^T\boldsymbol{x} + b_j$ is replaced by the nonlinear function $\Sigma_i\gamma_{j.i}K(\boldsymbol{x}^k, \boldsymbol{x}) + b_j$, where K is a kernel function. Thus, the SVR model is defined as

$$s(\boldsymbol{x}) = \begin{bmatrix} \sum_{i=1}^{N} \gamma_{1.i} K(\boldsymbol{x}^i, \boldsymbol{x}) + b_1 \\ \vdots \\ \sum_{i=1}^{N} \gamma_{m.i} K(\boldsymbol{x}^i, \boldsymbol{x}) + b_m \end{bmatrix} \tag{4.14}$$

with parameters $\gamma_{j.i}$ and b_j, $j = 1, \ldots, m$, $i = 1, \ldots, N$ obtained according to a general SVR methodology. In particular, Gaussian kernels of the form $K(\boldsymbol{x},\boldsymbol{y}) = \exp(-0.5 \cdot \|\boldsymbol{x} - \boldsymbol{y}\|^2/c^2)$ with $c > 0$ can be used, where c is the scaling parameter. Both c as well as parameters C_j and ε can be adjusted to minimize the generalization error calculated using a cross-validation method (Queipo et al. 2005).

4.2.3.6 Fuzzy Systems

Fuzzy systems are commonly used in machine control (Passino and Yurkovich 1998) where the expert knowledge and a set of sampled input-output (state-control) pairs recorded from successful control are translated into the set of "IF-THEN" rules that state in what situations which actions should be taken (Wang and Mendel 1992). Because of the incomplete and qualitative character of such information, it is represented using a fuzzy set theory (Zadeh 1965), where given piece of information (element) belongs to a given (fuzzy) subset of an input space with a certain degree, according to so-called membership function (Li and Lan 1989). The process of converting a crisp input value to a fuzzy value is called "fuzzification." Given the specific input state, the "IF-THEN" rules that apply are invoked, using the membership functions and truth values obtained from the inputs, to determine the result of the rule. This result in turn will be mapped into a membership function and truth value controlling the output variable. These results are combined to give a specific answer, using a procedure known as "defuzzification." The "centroid" defuzzification method is very popular, in which the "center of mass" of the result provides the crisp output value.

Fuzzy systems can also be used as universal function approximators (Wang and Mendel 1992). In particular, given a set of numerical data pairs, it is possible to obtain a fuzzy-rule-based mapping from the input space (here, design variables) to the output space (here, surrogate model response). The mapping is realized by dividing the input and output spaces into fuzzy regions, generating fuzzy rules from given desired input-output data pairs, assigning a degree to each generated rule and forming a combined fuzzy rule base, and, finally, performing defuzzification (Wang and Mendel 1992). It can be shown that under certain conditions, such a mapping is capable of approximating any real continuous function over the compact domain to arbitrary accuracy (Wang and Mendel 1992).

Assume that we have data pairs $(\boldsymbol{x}^k, \boldsymbol{f}^k)$, where $\boldsymbol{x}^k \in X_B$ and $k = 1, 2, \ldots, N$. Membership functions for the ith variable are defined as shown in Fig. 4.5. Each interval $[x_{0.i} - \delta_{0.i}, x_{0.i} + \delta_{0.i}]$, $i = 1, 2, \ldots, n$, is divided into K subintervals (fuzzy regions). The number K corresponds to the number of base points N and is given by

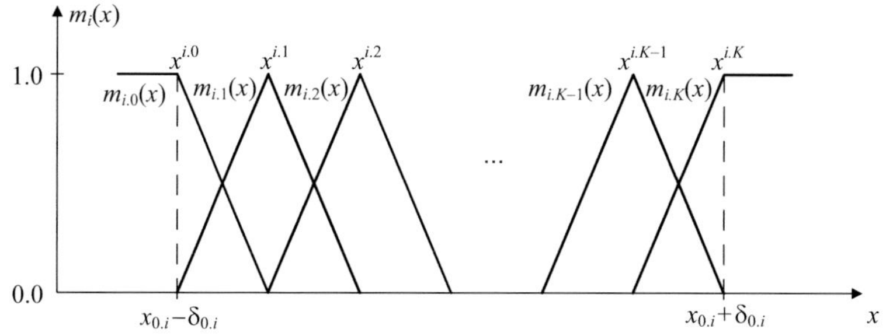

Fig. 4.5 Division of the input interval $[x_{0.i}-\delta_{0.i},\ x_{0.i}+\delta_{0.i}]$ into fuzzy regions and the corresponding membership functions

$K = \lfloor N^{1/n} \rfloor - 1$. In particular, if X_B consists of base points uniformly distributed in the region of interest X_R then $K+1$ is exactly the number of points of this uniform grid along any of the design variable axes. In general, K is chosen in such a way that the number of n-dimensional subintervals (and, consequently, the maximum number of rules) is not larger than the number of base points. Division of $[x_{0.i}-\delta_{0.i}, x_{0.i}+\delta_{0.i}]$ into K subintervals creates $K+1$ values $x^{i.k}$, $k=0, 1, \ldots, K$. In the case of a uniform base set, points $\boldsymbol{x}^{\mathbf{q}} = [x^{1.q_1} \ \ldots \ x^{n.q_n}]^T$, $\boldsymbol{q} \in \{0, 1, \ldots, K\}^n$ coincide with the base points. Value $x^{i.k}$ corresponds to the fuzzy region $[x^{i.k-1}, x^{i.k+1}]$ for $k=1, \ldots, K-1$ ($[x^{i.0}, x^{i.1}]$ for $k=0$, and $[x^{i.K-1}, x^{i.K}]$ for $k=K$). We also use the symbol $\boldsymbol{x}^{\boldsymbol{q}}$ to denote the n-dimensional fuzzy region $[x^{1.q_1} \ \ldots \ x^{n.q_n}]^T$. For any given x, the value of membership function $m_{i.k}(x)$ determines the degree of x in the fuzzy region $x^{i.k}$. Triangular membership functions are widely used. Here, one vertex lies at the center of the region and has membership value unity, and the other two vertices lie at the centers of the two neighboring regions, respectively, and have membership values equal to zero.

Having defined the membership functions we need to generate the fuzzy rules from given data pairs. We use if-then rules of the form IF $\boldsymbol{x}^k$ is in $\boldsymbol{x}^{\boldsymbol{q}}$ THEN $\boldsymbol{y}=\boldsymbol{f}^k$, where $\boldsymbol{y}$ is the response of the rule. At the level of vector components it means

$$\begin{aligned} &\text{IF } x_{k.1} \text{ is in } x^{1.q_1} \text{ AND } x_{k.2} \text{ is in } x^{2.q_2} \text{ AND } \ldots \text{ AND } x_{k.n} \text{ is in } x^{n.q_n} \\ &\text{THEN } \boldsymbol{y} = \boldsymbol{f}^{\mathrm{k}} \end{aligned} \tag{4.15}$$

where $x_{k.i}$, $i=1, \ldots, n$ are components of vector $\boldsymbol{x}^k$. In general, it may happen that there are some conflicting rules, i.e., rules that have the same IF part but a different THEN part. Such conflicts are resolved by assigning a degree to each rule and accepting only the rule from a conflict group that has a maximum degree. A degree is assigned to a rule in the following way. For the rule "IF $\boldsymbol{x}_{k.1}$ is in $\boldsymbol{x}^{1.q_1}$ AND $\boldsymbol{x}_{k.2}$ is in $\boldsymbol{x}^{2.q_2}$ AND … AND $\boldsymbol{x}_{k.n}$ is in $\boldsymbol{x}^{n.q_n}$ THEN $\boldsymbol{y}=\boldsymbol{f}^k$," the degree of this rule, denoted by $D(\boldsymbol{x}^k)$ is defined as

$$D(\boldsymbol{x}^k, \boldsymbol{x}^q) = \prod_{i=1}^{n} m_{i.q_i}(x_{k.i}), \tag{4.16}$$

Having resolved the conflicts we have a set of nonconflicting rules, which we denote as s_i, $i = 1, 2, \ldots, L$. We denote $\boldsymbol{s}$: $X_R \rightarrow R^m$ as the output of the fuzzy system, which is determined using a centroid defuzzification

$$\boldsymbol{s}(\boldsymbol{x}) = \frac{\sum_{i=1}^{L} D(\boldsymbol{x}, \boldsymbol{x}^i)\boldsymbol{y}_i}{\sum_{i=1}^{L} D(\boldsymbol{x}, \boldsymbol{x}^i)}, \tag{4.17}$$

where $\boldsymbol{x}^i$ is an n-dimensional fuzzy region corresponding to the ith rule, and $\boldsymbol{y}_i$ is the output of the ith rule.

4.2.3.7 Other Approximation Methods

There are several other popular approximation techniques that should be briefly mentioned in this chapter. One of these is moving least squares (Levin 1998), where the error contribution from each training point $\boldsymbol{x}^{(i)}$ is multiplied by a weight ω_i that depends on the distance between $\boldsymbol{x}$ and $\boldsymbol{x}^{(i)}$. A typical choice for the weights is

$$\omega_i\left(||\boldsymbol{x} - \boldsymbol{x}^{(i)}||\right) = \exp\left(-||\boldsymbol{x} - \boldsymbol{x}^{(i)}||^2\right). \tag{4.18}$$

Adding weights improves the flexibility of the model, however, at the expense of increased computational complexity, since computing the approximation for each point $\boldsymbol{x}$ requires solving a new optimization problem.

GPR (Rasmussen and Williams 2006) is another surrogate modeling technique that, as kriging, addresses the approximation problem from a stochastic point view. From this perspective, and since Gaussian processes are mathematically tractable, it is relatively easy to compute error estimations for GPR-based surrogates in the form of uncertainty distributions. Under certain conditions, GPR models can be shown to be equivalent to large neural networks (Rasmussen and Williams 2006) while requiring much less regression parameters than NNs.

Cokriging (Forrester et al. 2007; Toal and Keane 2011) is an interesting variation of the standard kriging interpolation (cf. Sect. 4.2.3.3). It allows for a fusion of information from computational models of various fidelity. The major advantage of cokriging is that—by exploiting knowledge embedded in the low-fidelity model—the surrogate can be created at much lower computational cost than for the models exclusively based on high-fidelity data. Cokriging is a rather recent method with relatively few applications in engineering (Toal and Keane 2011; Huang and Gao 2012; Laurenceau and Sagaut 2008; Koziel et al. 2013b).

4.2.3.8 Surrogate Model Validation

Model validation is an important stage of the modeling process in which one assesses the quality of the surrogate, among others, its predictive power. Clearly, it is not a good idea to evaluate the model merely based on its performance on the training set where, in particular, interpolative models exhibit zero error by definition. Some of the techniques described above identify a surrogate model together with some estimation of the attendant approximation error (e.g., kriging or GPR). Alternatively, there are procedures that can be used in a stand-alone manner to validate the prediction capability of a given model beyond the set of training points. A simple and probably the most popular way for validating a model is the split-sample method (Queipo et al. 2005), where part of the available data set (the training set) is used to construct the surrogate, whereas the second part (the test set) serves purely for model validation. However, the error estimated by a split-sample method depends strongly on how the set of data samples is partitioned.

More accurate estimation of the model generalization error can be obtained using cross-validation (Queipo et al. 2005; Geisser 1993). In cross-validation the data set is divided into K subsets, and each of these subsets is sequentially used as testing set for a surrogate constructed on the other $K - 1$ subsets. The prediction error can be estimated with all the K error measures obtained in this process (for example, as an average value). Cross-validation provides an error estimation that is less biased than with the split-sample method. The disadvantage of this method is that the surrogate has to be constructed more than once.

The surrogate modeling flow, i.e., the procedure of allocating samples, acquiring data, model identification, and validation can be repeated until the prescribed surrogate accuracy level is reached. In each repetition, a new set of training samples is added to the existing ones. The strategies of allocating the new samples (so-called infill criteria; Forrester and Keane 2009) usually aim at improving the global accuracy of the model, i.e., inserting new samples at the locations where the estimated modeling error is the highest.

From the point of view of simulation-driven design, the main advantage of approximation surrogates is that they are very fast. Unfortunately, the high computational cost of setting up such models (related to training data acquisition) is a significant disadvantage. In order to ensure decent accuracy, hundreds or even thousands of data samples are required and the number of training points quickly grows with dimensionality of the design space. Therefore, approximation surrogates are mostly suitable to build multiple-use library models. Their use for ad-hoc optimization of expensive computational models is rather limited.

4.3 Surrogate Modeling: Physics-Based Surrogates

As mentioned in the previous section, applicability of data-only-driven models may be limited for handling engineering problems where the involved computational models are very expensive. In short, the construction of a reliable surrogate may simply require too much computational effort in the form of training data acquisition. A second group of surrogate models, the physics-based ones, is more attractive with this respect. The physics-based surrogates are created by correcting (or enhancing) the underlying low-fidelity model which is typically a simplified representation of the system/component of interest. Very often, it is also a simulation model, however, featuring coarser discretization of the structure or relaxed convergence criteria. More detailed discussion about setting up the low-fidelity models can be found in Sect. 5.3 of this book.

Perhaps the major advantage of physics-based surrogates is the fact that only a limited amount of high-fidelity data is normally sufficient to ensure reasonable accuracy of the model. This is achieved by exploiting the knowledge about the systems at hand, embedded in the low-fidelity model. For the same reason, physics-based surrogates are characterized by good generalization capability, i.e., they can provide reliable prediction of the high-fidelity model response at the designs not seen in the training process. Clearly, improved (compared to data-driven models) generalization results in better computational efficiency of the SBO process (Koziel et al. 2011).

SBO techniques, particularly methods exploiting physics-based models, are the main topic of this book. We are mostly interested in so-called response correction techniques to construct the surrogates. More detailed exposition of these techniques is presented in Chaps. 5 through 7 with the related methods discussed in Chaps. 8 through 10. In this section, we only present motivation and a few concepts related to the surrogate model construction using underlying low-fidelity models.

Let $\boldsymbol{c}(\boldsymbol{x})$ denotes a low-fidelity model of the device or system of interest. In order to grasp the basic idea behind constructing the physics-based surrogate of a high-fidelity model f, we will discuss a simple case of multiplicative response correction, considered in the context of the SBO (4.1). The algorithm (4.1) produces a sequence $\{\boldsymbol{x}^{(i)}\}$ of approximate solutions to the original problem (3.1). From the algorithm convergence standpoint (particularly if the algorithm is embedded in the trust-region framework, cf. (4.2)), a local alignment between the surrogate and the high-fidelity model is of fundamental importance. The surrogate $\boldsymbol{s}^{(i)}(\boldsymbol{x})$ at iteration i can be constructed as

$$\boldsymbol{s}^{(i)}(\boldsymbol{x}) = \beta_k(\boldsymbol{x})\boldsymbol{c}(\boldsymbol{x}), \tag{4.19}$$

where $\beta_k(\boldsymbol{x}) = \beta_k(\boldsymbol{x}^{(i)}) + \nabla\beta(\boldsymbol{x}^{(i)})^T(\boldsymbol{x} - \boldsymbol{x}^{(i)})$, where $\beta(\boldsymbol{x}) = \boldsymbol{f}(\boldsymbol{x})/\boldsymbol{c}(\boldsymbol{x})$. This ensures a so-called zero- and first-order consistency between $\boldsymbol{s}$ and $\boldsymbol{f}$, i.e., agreement of function values and their gradients at $\boldsymbol{x}^{(i)}$ (Alexandrov and Lewis 2001). Figure 4.6 illustrates correction (4.19) for analytical functions.

A low-fidelity model correction can be applied at the level of the model domain. An example is the so-called input space mapping (ISM) (Bandler et al. 2004a), where the surrogate is created as (here, we use vector notation for the models, i.e., $\boldsymbol{c}$ and $\boldsymbol{s}$ for the low-fidelity and the surrogate models, respectively)

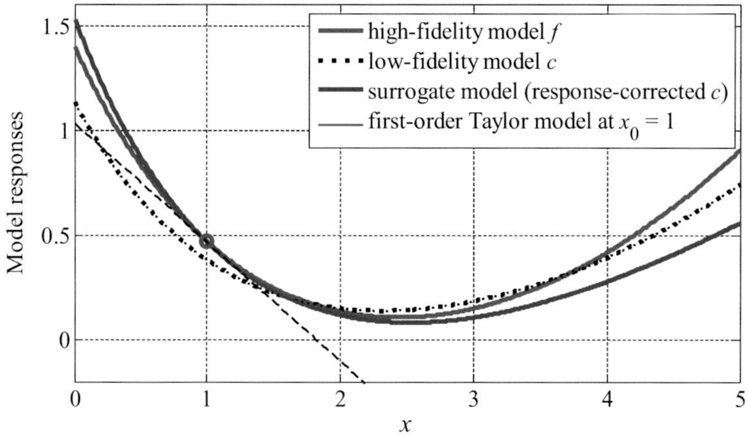

Fig. 4.6 Visualization of the response correction (4.19) for the analytical functions c (low-fidelity model) and f (high-fidelity model). The correction established at $\boldsymbol{x}_0 = 1$. It can be observed that corrected model (surrogate s) exhibits good alignment with the high-fidelity model in relatively wide vicinity of $\boldsymbol{x}_0$, especially compared to the first-order Taylor model setup using the same data from f (the value and the gradient at $\boldsymbol{x}_0$)

$$\boldsymbol{s}^{(i)}(\boldsymbol{x}) = \boldsymbol{c}\left(\boldsymbol{x} + \boldsymbol{q}^{(i)}\right). \tag{4.20}$$

The parameter vector $\boldsymbol{q}^{(i)}$ is obtained by minimizing $||\boldsymbol{f}(\boldsymbol{x}^{(i)}) - \boldsymbol{c}(\boldsymbol{x}^{(i)} + \boldsymbol{q}^{(i)})||$, which represents the misalignment between the surrogate and the high-fidelity model; $\boldsymbol{x}^{(i)}$ is a reference design at which the surrogate model is established (e.g., a current design in the optimization context). Figure 4.7 shows an example of a microwave filter structure (Salleh et al. 2008) evaluated using electromagnetic simulation (high-fidelity model), its equivalent circuit (low-fidelity model) and the corresponding responses (here, so-called transmission characteristics $|S_{21}|$ versus frequency) before and after applying the ISM correction. Figure 4.7d indicates excellent generalization capability of the surrogate (recall that a single high-fidelity training point was used to set up the model).

Another type of the model correction may involve additional parameters that are normally fixed in the high-fidelity model but can be adjusted in the low-fidelity model. The latter is just an auxiliary design tool after all: it is not supposed to be built or measured.

This concept is utilized, among others, in the so-called implicit space-mapping technique (Bandler et al. 2003, 2004a; Koziel et al. 2010b), where the surrogate is created as

$$\boldsymbol{s}^{(i)}(\boldsymbol{x}) = \boldsymbol{c}_I\left(\boldsymbol{x}, \boldsymbol{p}^{(i)}\right). \tag{4.21}$$

Here, $\boldsymbol{c}_I$ is the low-fidelity model with the explicit dependence on additional parameters $\boldsymbol{p}$. The vector $\boldsymbol{p}^{(i)}$ is obtained by minimizing $||\boldsymbol{f}(\boldsymbol{x}^{(i)}) - \boldsymbol{c}_I(\boldsymbol{x}^{(i)}, \boldsymbol{p})||$.

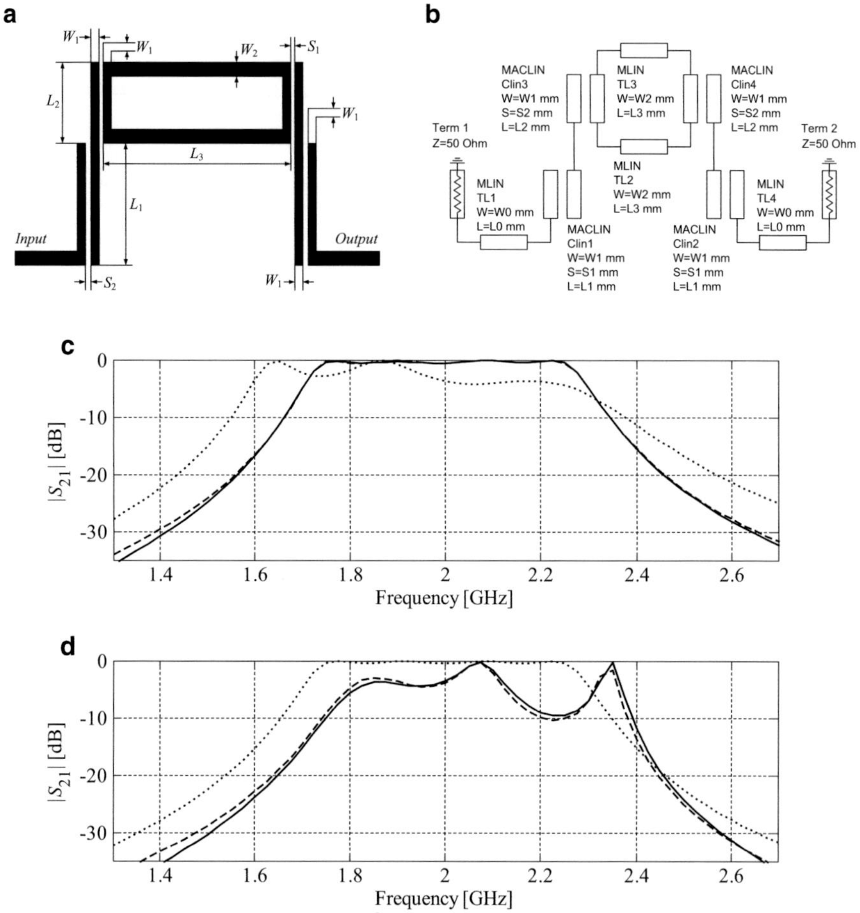

Fig. 4.7 Low-fidelity model correction through parameter shift (input space mapping): (**a**) microstrip filter geometry (high-fidelity model $\boldsymbol{f}$ evaluated using EM simulation); (**b**) low-fidelity model $\boldsymbol{c}$ (equivalent circuit); (**c**) response of $\boldsymbol{f}$ (—) and $\boldsymbol{c}$ (····), as well as response of the surrogate model $\boldsymbol{s}$ (- - -) created using input space mapping; and (**d**) surrogate model verification at a different design (other than that at which the model was created) $\boldsymbol{f}$ (—), $\boldsymbol{c}$ (····), and $\boldsymbol{s}$ (- - -). Good alignment indicates excellent generalization of the model

For the sake of illustration, consider again the filter of Fig. 4.7a, b with the implicit space-mapping parameters being dielectric permittivities of the microstrip-line component substrates (rectangle elements in Fig. 4.7b). Normally, the entire circuit is fabricated on a dielectric substrate with a specified (and fixed) characteristics. For the sake of correcting the low-fidelity model, the designer is free to adjust these characteristics (in particular, the value of dielectric permittivity) individually for each component of the equivalent circuit. Figure 4.8 the responses before and after applying the implicit space-mapping correction. Again, good generalization of the surrogate model can be observed.

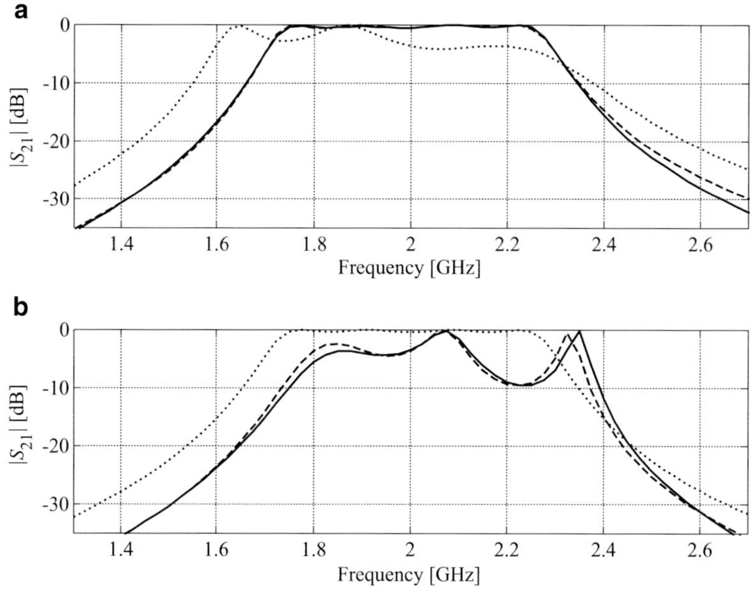

Fig. 4.8 Low-fidelity model correction through implicit space mapping applied to a microstrip filter of Fig. 4.7: (**a**) response of ***f*** (—) and ***c*** (····), as well as response of the surrogate model ***s*** (- - -) created using implicit space mapping and (**b**) surrogate model verification at a different design (other than that at which the model was created) ***f*** (—), ***c*** (····), and ***s*** (- - -)

In many cases, the vector-valued responses of the system are actually evaluations of the same design but at different values of certain parameters such as the time, frequency (e.g., for microwave structures) or a specific geometry parameter (e.g., chord line coordinate for the airfoil profiles). In such situations, it might be convenient to apply a linear or nonlinear scaling to this parameter so as to shape the response accordingly. A good example of such a correction procedure is frequency scaling often utilized in electrical engineering (Koziel et al. 2006a).

As a matter of fact, for many components simulated using electromagnetic solvers, a frequency shift is the major type of discrepancy between the low- and high-fidelity models. Let us consider a simple frequency scaling procedure, also referred to as frequency SM (Koziel et al. 2006a). We assume that $\boldsymbol{f}(\boldsymbol{x}) = [f(\boldsymbol{x},\omega_1)\ f(\boldsymbol{x},\omega_2)\ \ldots\ f(\boldsymbol{x},\omega_m)]^T$, where $f(\boldsymbol{x},\omega_k)$ is the evaluation of the high-fidelity model at a frequency ω_k, whereas ω_1 through ω_m represent the entire discrete set of frequencies at which the model is evaluated. A similar convention is used for the low-fidelity model. The frequency-scaled model $\boldsymbol{s}_F(\boldsymbol{x})$ is defined as

$$\boldsymbol{s}_F(\boldsymbol{x}, [F_0\ \ F_1]) = [\boldsymbol{c}(\boldsymbol{x}, F_0 + F_1\omega_1)\ \ \ldots\ \ \boldsymbol{c}(\boldsymbol{x}, F_0 + F_1\omega_m)]^T, \tag{4.22}$$

where F_0 and F_1 are scaling parameters obtained to minimize misalignment between $\boldsymbol{s}$ and $\boldsymbol{f}$ at a certain reference design $\boldsymbol{x}^{(i)}$ as

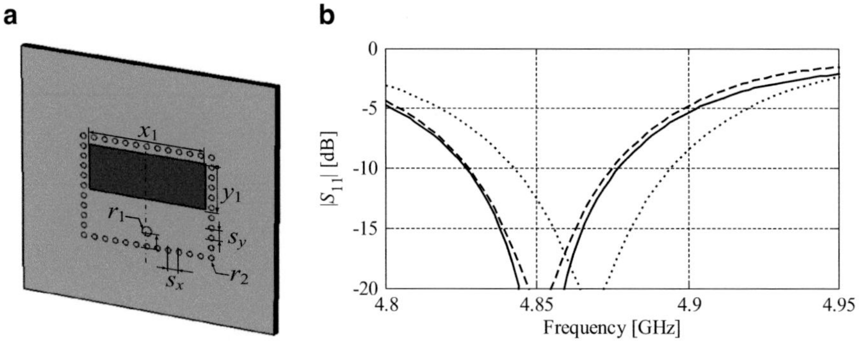

Fig. 4.9 Low-fidelity model correction through frequency scaling: (**a**) antenna geometry (both $\boldsymbol{f}$ and $\boldsymbol{c}$ evaluated using EM simulation, coarse discretization used for $\boldsymbol{c}$) and (**b**) response of $\boldsymbol{f}$ (—) and $\boldsymbol{c}$ (····), as well as response of the surrogate model $\boldsymbol{s}$ (- - -) created using frequency scaling as in Eqs. (4.21)–(4.23)

$$[F_0, F_1] = \arg \min_{[F_0, F_1]} \left|\left|\boldsymbol{f}\left(\boldsymbol{x}^{(j)}\right) - \boldsymbol{s}_F\left(\boldsymbol{x}^{(i)}, [F_0 \ \ F_1]\right)\right|\right|. \tag{4.23}$$

An example of frequency scaling applied to the low-fidelity model of a substrate-integrated cavity antenna (Koziel and Ogurtsov 2013b) is shown in Fig. 4.9. Here, both the low- and high-fidelity models are evaluated using EM simulation ($\boldsymbol{c}$ with coarse discretization).

More information about physics-based surrogates can be found in the literature (e.g., Bandler et al. 2004a; Koziel et al. 2011, 2013a; Leifsson and Koziel 2015a, b, c). The purpose of this section was merely to give the reader an idea of possible ways of correcting the low-fidelity model in order to align it with the high-fidelity one and construct the surrogate model that exhibits good generalization. On the other hand, the remaining part of this book contains a detailed exposition of a specific type of model enhancement methods, i.e., response correction techniques.

4.4 SBO with Approximation Surrogates

In this section, we provide an outline of optimization using approximation-based surrogates. In particular, we discuss the response surface design methodology as well as SBO techniques exploiting kriging interpolation. One of the important issues discussed here is a proper balance between exploration and exploitation of the design space.

4.4.1 Response Surface Methodologies

SBO allows for efficient handling (optimization) of the expensive objective functions through iterative enhancement and reoptimization of a suitably constructed surrogate model (Forrester and Keane 2009; Koziel et al. 2011). Another important advantage of SBO is a possibility of handling nonsmooth or noisy responses (Queipo et al. 2005), as well as carrying out a low-cost sensitivity analysis or even tolerance-aware design (Bandler et al. 2004a).

In this section, we briefly outline the optimization methods exploiting data-driven surrogates (also referred to as response surface approximations). One can distinguish the two main steps of the optimization process, i.e., construction of the initial surrogate model (using the initial set of training samples), and iterative prediction–correction scheme in which the surrogate is used to identify the promising locations in the design space, and then updated using the high-fidelity model data evaluated at these candidate designs. The overall flow of the SBO process can be summarized as follows (Queipo et al. 2005):

1. Construct an initial surrogate model.
2. Estimate the high-fidelity optimum design using the surrogate model.
3. Evaluate the high-fidelity model at the design found in Step 2.
4. Update the surrogate model using new high-fidelity data.
5. If the termination condition is not satisfied, go to 2.

The above SBO process is illustrated in Fig. 4.10 using a simple one-dimensional function minimization problem.

The infill criteria, i.e., the strategy for allocating the new samples, do not have to be exclusively based on optimizing the surrogate model. In practice, other strategies are possible and often utilized, where one of important objectives is improvement of global modeling accuracy of the surrogate (Couckuyt 2013) or balanced exploration and exploitation of the design space (Forrester and Keane 2009), see Sect. 4.4.3 for more details. If the primary objective is local optimization, the entire SBO algorithm may be embedded in the trust-region-like framework (Forrester and Keane 2009) for improved convergence.

It should also be mentioned that SBO may be implemented as a one-shot approach, in which the surrogate model is constructed using sufficient amount of training data (so as to ensure global accuracy), and then optimized without further updates. Clearly, construction of a globally accurate surrogate requires considerable computational effort related to training data acquisition, however, it may be beneficial in case the surrogate model is intended for multiple use (Kabir et al. 2008).

4.4.2 Sequential Approximate Optimization

Sequential approximate optimization (SAO) is a simple variation of SBO with approximation-based surrogates (Jacobs et al. 2004; Giunta and Eldred 2000;

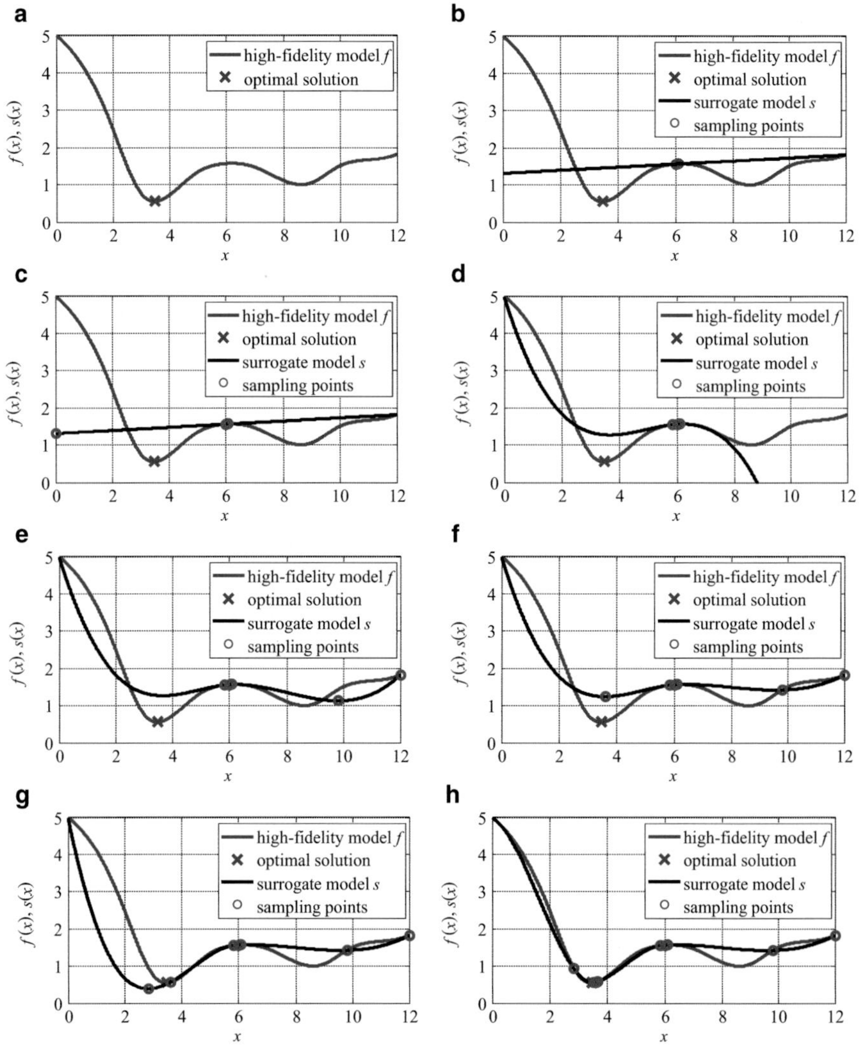

Fig. 4.10 SBO illustration: (**a**) high-fidelity model f (····) to be minimized and the minimum of f (×); (**b**) initial surrogate model (here, cubic splines) constructed using two data points; (**c**) a new data point obtained by optimizing the surrogate (evaluation of f at this point is used to update the surrogate); and (**d**)–(**h**) after a few iterations, the surrogate model optimum adequately predicts the high-fidelity minimizer; global accuracy of the surrogate is not of primary concern

Pérez et al. 2002; Hosder et al. 2001; Roux et al. 1998; Sobieszczanski-Sobieski and Haftka 1997; Giunta 1997). The main principle of SAO is to restrict the search to a subregion of the design space (usually, a multidimensional interval) using a local approximation surrogate (such as low-order polynomial) as a prediction tool. Upon optimizing the surrogate, a new subregion is defined according to the chosen

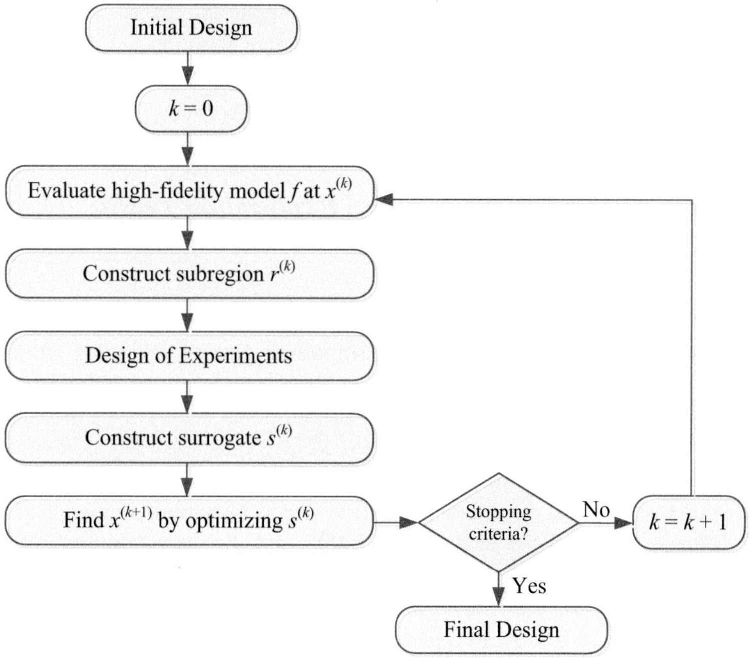

Fig. 4.11 Sequential approximate optimization workflow

relocation strategy. The SAO flow has the following steps (see also Fig. 4.11; Jacobs et al. 2004):

1. Set iteration counter $k = 0$.
2. Evaluate the high-fidelity model $\boldsymbol{f}$ at the design $\boldsymbol{x}^{(k)}$.
3. Construct a search region $\boldsymbol{r}^{(k)}$ based on a relocation strategy.
4. Sample $\boldsymbol{r}^{(k)}$ using a selected DOE plan.
5. Acquire high-fidelity data at the designs selected in Step 4.
6. Construct the surrogate $\boldsymbol{s}^{(k)}$ using the data set of Step 5.
7. Find a new design $\boldsymbol{x}^{(k+1)}$ by optimizing the surrogate $\boldsymbol{s}^{(k)}$ within the region $\boldsymbol{r}^{(k)}$.
8. Terminate the algorithm if the stopping criteria are met; otherwise set $k = k + 1$ and go to 2.

The advantage of SAO is that restriction of the search to one small subset of the design space at a time allows us to construct a simple surrogate (e.g., low-order polynomial) using a limited number of training samples.

Both the location and the size of the new region have to be selected at the beginning of each iteration (Step 3). A typical relocation strategy is to move toward the most current optimum (Jacobs et al. 2004) so that the last optimum becomes a center of the new region. In multipoint strategies, the last iteration optimum may become the corner point for the next search subregion (Toropov et al. 1993). Adjustment of the subregion size may be based on the quality of the designs obtained by optimizing the surrogate, e.g., comparison of the actual versus predicted improvement of the high-fidelity objective function as in trust-region-like framework (Giunta and Eldred 2000; Koziel and Ogurtsov 2012b).

4.4.3 *Optimization Using Kriging: Exploration Versus Exploitation*

Selecting the strategy for allocating the new training samples (so-called infill points, Forrester and Keane 2009) is an important part of designing SBO algorithms. The two main criteria for selecting these points are: a reduction of the objective function value and improvement of the global accuracy of the surrogate model (Couckuyt 2013).

One of the simplest infill strategies is to allocate a single sample at the surrogate model optimum. If the surrogate optimization step is constrained using, e.g., trust-region framework (Forrester and Keane 2009), and the surrogate model is first-order consistent with the high-fidelity model (Alexandrov et al. 1998), this strategy is capable of finding at least a local minimum of the high-fidelity model. In general, an infill strategy may be oriented toward global search as well as constructing a globally accurate surrogate. For that purpose, the most convenient type of approximation-based surrogates is kriging because, on the top of the predicted response, it also provides information about the expected model error (Kleijnen 2008; Jones et al. 1998; Gorissen et al. 2010).

Here is the list of the most commonly used infill criteria:

1. Maximization of the expected improvement, i.e., the improvement one expects to achieve at an untried point $\boldsymbol{x}$ (Jones et al. 1998).
2. Minimization of the predicted objective function $\hat{y}(\boldsymbol{x})$, i.e., surrogate optimization already mentioned above. Here, a reasonable global accuracy level of the surrogate is assumed (Liu et al. 2012).
3. Minimization of the statistical lower bound, i.e., $LB(\boldsymbol{x}) = \hat{y}(\boldsymbol{x}) - As(\boldsymbol{x})$ (Forrester and Keane 2009), where $\hat{y}(\boldsymbol{x})$ is the surrogate model prediction and $s^2(\boldsymbol{x})$ is the variance; A is a user-defined coefficient.
4. Maximization of the probability of improvement, i.e., identifying locations that give the highest change of improving the objective function value (Forrester and Keane 2009).
5. Maximization of the mean square error, i.e., finding locations where the mean square error (as predicted by the kriging model) is the highest (Liu et al. 2012).

Normally, finding the new samples according to the above infill criteria requires global optimization (Couckuyt 2013). SBO for constrained optimization problems using the above infill strategies is more involved, see, e.g., Forrester and Keane 2009; Queipo et al. 2005; Liu et al. 2012.

The adaptive sampling techniques outlined above may target the construction of globally accurate surrogates or finding the optimum design of the high-fidelity model (either in a local or a global sense). Methods focused on exploitation of the design space are usually more efficient in the computational sense, however, they may not be able to find the best possible design. On the other hand, the exploration of the design space implies—in most cases—a global search. For non-convex objective functions this usually requires performing a global sampling of the search space, e.g., by selecting the points that maximize the estimated error associated with the surrogate considered (Forrester and Keane 2009). However, global exploration is often impractical, especially for expensive functions with a medium/large number of optimization variables (more than a few tens). In an optimization context, there should be a balance between exploitation and exploration. An example of an infill criterion that realizes such a balanced search is criterion number 3 in the list above. The constant A in that criterion controls the exploitation/exploration balance (from pure exploitation, i.e., $\mathrm{LB}(\boldsymbol{x}) \to \hat{y}(\boldsymbol{x})$, for $A \to 0$ to pure exploration for $A \to \infty$). Choosing a good value of A is a nontrivial task though (Forrester and Keane 2009).

SBO methods with adaptive sampling are often referred to as efficient global optimization (EGO) techniques. A detailed discussion of EGO can be found in Jones et al. (1998) and in Forrester and Keane (2009).

4.4.4 Surrogate Management Framework

Surrogate management framework (SMF) (Booker et al. 1999) is a pattern-search-based optimization technique exploiting kriging interpolation models. Each iteration of the search procedure consists of the two main steps: SEARCH and POLL. Each iteration starts with a pattern of size Δ centered at $\boldsymbol{x}^{(i)}$. The SEARCH step is optional and is always performed before the POLL step. In the SEARCH step, a (small) number of points are selected from the search space (typically by means of a surrogate), and the cost function $\boldsymbol{f}(\boldsymbol{x})$ is evaluated at these points. If the cost function for some of them improves on $\boldsymbol{f}(\boldsymbol{x}^{(i)})$, the current pattern is centered at this new point, and a new SEARCH step is started. Otherwise a POLL (or a local search) step is taken. Polling requires computing $\boldsymbol{f}(\boldsymbol{x})$ for neighboring points in the pattern. The POLL step is successful if one of these points improves $\boldsymbol{f}(\boldsymbol{x}^{(i)})$, in which case the pattern is translated to this new point, and a new SEARCH step is performed. Otherwise, the entire pattern search iteration is considered unsuccessful and the termination condition is checked. This stopping criterion is typically based on the pattern size Δ (Booker et al. 1999; Marsden et al. 2004). If, after the unsuccessful pattern search iteration another iteration is needed, the pattern size Δ is decreased, and a new SEARCH step is taken with the pattern centered again at $\boldsymbol{x}^{(i)}$. Surrogates are incorporated in the SMF through the SEARCH step. For example, kriging (with Latin hypercube sampling) is considered in the SMF application studied in Marsden et al. (2004). The flow of the algorithm is shown in Fig. 4.12.

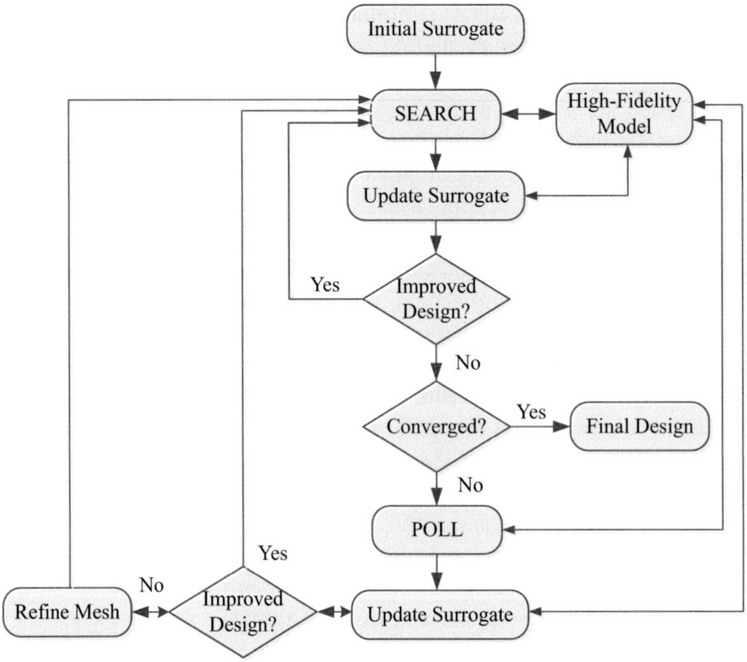

Fig. 4.12 Surrogate management framework (SMF) flowchart

In order to guarantee convergence to (at least local) optimum of the high-fidelity model, the set of vectors formed by each pattern point and the pattern center should be a generating (or positive spanning) set (Kolda et al. 2003; Marsden et al. 2004). A generating set for R^n consists of a set of vectors whose nonnegative linear combinations span R^n. Generating sets are crucial in proving convergence (for smooth objective functions) due to the following property: if a generating set is centered at $\boldsymbol{x}^{(i)}$ and $\nabla \boldsymbol{f}(\boldsymbol{x}^{(i)}) \neq 0$, then at least one of the vectors in the generating set defines a descent direction (Kolda et al. 2003). Consequently, if $\boldsymbol{f}(\boldsymbol{x})$ is smooth and $\nabla \boldsymbol{f}(\boldsymbol{x}^{(i)}) \neq 0$, some of the points in the associated stencil will improve on $\boldsymbol{f}(\boldsymbol{x}^{(i)})$ for sufficiently small pattern size Δ.

4.4.5 Summary

The essence of the SBO methods exploiting approximation-based surrogates is in the iterative construction of the computationally cheap and smooth data-driven surrogate, and using it to find promising locations in the design space, in particular, predicting the high-fidelity model optimum. One of the issues here is finding a proper balance between exploration of the design space and exploitation of the promising regions. While exploration usually leads to improving global accuracy of the surrogate, it might be computationally expensive and not really necessary if the goal is to optimize the high-fidelity model. In general, approximation-based surrogates may be appropriate solutions for problems with small or medium number of designable parameters. For more complex problems the computational cost of creating the surrogate may be prohibitive.

4.5 SBO with Physics-Based Surrogates

Optimization techniques exploiting physics-based surrogates is the main theme of this book. Perhaps the most attractive feature of physics-based models is their good generalization properties, which result from the knowledge about the system of interest embedded in the underlying low-fidelity models. Consequently, construction of a reliable surrogate may require significantly less high-fidelity training points than for approximation-based models (Koziel et al. 2011); see also Sect. 4.3 and the examples therein. The low-fidelity models are computationally cheap representations of the system/component under design. An important class of such models are variable-fidelity (or variable-resolution) ones that are evaluated using the same simulation tools as those used for high-fidelity models but with coarser discretization of the structure and/or relaxed termination criteria for the simulation process.

In this section, we briefly outline a few selected physics-based SBO techniques, in particular, some variations of SM (Bandler et al. 2004a; Koziel et al. 2006a) and approximation model management optimization (AMMO, Alexandrov and Lewis 2001). A detailed exposition of a large class of SBO methods involving response correction of the low-fidelity model will be given in the remaining chapters of this book.

4.5.1 Space Mapping

SM (Bandler et al. 1994, 2004a; Koziel et al. 2006a) was originally developed for solving expensive problems in computational electromagnetics, in particular, microwave engineering. SM has evolved over the last twenty years, and, nowadays, its popularity has been spreading across several engineering disciplines (Redhe and Nilsson 2004; Bandler et al. 2004a; Priess et al. 2011; Marheineke et al. 2012; Koziel and Leifsson 2012b; Tu et al., 2013).

The original space-mapping optimization methodologies were based on transforming the domain of the low-fidelity model (aggressive SM, input SM; Bandler et al. 2004a), which works well for many engineering problems, especially in electrical engineering, where the ranges of both the low- and high-fidelity model responses are similar and model discrepancy is mostly due to certain second-order effects (e.g., the presence of parasitic components). If the ranges of the low- and high-fidelity models are severely misaligned, input-like SM is not sufficient and response correction (output SM; Bandler et al. 2003) is usually necessary. Other types of SM have also been developed (Bandler et al. 2004a; Koziel et al. 2006a, 2008a, b; Robinson et al. 2008).

4.5.1.1 SM Concept

The original SM idea assumes the existence of a mapping $\boldsymbol{P}$ that relates the high- and low-fidelity model parameters (Bandler et al. 1994)

$$\boldsymbol{x}_c = \boldsymbol{P}(\boldsymbol{x}_f) \tag{4.24}$$

so that $\boldsymbol{c}(\boldsymbol{P}(\boldsymbol{x}_f)) \approx \boldsymbol{f}(\boldsymbol{x}_f)$ at least in some region of the high-fidelity parameter space. Given $\boldsymbol{P}$, the direct solution of the original problem (3.1), can be replaced by finding $\boldsymbol{x}_f^{\#} = \boldsymbol{P}^{-1}(\boldsymbol{x}_c^{*})$. Here, $\boldsymbol{x}_c^{*}$ is the optimal design of $\boldsymbol{c}$ defined as $\boldsymbol{x}_c^{*} = \mathrm{argmin}\{\boldsymbol{x}_c: U(\boldsymbol{c}(\boldsymbol{x}_c))\}$, whereas $\boldsymbol{x}_f^{\#}$ can be considered as a reasonable estimate of $\boldsymbol{x}_f^{*}$. Thus, the problem (3.1) can be reformulated as

$$\boldsymbol{x}_f^{\#} = \arg\min_{\boldsymbol{x}_f} U\left(\boldsymbol{c}\left(\boldsymbol{P}\left(\boldsymbol{x}_f\right)\right)\right), \tag{4.25}$$

where $\boldsymbol{c}(\boldsymbol{P}(\boldsymbol{x}_f))$ is an enhanced low-fidelity model (or, the surrogate). However, $\boldsymbol{P}$ is not given explicitly: it can only be evaluated at any $\boldsymbol{x}_f$ by means of a parameter extraction (PE) procedure

$$\boldsymbol{P}(\boldsymbol{x}_f) = \arg\min_{\boldsymbol{x}_c} \left|\left|\boldsymbol{f}(\boldsymbol{x}_f) - \boldsymbol{c}(\boldsymbol{x}_c)\right|\right|. \tag{4.26}$$

One of the issues here is possible nonuniqueness of the solution to (4.26) (Bandler et al. 1995). Another issue is the assumption on the similarity of high- and low-fidelity model ranges, which is a very strong one (Alexandrov and Lewis 2001). These and other issues led to numerous improvements, including parametric SM (cf. Sect. 4.5.1.3).

4.5.1.2 Aggressive SM

Aggressive SM (ASM) (Bandler et al. 1995) is one of the first versions of SM, and it is still popular in microwave engineering (Sans et al. 2014). Assuming uniqueness of the low-fidelity model optimum $\boldsymbol{x}_c^{*}$, the solution to Eq. (4.26) is equivalent to reducing the residual vector $\boldsymbol{f} = \boldsymbol{f}(\boldsymbol{x}_f) = \boldsymbol{P}(\boldsymbol{x}_f) - \boldsymbol{x}_c^{*}$ to zero. The first step of the ASM algorithm is to find $\boldsymbol{x}_c^{*}$. Next, ASM iteratively solves the nonlinear system

$$f(x_f) = 0 \tag{4.27}$$

for x_f. At the jth iteration, the calculation of the error vector $f^{(j)}$ requires an evaluation of $P^{(j)}(x_f^{(j)})$. This is realized by executing (4.26), i.e., $P(x_f^{(j)}) = \arg\min\{x_c: f(x_f^{(j)}) - c(x_c)\|\}$. The quasi-Newton step in the high-fidelity model space is given by

$$B^{(j)} h^{(j)} = -f^{(j)}, \tag{4.28}$$

where $B^{(j)}$ is the approximation of the SM Jacobian $J_P = J_P(x_f) = [\partial P^T/\partial x_f]^T = [\partial (x_c^T)/\partial x_f]^T$. Solving Eq. (4.28) for $h^{(j)}$ gives the next iterate $x_f^{(j+1)}$

$$x_f^{(j+1)} = x_f^{(j)} + h^{(j)}. \tag{4.29}$$

The algorithm terminates if $\|f^{(j)}\|$ is sufficiently small. The output of the algorithm is an approximation to $x_f^{\#} = P^{-1}(x_c^{*})$. A popular way of obtaining the matrix B is through a rank one Broyden update (Broyden 1965) of the form $B^{(j+1)} = B^{(j)} + (f^{(j+1)} - f^{(j)} - B^{(j)}h^{(j)})h^{(j)T}/\|h^{(j)}\|^2$. Several improvement of the original ASM algorithm have been proposed in the literature, including hybrid ASM (Bakr et al. 1999a) and trust-region ASM (Bakr et al. 1998).

4.5.1.3 Parametric SM

In parametric SM, the optimization algorithm is an iterative process (4.1), whereas the transformation between the low- and high-fidelity models (model correction) assumes certain analytical form. Simple examples of input and implicit SM were shown in Sect. 4.3. More generally, the input SM surrogate model can take the form (Koziel et al. 2006a)

$$s^{(i)}(x) = c\left(B^{(i)} \cdot x + q^{(i)}\right), \tag{4.30}$$

where $B^{(i)}$ and $q^{(i)}$ are matrices obtained in the PE process

$$\left[B^{(i)}, q^{(i)}\right] = \arg\min_{[B,q]} \sum_{k=0}^{i} w_{i.k} \left\| f\left(x^{(k)}\right) - c\left(B \cdot x^{(k)} + q\right)\right\|. \tag{4.31}$$

Here, $w_{i.k}$ are weighting factors; a common choice for $w_{i.k}$ is $w_{i.k} = 1$ for all i and all k (all previous designs contribute to the PE process) or $w_{i.1} = 1$ and $w_{i.k} = 0$ for $k < i$ (the surrogate model depends on the most recent design only).

In general, the SM surrogate model is constructed as follows:

$$s^{(i)}(x) = \bar{s}\left(x, p^{(i)}\right), \tag{4.32}$$

where $\bar{s}$ is a generic SM surrogate model, i.e., the low-fidelity model $\boldsymbol{c}$ composed with suitable (usually linear) transformations. The parameters $\boldsymbol{p}$ are obtained in the extraction process similar to (4.31).

A number of SM surrogate models have been developed, some of them rather problem specific such as frequency SM (cf. Sect. 4.3). Interested readers are referred to the literature (e.g., Bandler et al. 2004a; Koziel et al. 2006a).

The robustness of SM and the computational cost of the optimization process depend on the quality of the low-fidelity model but also on the proper selection of the SM transformations (Koziel and Bandler 2007a, b). Although output SM can be used to obtain both zero- and first-order consistency conditions with $\boldsymbol{f}(\boldsymbol{x})$, many other SM-based optimization algorithms that have been applied in practice do not satisfy those conditions, and on some occasions convergence problems have been identified (Koziel et al. 2008b). Also, the choice of an adequate SM correction may not be trivial (Koziel et al. 2008b). On the other hand, SM has been found to be a very efficient tool for yielding satisfactory designs in various engineering disciplines (e.g., Redhe and Nilsson 2004; Bandler et al. 2004a; Marheineke et al. 2012). At the same time, a number of enhancements of SM algorithms have been suggested to alleviate the difficulties mentioned above (e.g., Koziel et al. 2010a, c).

4.5.2 Approximation Model Management Optimization

The AMMO algorithm (Alexandrov and Lewis 2001) is essentially a response correction type of approach which aims at enhancing the low-fidelity model at the current iteration point so as to ensure first-order consistency between the surrogate and the high-fidelity model, i.e., $\boldsymbol{s}^{(i)}(\boldsymbol{x}^{(i)}) = \boldsymbol{f}(\boldsymbol{x}^{(i)})$, and $\nabla \boldsymbol{s}^{(i)}(\boldsymbol{x}^{(i)}) = \nabla \boldsymbol{f}(\boldsymbol{x}^{(i)})$. Additionally, AMMO exploits the trust-region methodology (Conn et al. 2000) to guarantee convergence of the optimization process to the high-fidelity model optimum.

The model correction utilized in AMMO works as follows. Let $\beta(\boldsymbol{x}) = \boldsymbol{f}(\boldsymbol{x})/\boldsymbol{c}(\boldsymbol{x})$ be the correction function. We define

$$\beta_i(\boldsymbol{x}) = \beta\left(\boldsymbol{x}^{(i)}\right) + \nabla\beta\left(\boldsymbol{x}^{(i)}\right)^T\left(\boldsymbol{x} - \boldsymbol{x}^{(i)}\right). \tag{4.33}$$

The surrogate model is then defined as

$$\boldsymbol{s}^{(i)}(\boldsymbol{x}) = \beta_i(\boldsymbol{x})\boldsymbol{c}(\boldsymbol{x}) \tag{4.34}$$

and can be shown to satisfy the aforementioned consistency conditions. Obviously, correction (4.33) requires both low- and high-fidelity derivatives. The flow of the algorithm is given in Fig. 4.13.

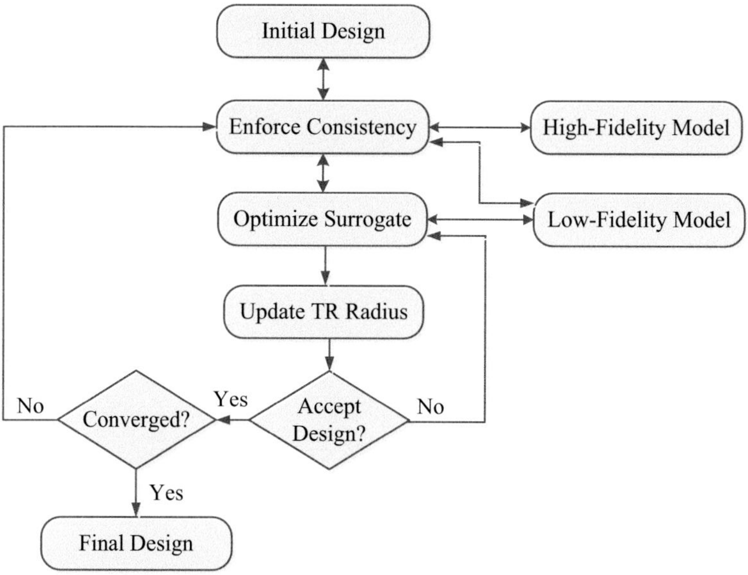

Fig. 4.13 A flowchart of the approximation model management optimization (AMMO) algorithm

Chapter 5
Design Optimization Using Response Correction Techniques

The material presented in Chap. 4 was intended to provide the reader with a general introduction to surrogate modeling and surrogate-based optimization. Among the methods presented in there, optimization using physics-based models belongs to the most promising approaches in terms of computational efficiency. In Sects. 4.3 and 4.5, we outlined basic methods of constructing the physics-based surrogates by enhancing the underlying low-fidelity models, as well as utilizing such models in the design optimization process. At that point, it was mentioned that response correction is perhaps the simplest way of implementing such a model enhancement. In this chapter, we formulate the response correction task in a more rigorous manner, and then broadly discuss available response correction techniques while categorizing them into parametric and non-parametric ones. We also discuss in more depth the issue of low-fidelity modeling, including the low-fidelity model setup and finding a proper balance between the model speed and accuracy. More detailed exposition of various response correction methods and their applications for solving computationally expensive design problems in various engineering fields are given in the remaining chapters of this book.

5.1 Introduction: Parametric and Non-parametric Response Correction

As mentioned on several occasions, correcting the response of a low-fidelity model is not the only way, but perhaps the most straightforward one, of constructing a physics-based surrogate model. On a generic level, the surrogate, $\boldsymbol{s}(\boldsymbol{x})$, obtained this way can be described as

$$\boldsymbol{s}(\boldsymbol{x}) = \boldsymbol{C}(\boldsymbol{c}(\boldsymbol{x})), \tag{5.1}$$

S. Koziel, L. Leifsson, *Simulation-Driven Design by Knowledge-Based Response Correction Techniques*, DOI 10.1007/978-3-319-30115-0_5

where $\boldsymbol{C}$: $R^m \rightarrow R^m$ is a response correction function, $\boldsymbol{c}(\boldsymbol{x})$ is a vector of the low-fidelity model responses, and $\boldsymbol{x}$ is a vector of design variables.

Typically (cf. Sect. 4.1), an optimization process involving surrogates is performed in an iterative manner so that the sequence $\boldsymbol{x}^{(1)}, \boldsymbol{x}^{(2)}, \ldots$, of (hopefully) better and better approximations to $\boldsymbol{x}^*$ (the solution to the original problem (3.1)) is obtained as in (4.1), i.e., $\boldsymbol{x}^{(i+1)} = \mathrm{argmin}\{\boldsymbol{x} : U(\boldsymbol{s}^{(i)}(\boldsymbol{x}))\}$ with $\boldsymbol{s}^{(i)}$ being the surrogate model at iteration i. Here, $\boldsymbol{s}^{(i)}(\boldsymbol{x}) = \boldsymbol{C}^{(i)}(\boldsymbol{c}(\boldsymbol{x}))$, where $\boldsymbol{C}^{(i)}$ is the correction function at iteration i. For surrogates constructed using a response correction, we typically require that, at least, zero-order consistency between the surrogate and the high-fidelity model, $\boldsymbol{f}(\boldsymbol{x})$, is satisfied, i.e., $\boldsymbol{s}^{(i)}(\boldsymbol{x}^{(i)}) = \boldsymbol{f}(\boldsymbol{x}^{(i)})$. It can be shown (Alexandrov et al. 1998) that satisfaction of first-order consistency, i.e., $\boldsymbol{J}[\boldsymbol{s}^{(i)}(\boldsymbol{x}^{(i)})] = \boldsymbol{J}[\boldsymbol{f}(\boldsymbol{x}^{(i)})]$ (here, $\boldsymbol{J}[\cdot]$ denotes the Jacobian of the respective model), guarantees convergence of $\{\boldsymbol{x}^{(i)}\}$ to a local optimum of $\boldsymbol{f}$ assuming that the algorithm is enhanced by the trust region mechanism (Conn et al. 2000; see also (4.2)) and the functions involved are sufficiently smooth.

Response correction approaches can be roughly categorized into parametric techniques, where the response correction function (5.1) can be represented using explicit formulas, and non-parametric ones, where the response correction is defined implicitly through certain relations between the low- and high-fidelity models which may not have any compact analytical form.

There are several parametric response correction methods covered in this book, including output space mapping (Bandler et al. 2004a), manifold mapping (Echeverria and Hemker 2005), and multipoint response correction (Koziel 2011a). They will be discussed in detail in Chap. 6. Also, the approximation model management optimization technique discussed in Sect. 4.5.2 belongs to this group. Constructing a parametric surrogate is usually straightforward and it boils down to applying a specific formula (that may involve the low- and high-fidelity model responses at one or more points, as well as their derivatives), or solving a suitably formulated (usually linear) regression problem (Leifsson and Koziel 2015a).

Non-parametric correction methods will be described in detail in Chap. 7. Constructing the surrogate model is more involved here because it normally requires a careful analysis of the model responses in order to identify relevant (method dependent) features that allow exploring the correlations between the models of various fidelities. On one hand, this makes the non-parametric methods less general and more complex in implementation. On the other hand, non-parametric surrogate models typically exhibit a better generalization capability, which translates into improved computational efficiency of the related SBO algorithms. The specific non-parametric methods covered in this book include adaptive response correction (Koziel et al. 2009a), adaptive response prediction (Koziel and Leifsson 2012a), and shape-preserving response prediction (Koziel 2010a). We also discuss some of related techniques, specifically, adaptively adjusted design specifications (Koziel 2010b) and feature-based optimization (Koziel and Bandler 2015).

It should also be mentioned that—in practice—response correction methods are often used in composition with other types of low-fidelity model enhancement techniques. In such setups, the initial low-fidelity model preconditioning might be realized by appropriate transformations of the parameter space (e.g., by means of input space mapping; Koziel and Echeverría Ciaurri 2010) to ensure improved global matching between the models, whereas response correction might be applied to provide zero- (and, if possible, first-) order consistency between the models at a current iteration point of the SBO algorithm (Koziel et al. 2013c).

5.2 Function Correction

In this section, to give the reader a flavor of response correction methods, we briefly outline simple one-point function correction techniques, i.e., those that utilize the model responses (and, in some cases, derivatives) at a single point. There are three main groups of the correction techniques: compositional, additive, or multiplicative ones. We will briefly illustrate each of these categories for correcting the low-fidelity model $c(\boldsymbol{x})$, as well as discuss whether zero- and first-order consistency conditions with $f(\boldsymbol{x})$ (Alexandrov and Lewis 2001) can be satisfied. Here, we assume that the models, $c(\boldsymbol{x})$ and $f(\boldsymbol{x})$, are scalar functions; however, generalizations for vector-valued functions are straightforward in some cases.

The compositional correction (Søndergaard 2003)

$$s^{(i+1)}(\boldsymbol{x}) = g(c(\boldsymbol{x})) \tag{5.2}$$

is a simple scaling of c. Since the mapping g is a real-valued function of a real variable, a compositional correction will not in general yield first-order consistency conditions. By selecting a mapping g that satisfies

$$g'\left(c\left(\boldsymbol{x}^{(i)}\right)\right) = \frac{\nabla f\left(\boldsymbol{x}^{(i)}\right)\nabla c\left(\boldsymbol{x}^{(i)}\right)^T}{\nabla c\left(\boldsymbol{x}^{(i)}\right)\nabla c\left(\boldsymbol{x}^{(i)}\right)^T}, \tag{5.3}$$

the discrepancy between $\nabla f(\boldsymbol{x}^{(i)})$ and $\nabla s^{(i+1)}(\boldsymbol{x}^{(i)})$ (in a norm sense) is minimized. Note that the correction in (5.3), as many transformations that ensure first-order consistency, requires a high-fidelity gradient. Still, numerical estimates of $\nabla f(\boldsymbol{x}^{(i)})$ may yield acceptable results.

The compositional correction can also be introduced in the parameter space (Bandler et al. 2004a)

$$s^{(i+1)}(\boldsymbol{x}) = c(\boldsymbol{p}(\boldsymbol{x})). \tag{5.4}$$

If the ranges of $f(\boldsymbol{x})$ and $c(\boldsymbol{x})$ are different, then the condition $c(\boldsymbol{p}(\boldsymbol{x}^{(i)}))=f(\boldsymbol{x}^{(i)})$ cannot be satisfied. This difficulty can be alleviated by combining both types of compositional corrections with g and $\boldsymbol{p}$ taking the forms

$$g(t) = t - c\left(\boldsymbol{x}^{(i)}\right) + f\left(\boldsymbol{x}^{(i)}\right), \tag{5.5}$$

$$\boldsymbol{p}(\boldsymbol{x}) = \boldsymbol{x}^{(i)} + \boldsymbol{J}_p\left(\boldsymbol{x} - \boldsymbol{x}^{(i)}\right), \tag{5.6}$$

where $\boldsymbol{J}_p$ is an $n \times n$ matrix for which $\boldsymbol{J}_p{}^T\, \nabla c = \nabla f(\boldsymbol{x}^{(i)})$ guarantees consistency.

Additive and multiplicative corrections may ensure first-order consistency conditions. For the additive case one can generally express the correction as

$$s^{(i+1)}(\boldsymbol{x}) = \lambda(\boldsymbol{x}) + s^{(i)}(\boldsymbol{x}). \tag{5.7}$$

The associated consistency conditions require that $\lambda(\boldsymbol{x})$ satisfies $\lambda\left(\boldsymbol{x}^{(i)}\right) = f\left(\boldsymbol{x}^{(i)}\right) - c\left(\boldsymbol{x}^{(i)}\right)$, and $\nabla\lambda\left(\boldsymbol{x}^{(i)}\right) = \nabla f\left(\boldsymbol{x}^{(i)}\right) - \nabla c\left(\boldsymbol{x}^{(i)}\right)$. These can be achieved by means of the following linear additive correction:

$$s^{(i+1)}(\boldsymbol{x}) = f\left(\boldsymbol{x}^{(i)}\right) - c\left(\boldsymbol{x}^{(i)}\right) + \left(\nabla f\left(\boldsymbol{x}^{(i)}\right) - \nabla c\left(\boldsymbol{x}^{(i)}\right)\right)\left(\boldsymbol{x} - \boldsymbol{x}^{(i)}\right) + c(\boldsymbol{x}). \tag{5.8}$$

Multiplicative corrections, also known as the β-correlation method (Søndergaard 2003), can be generally represented as

$$s^{(i+1)}(\boldsymbol{x}) = \alpha(\boldsymbol{x})s^{(i)}(\boldsymbol{x}). \tag{5.9}$$

If $c(\boldsymbol{x}^{(i)}) \neq 0$, zero- and first-order consistency can be achieved if $\alpha(\boldsymbol{x}^{(i)}) = f(\boldsymbol{x}^{(i)})/c(\boldsymbol{x}^{(i)})$ and $\nabla\alpha(\boldsymbol{x}^{(i)}) = [\nabla f(\boldsymbol{x}^{(i)}) - f(\boldsymbol{x}^{(i)})/c(\boldsymbol{x}^{(i)})\nabla c(\boldsymbol{x}^{(i)})]/c(\boldsymbol{x}^{(i)})$.

In practice, the requirement $c(\boldsymbol{x}^{(i)}) \neq 0$ is not very restrictive because the range of $f(\boldsymbol{x})$ (and thus, of the surrogate $c(\boldsymbol{x})$) is often known beforehand, and hence a bias can be introduced both for $f(\boldsymbol{x})$ and $c(\boldsymbol{x})$ to avoid cost function values equal to zero. In these circumstances, the following multiplicative correction

$$s^{(i+1)}(\boldsymbol{x}) = \left[\frac{f\left(\boldsymbol{x}^{(i)}\right)}{c\left(\boldsymbol{x}^{(i)}\right)} + \frac{\nabla f\left(\boldsymbol{x}^{(i)}\right)c\left(\boldsymbol{x}^{(i)}\right) - f\left(\boldsymbol{x}^{(i)}\right)\nabla c\left(\boldsymbol{x}^{(i)}\right)}{c\left(\boldsymbol{x}^{(i)}\right)^2}\left(\boldsymbol{x} - \boldsymbol{x}^{(i)}\right)\right]c(\boldsymbol{x}) \tag{5.10}$$

ensures both zero-and first-order consistency between $s^{(i+1)}$ and f.

5.3 Low-Fidelity Modeling

Fundamental components of any surrogate-based process utilizing physics-based surrogates, be it modeling or optimization, are the low-fidelity models which are simplified (consequently, of limited accuracy but fast) representations of the original, high-fidelity computational model of the system at hand. The success and computational cost of the surrogate-based process depend on the quality and the speed of the low-fidelity model. This section provides a general overview of low-fidelity modeling, providing illustrations of their development through examples from the areas of microwave engineering, antenna design, and aerodynamics.

5.3.1 Principal Properties and Methods

In design optimization, information from the surrogate is used to navigate the high-fidelity design space and search for the optimum. Therefore, the surrogate needs to be able to capture the changes in the measures of merit with respect to the design parameters in the same way as the high-fidelity model. If, for example, a measure of merit of the high-fidelity model decreases due to design variable changes, the measure of merit at the surrogate model level has to decrease as well for the same design variable changes. However, the amount by which the measure of merit changes does not need to be the same for both models. In other words, the surrogate and the high-fidelity model have to be sufficiently well correlated. The purpose of the response correction, in general, is not to replace any missing physics of the low-fidelity model. Hence, it is up to the low-fidelity model to be capable of following the high-fidelity model within the design space. Moreover, to accelerate the design process the low-fidelity model has to be fast.

The following general approaches are commonly used to develop low-fidelity models (Alexandrov et al. 1998):

1. *Simplified physics*: Replace the set of governing equations by a set of simplified equations. These are often referred to as *variable-fidelity physics models.*
2. *Coarse discretization*: Use the same governing equations as in the high-fidelity model, but with a coarser computational grid discretization. Often referred to as *variable-resolution models.*
3. *Relaxed convergence criteria*: Reduce/relax the solver convergence criteria. Sometimes referred to as *variable-accuracy models.*
4. A combination of (1)–(3).

The above list is a general compilation of the low-fidelity modeling approaches. Of course, implementation of (1)–(4) can be domain specific. For the sake of example, in the modeling of microwave devices the following computational simplifications, aside from coarsening the mesh, are included:

1. Reducing the computational domain and applying simple absorbing boundaries with the finite-volume methods implemented in full-wave EM simulations.
2. Using low-order basis functions with the finite-element and moment method solvers.
3. Using relaxed solution termination criteria such as the S-parameter error for the frequency domain methods with adaptive meshing, and residue energy for the time-domain solvers.

Moreover, simplifications of the physics for modeling of microwave devices, aside from using different governing physics, include:

4. Ignoring dielectric and metal losses as well as material dispersion if their impact to the simulated response is not significant.
5. Setting metallization thickness to zero for traces, strips, and patches.
6. Ignoring moderate anisotropy of substrates.
7. Energizing the antenna with discrete sources rather than waveguide ports.

Clearly, the particular simplification approaches depend on the engineering discipline and the specific type of simulation utilized in there.

In the following two sections, we demonstrate the above low-fidelity modeling approaches using the examples from the areas of computational electromagnetics and aerodynamics.

5.3.2 Variable-Resolution and Variable-Accuracy Modeling

Let us consider the microstrip antenna shown in Fig. 5.1. The two computational models of the antenna are shown in Fig. 5.2. Both models are defined, discretized, and simulated using CST MWS (CST Microwave Studio 2013). The finer model is discretized with about 410,000 mesh cells and evaluates in about 70 min, whereas the coarser model contains about 20,000 mesh cells and its simulation time is only 26 s.

The responses of the models are shown in Fig. 5.3a, along with responses of other models with various discretization levels. One can infer from this figure that the two "finest" coarse-discretization models (with ~400,000 and ~740,000 mesh cells) represent the high-fidelity model response (the model with ~1,600,000 mesh cells, shown as a thick solid line) quite properly. The model with ~270,000 cells can be considered as a borderline one. The two remaining models could be considered as poor ones, particularly the model with ~20,000 cells; its response is essentially unreliable.

The simulation time increases rapidly with the number of mesh cells, as can be seen in Fig. 5.3b. It takes around 120 min to simulate the high-fidelity model, whereas the second finest model (~740,000 cells) takes 42 min, and the coarsest model needs roughly 1 min.

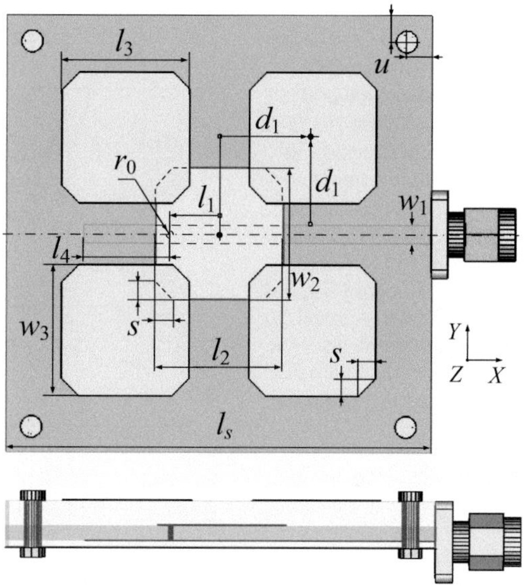

Fig. 5.1 Microstrip antenna: top/side views, substrates shown transparent (Koziel and Ogurtsov 2012a)

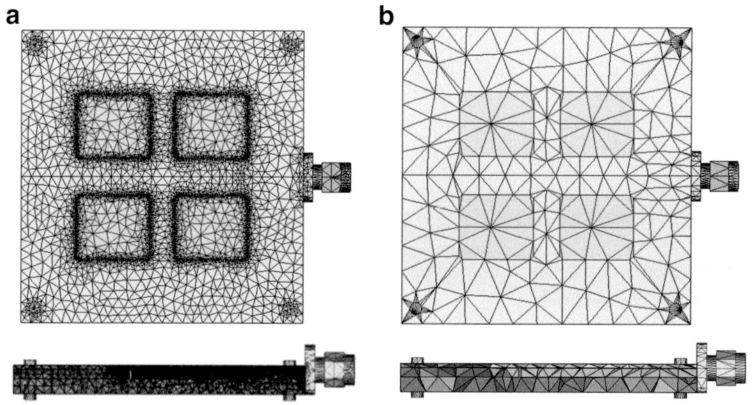

Fig. 5.2 Variable-resolution models of the microstrip antenna in Fig. 5.1: (**a**) High-fidelity model shown with a fine tetrahedral mesh (~410,000 cells); and (**b**) low-fidelity model shown with a much coarser mesh (~20,000 cells). Both models are developed in CST Microwave Studio (Koziel and Ogurtsov 2012a)

Now, let us consider two-dimensional transonic airflow past an airfoil (such as sections of a typical transport aircraft wing). Typical computational meshes for the airfoil flow are shown in Fig. 5.4. Variations of the measures of merit (the lift (C_l) and drag (C_d) coefficients) with the mesh discretization are shown in Fig. 5.5a. Mesh f is the high-fidelity mesh (~408,000 cells) and meshes $\{c_i\}$, $i = 1,2,3,4$, are the coarse meshes. Simulation time of the high-fidelity mesh is approximately 33 min (Fig. 5.5b).

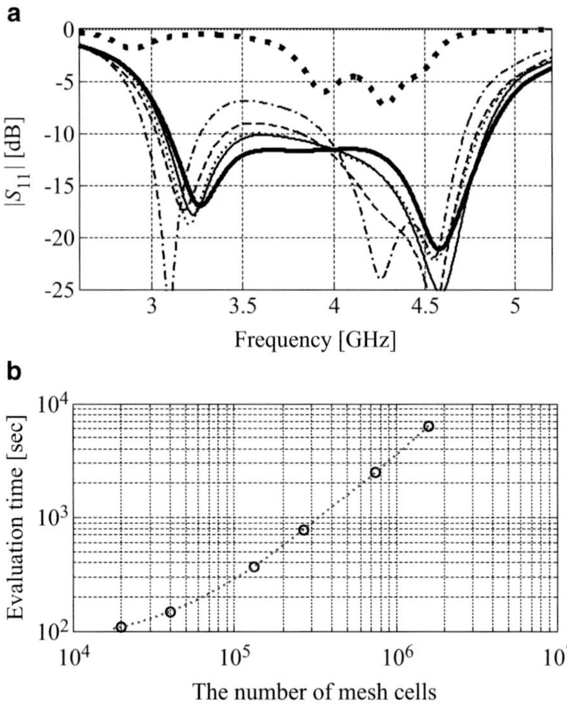

Fig. 5.3 Antenna of Fig. 5.1 at a selected design simulated with the CST MWS transient solver (CST Microwave Studio 2013): (**a**) reflection responses at different discretization densities, 19,866 cells (▪▪▪), 40,068 cells (· — ·), 266,396 cells (– –), 413,946 cells (···), 740,740 cells (—), and 1,588,608 cells (—); and (**b**) the antenna simulation time versus the number of mesh cells (Koziel and Ogurtsov 2012a)

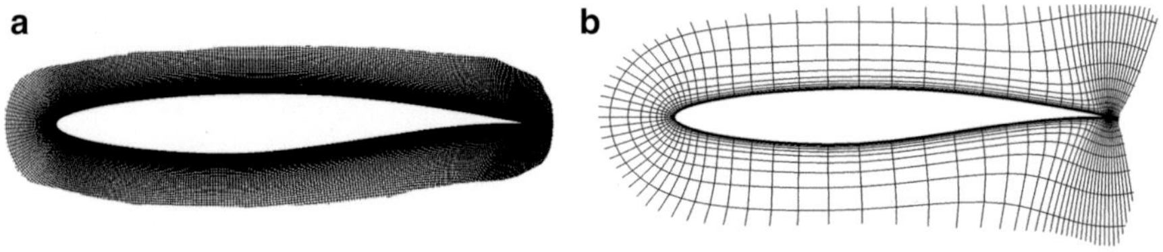

Fig. 5.4 Structured computational mesh close to the surface of a supercritical airfoil: (**a**) fine mesh, and (**b**) coarse mesh

By visual inspection of Fig. 5.5a, one can see that mesh c_4 is the most inaccurate model and only slightly faster than model c_3. In this case, one can select model c_3 as the low-fidelity model. This particular grid has roughly 32,000 cells and takes about 2 min to run.

The flow solution history for the low-fidelity model is shown in Fig. 5.6a and it indicates the lift and drag coefficients being nearly converged after 80–100 iterations. The maximum number of iterations can, therefore, be set to 100 for the low-fidelity model. This reduces the overall simulation time to 45 s.

A comparison of the pressure distributions of the high- and low-fidelity models, shown in Fig. 5.6b, indicates that the low-fidelity model, in spite of being based on much coarser mesh and reduced flow solver iterations, captures the main features of the high-fidelity model pressure distribution quite well. The biggest discrepancy in

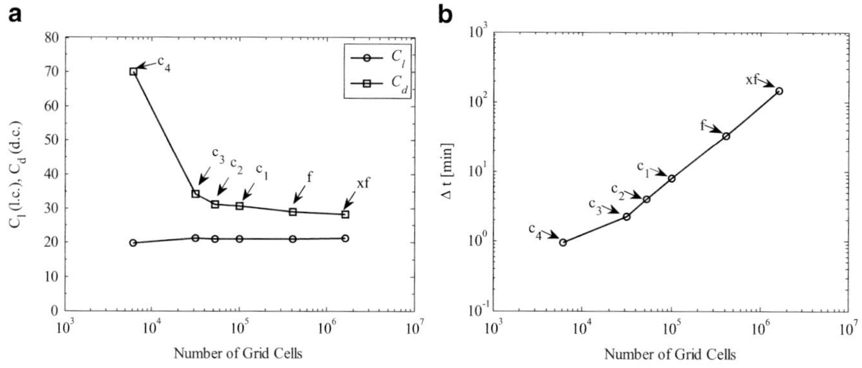

Fig. 5.5 Variation of the measures of merit with the number of grid cells: (**a**) lift (C_l) and drag (C_d) coefficients, and (**b**) simulation time

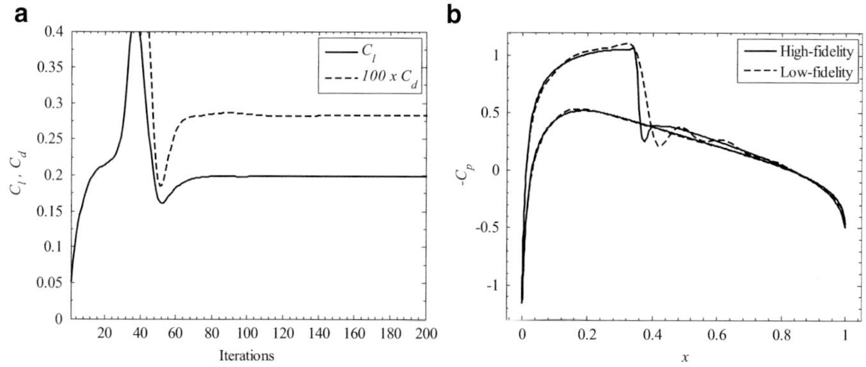

Fig. 5.6 Low-fidelity model (model c_3 in Fig. 5.5) responses: (**a**) evolution of the lift and drag coefficients during flow solver solution, and (**b**) comparison of the pressure distribution with the high-fidelity model

the distributions is around the shock on the upper surface, leading to an overestimation of both lift and drag (see Fig. 5.5a).

The ratio of the simulation times of the high- and low-fidelity model in this case is 44. However, in many cases, the solver does not fully converge with respect to the residuals and goes on up to the (user-specified) maximum allowable iterations. In those cases the ratio of simulation times of the high- and low-fidelity models is over 100; that is, the low-fidelity model is over two orders of magnitude faster than the high-fidelity one.

5.3.3 *Variable-Fidelity Physics Modeling*

Variable-fidelity physics models are developed by replacing the set of high-fidelity governing equations by a set of simplified or modified equations. In the case of

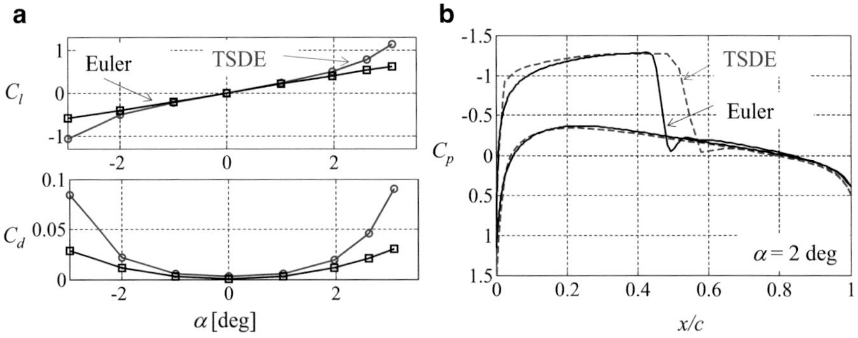

Fig. 5.7 Responses of the Euler equations and the transonic small disturbance equation (TSDE) for flow past the NACA 0012 airfoil in transonic flow: (**a**) lift (C_l) and drag (C_d) coefficients, and (**b**) pressure coefficient (C_p) distributions

aerodynamics, models with lower fidelity governing equations are generated by making certain assumptions about the flow field. As an example, the Euler equations can be used as the high-fidelity model, and by assuming weak shocks and small disturbances, the transonic small disturbance equation (TSDE) can be used as the low-fidelity one. Figure 5.7 gives a comparison of these models for transonic flow past at a given airfoil shape and operating condition. The measures of merit compare well for small perturbations (small angles of attack), as can be seen from Fig. 5.7a. The airfoil surface responses also compare well, aside from an area around the upper surface shock (Fig. 5.7b).

Now, the main question is whether the low-fidelity model exhibits sufficient correlations with the high-fidelity one. Again, let us consider two-dimensional transonic flow past a supercritical airfoil. We use the compressible RANS equations with the Spalart-Allmaras turbulence model as the high-fidelity model. The low-fidelity model solves the compressible Euler equations coupled with a two-equation integral boundary-layer formulation using the displacement thickness concept. On a typical desktop computer, the high-fidelity model evaluates in roughly 40 min, whereas the low-fidelity model in less than 15 s, or 160 times faster. Figure 5.8 shows a comparison of the output of these models for several (locally) perturbed shapes of the original airfoil, indicating that the lower fidelity model captures (at least locally) the main physics of the higher fidelity model.

Similarly, in microwave engineering, variable-fidelity physics models can be developed using simplified models. Consider, for example, the dual-band microstrip filter (Guan et al. 2008) shown in Fig. 5.9. The high-fidelity model (Fig. 5.9a) determines the scattering parameters of the device using a commercial planar-3D solver Sonnet ***em*** (Sonnet 2012). Simulation time is around 20 min per design. The low-fidelity model (Fig. 5.9b) is an equivalent circuit implemented in Agilent ADS (ADS 2011). The low-fidelity model is a simplified representation of the filter, where the lumped microstrip line models are connected together using circuit theory rules. The low-fidelity model is very fast (simulation time is in the

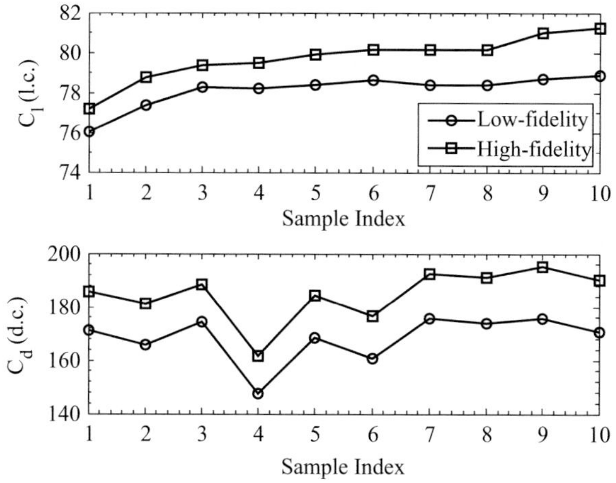

Fig. 5.8 Measures of merit for several (locally) perturbed shapes of a supercritical airfoil at a transonic flow condition computed using variable-fidelity physics models

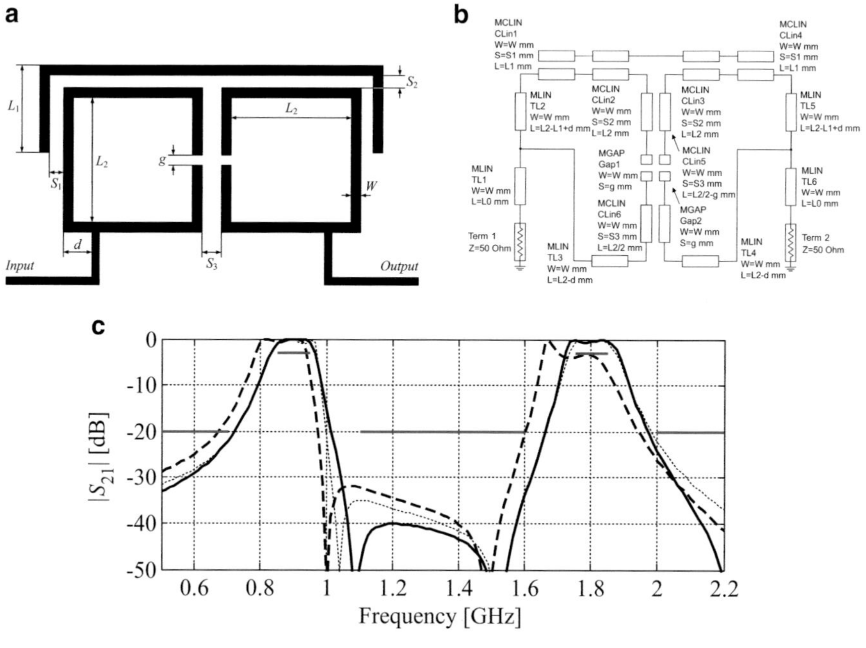

Fig. 5.9 Dual-band band-pass filter (Guan et al. 2008): (**a**) high-fidelity model, (**b**) low-fidelity model, (**c**) responses at two designs: *dashed line* = high-fidelity model at design 1, *thin dashed line* = low-fidelity model at design 2, *solid line* = high-fidelity model at design 2, *thick horizontal lines* = minimax constraints

range of milliseconds) so that all the operations on the low-fidelity model can be neglected in the design process. A comparison of the model responses is given in Fig. 5.9c. In should be emphasized here that while equivalent circuit models are generally acceptable (from the SBO standpoint) for relatively simple structures such as the filter in Fig. 5.9, they are usually not sufficiently accurate for complex circuits featuring considerable electromagnetic cross-couplings (e.g., highly miniaturized microstrip passives; Koziel and Kurgan 2015).

5.3.4 Practical Issues of Low-Fidelity Model Selection

In practice, the selection of an appropriate low-fidelity model is done by the designer. As it can be concluded from the examples presented in Sects. 5.3.2 and 5.3.3, the selection of the model coarseness strongly affects the simulation time, and, therefore, the performance of the design optimization process. Coarser models are faster, which turns into a lower cost per design iteration of the SBO process. The coarser models, however, are less accurate, which may result in a larger number of iterations necessary to find a satisfactory design. Also, there is an increased risk of failure for the optimization algorithm to find a good design. Finer models, on the other hand, are more expensive but they are more likely to produce a useful design with a smaller number of iteration. Experimental studies have been carried out to investigate the issues of the low-fidelity model selection and its impact on the performance of the SBO process (Leifsson and Koziel 2015a). At the same time, no general-purpose methods for automated low-fidelity model selection exist so far. The decision about the "right" low-fidelity model setup is normally made based on grid convergence studies such as those presented in Figs. 5.3 and 5.5, engineering experience, and visual inspection of the low- and high-fidelity model responses. On the other hand, preliminary results reported by Leifsson et al. (2014b) indicate that certain parameters of the low-fidelity model (e.g., those controlling grid density) may be automatically adjusted using optimization methods. Also, multi-fidelity methods (e.g., Koziel et al. 2011a; Leifsson and Koziel 2015a) where the design optimization process is conducted using an entire family of models of increasing accuracy may, to some extent, mitigate the risk of improper model setup by automatically switching to the higher fidelity model in case the lower fidelity one fails to improve the design (Koziel and Ogurtsov 2013a; Koziel and Leifsson 2013a, b, c; Tesfahunegn et al. 2015).

Chapter 6
Surrogate-Based Optimization Using Parametric Response Correction

In this chapter, we briefly describe several parametric response correction techniques and illustrate their application for solving design optimization problems in various engineering disciplines such as electrical engineering, antenna design, hydrodynamics, and aerodynamic shape optimization. Parametric response correction is simple to implement and it boils down to determining the surrogate model parameters using available data from both the low- and high-fidelity model (normally obtained from previous iterations of the optimization algorithm) and by evaluating specific (explicit) formulas and/or solving simple (usually linear) regression problems. A simple example of a parametric response correction is the AMMO algorithm (cf. Sect. 4.5.2), other examples can be found in Sect. 5.2. Here, we focus on the methods working with vector-valued responses, such as output space mapping, manifold mapping, and multi-point response correction.

6.1 Output Space Mapping

Output space mapping (OSM) (Bandler et al. 2003; Koziel et al. 2006a) is one of the simplest realizations of response correction. In this section, we consider a few variations of OSM and show their applications for solving design problems in microwave engineering and hydrodynamic shape optimization. A multi-point version of OSM is described in Sect. 6.2. It should be mentioned here that OSM is often used as an additional low-fidelity model correction layer, applied on the top of other types of corrections (such as input and/or implicit SM (Koziel et al. 2008a)).

S. Koziel, L. Leifsson, *Simulation-Driven Design by Knowledge-Based Response Correction Techniques*, DOI 10.1007/978-3-319-30115-0_6

6.1.1 Output Space Mapping Formulation

The additive OSM is defined as

$$s^{(i)}(x) = c(x) + \Delta_r^{(i)}. \tag{6.1}$$

Here, $\Delta_r^{(i)} = f(x^{(i)}) - c(x^{(i)})$, which ensures zero-order consistency; $x^{(i)}$ is the current iteration point generated by the SBO process (4.1). It should be noted that this way of creating the surrogate model is only based on the high-fidelity model response for a single design (here, at the most current iteration point $x^{(i)}$). Typically, output SM is used as a supplement for other SM techniques (Koziel et al. 2006a, b). Provided that the coarse model is sufficiently accurate, output SM can also work as a stand-alone technique (Koziel et al. 2006a).

The multiplicative OSM is defined as

$$s^{(i)}(x) = \Lambda^{(i)} \cdot c(x), \tag{6.2}$$

where $\Lambda^{(i)}$ is a diagonal $m \times m$ matrix $\Lambda^{(i)} = \mathrm{diag}([\lambda_1{}^{(i)}\ \lambda_2{}^{(i)}\ \ldots\ \lambda_m{}^{(i)}])$ with $\lambda_k{}^{(i)} = f_k(x^{(i)})/c_k(x^{(i)})$, with $f_k(x^{(i)})$ and $c_k(x^{(i)})$ being the kth components of the high- and low-fidelity model response vectors, respectively. The surrogate (6.2) satisfies the zero-order consistency condition.

The model (6.1) can be enhanced as follows

$$s^{(i)}(x) = c(x) + \Delta_r^{(i)} + \Delta_J^{(i)} \cdot \left(x - x^{(i)}\right), \tag{6.3}$$

where $\Delta_r^{(i)}$ is defined as in (6.1) and $\Delta_J^{(i)} = J[f(x^{(i)}) - c(x^{(i)})]$, which ensures first-order consistency. Alternatively, the Jacobian of $f(x^{(i)}) - c(x^{(i)})$ can be replaced by a suitable approximation obtained using, e.g., Broyden update (Broyden 1965). Similarly, as in the case the basic OSM, the model (6.3) is normally used as a supplement but it can, also, work as a stand-alone technique, particularly when the SBO algorithm (4.1) is embedded in the trust region framework (Conn et al. 2000; Koziel et al. 2010a). The multiplicative OSM surrogate (6.2) can be enhanced by sensitivity data in a similar manner.

6.1.2 Output Space Mapping for Microwave Filter Optimization

In this section, we illustrate the application of OSM for design optimization of microwave filters. We also demonstrate the importance of appropriate convergence safeguards, here, trust regions.

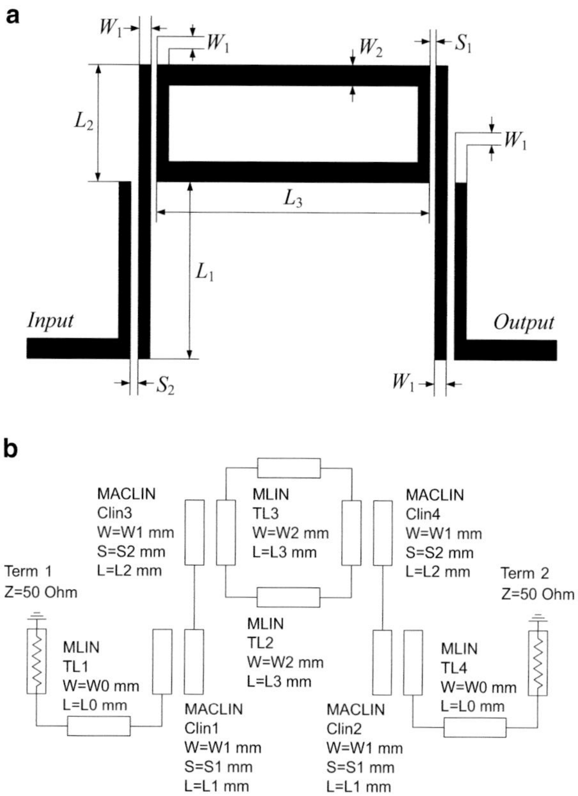

Fig. 6.1 Fourth-order ring resonator bandpass filter: (**a**) geometry (Salleh et al. 2008), (**b**) coarse model (Agilent ADS)

Consider the fourth-order ring resonator bandpass filter (Salleh et al. 2008) shown in Fig. 6.1a. The design parameters are $\boldsymbol{x} = [L_1\, L_2\, L_3\, S_1\, S_2\, W_1\, W_2]^T$ mm. The high-fidelity model $\boldsymbol{f}$ is implemented and simulated in an electromagnetic simulation software FEKO (FEKO 2008). The low-fidelity model, shown in Fig. 6.1b, is an equivalent circuit model implemented in Agilent ADS (Agilent 2011). The design specifications imposed on the transmission coefficient $|S_{21}|$ of the filter are $|S_{21}| \geq -1$ dB for 1.75 GHz $\leq \omega \leq$ 2.25 GHz, and $|S_{21}| \leq -20$ dB for 1.0 GHz $\leq \omega \leq$ 1.5 GHz and 2.5 GHz $\leq \omega \leq$ 3.0 GHz, where ω is a signal frequency. The filter was optimized using space mapping with the low-fidelity model $\boldsymbol{c}$ preconditioned in all iterations by the input SM (Koziel et al. 2010a). Two output SM models were considered: (1) model (6.1), and (2) model (6.3) with the linear term $\boldsymbol{\Delta}_J$ obtained using the Broyden update (Broyden 1965). The SBO algorithm with the model (6.1) was run twice: without and with embedding it in the trust-region (TR) framework (Conn et al. 2000). The algorithm with model (6.3) was only run with the TR enhancement. The results are presented in Fig. 6.2 and Table 6.1. It follows that the SBO algorithm using model (6.1) finds a design satisfying the specifications but it does not converge without TR safeguard. Using either TR or both TR and more involved model (6.3) improves performance.

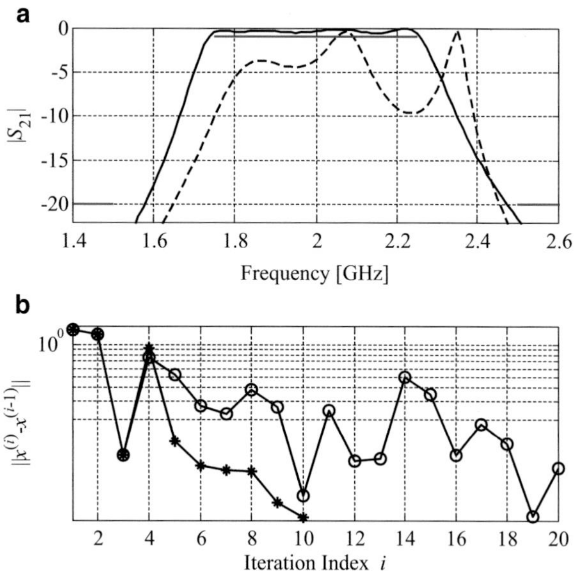

Fig. 6.2 Fourth-order ring resonator bandpass filter: (**a**) initial (*dashed line*) and optimized (*solid line*) fine model response obtained with trust-region enhanced SM algorithm exploiting model (6.3) with the linear term calculated through Broyden update; (**b**) convergence plots for SM using model (6.1) (*open circle*) and trust-region enhanced SM using the surrogate (6.3) (*asterisk*)

Table 6.1 Fourth-order ring resonator filter: optimization results

SM algorithm[a]	Specification error (dB) Final	Specification error (dB) Best found	Number of high-fidelity model evaluations
Output SM model (6.1)	−0.2	−0.3	21[b]
Output SM model (6.1) SBO algorithm (4.1) enhanced using TR	−0.5	−0.5	18
Output SM model (6.3) SBO algorithm (4.1) enhanced using TR	−0.4	−0.4	17

[a]Coarse model preconditioned in each iteration using input SM (Koziel et al. 2010a)
[b]Convergence not obtained; algorithm terminated after 20 iterations

6.1.3 *Hydrodynamic Shape Optimization of Axisymmetric Bodies Using OSM*

In this section, we discuss surrogate-based optimization of hydrodynamic shape exploiting the concept of multiplicative output space mapping similar to (6.2). We adopt the multi-fidelity approach with the high-fidelity model based on the Reynolds-Averaged Navier–Stokes (RANS) equations, and the low-fidelity model based on the same equations, but with a coarse discretization and relaxed convergence criteria. Here, we choose to focus on the clean hull design.

Autonomous underwater vehicles (AUVs) are becoming increasingly important in various marine applications, such as oceanography, pipeline inspection, and mine counter measures (Yamamoto 2007). Endurance (speed and range) is one of the more important attribute of AUVs (Allen et al. 2000). Vehicle drag reduction and/or an increase in the propulsion system efficiency will translate to a longer range for a given speed (or the same distance in a reduced time). A careful hydrodynamic design of the AUVs, including the hull shape, the protrusions, and the propulsion system, is therefore essential.

The goal of hydrodynamic shape optimization is to find an optimal—with respect to given objectives—hull shape, so that the given design constraints are satisfied. In this section, we consider both direct design (adjustment of the geometrical shape to maximize performance, e.g., minimize drag), and inverse design (obtaining a predefined flow behavior, see, e.g., Dalton and Zedan 1980). In case of the direct approach that involves the minimization the drag coefficient (C_D) and we set $\boldsymbol{f}(\boldsymbol{x}) = C_D(\boldsymbol{x})$. In the inverse approach, a target pressure distribution ($C_{p,t}$) is prescribed a priori and the function $\boldsymbol{f}(\boldsymbol{x}) = \|C_p(\boldsymbol{x}) - C_{p.t}\|^2$, where $C_p(\boldsymbol{x})$ is the hull surface pressure distribution, is minimized. No constraints are considered here aside from the design variable bounds.

Figure 6.3a shows a typical torpedo-shaped hull with a nose of length a, midsection of length b, overall length L, and maximum diameter of D. The nose and the tail are parameterized using Bézier curves (Lepine et al. 2001). We use five control points for the nose and four for the tail, as shown in Fig. 6.3b, c. Control points number three and eight are free (x- and y-coordinates), while the other points

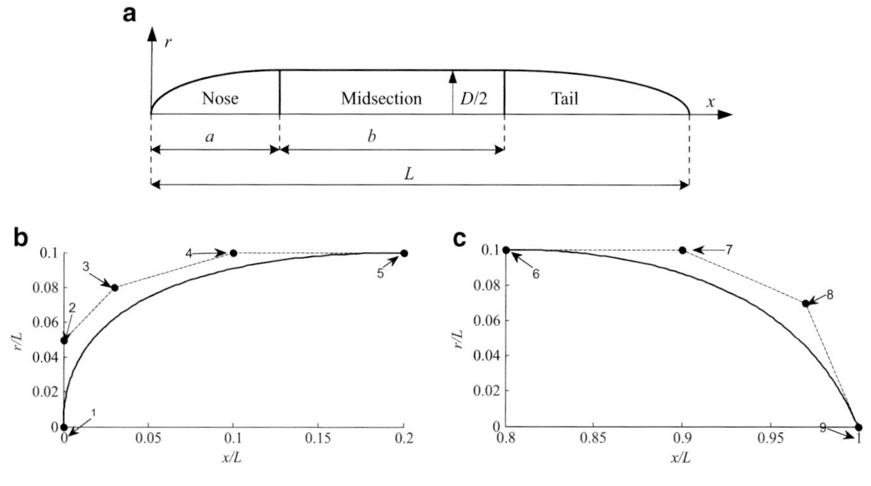

Fig. 6.3 (**a**) A sketch of a typical torpedo shaped hull form (axisymmetric with three sections: nose, middle, tail); typically, equipment such as the computer, sensors, electronics, batteries, and payload are housed in the nose and the midsection, whereas the propulsion system is in the tail; (**b**) Bézier curve representing the nose (5 control points); and (**c**) Bézier curve representing the tail (4 control points). Control points 3 and 8 are free, while the other points are essentially fixed (depend on L, a, b, and D)

are fixed. We, therefore, have two design variables for the nose and tail curves, a total of four design variables, aside from the hull dimensions a, b, L, and D. The flow past the hull is considered to be steady and incompressible. The Reynolds-Averaged Navier–Stokes (RANS) equations are assumed as the governing flow equations with the two-equation k-ε turbulence model with standard wall functions (Tannehill et al. 1997). The CFD computer code FLUENT (2015) is used for numerical simulations of the fluid flow. The computational grid is structured with quadrilateral elements. The elements are clustered around the body and grow in size with distance from the body. The grids are generated using ICEM CFD (ICEM 2013). The high-fidelity model has ~42,000 grid cells (elements), the low-fidelity model has 504 elements.

For a body in incompressible flow, the total drag is due to pressure and friction forces, which are calculated by integrating the pressure (C_p) and skin friction (C_f) distributions over the hull surface. The pressure coefficient is defined as $C_p \equiv (p - p_\infty)/q_\infty$, where p is the local static pressure, p_∞ is free-stream static pressure, and $q_\infty = (1/2\rho_\infty V_\infty{}^2)$ is the dynamic pressure, with ρ_∞ as the free-stream density, and the V_∞ free-stream velocity. The skin friction coefficient is defined as $C_f \equiv \tau/q_\infty$, where τ is the shear stress. Typical C_p and C_f distributions are shown in Fig. 6.4. The total drag coefficient is defined as $C_D \equiv d/(q_\infty S)$, where d is the total drag force, and S is the reference area. The drag coefficient is the sum of the pressure and friction drag, $C_D = C_{Dp} + C_{Df}$. Both can be calculated from C_p and C_f (Leifsson and Koziel 2015a).

The design problem is solved here using the SBO scheme (4.1) with the surrogate model using multiplicative OSM (6.2) which turns out to be sufficient for our purposes. The specific formulation of the surrogate model is given below. Let $C_{p.f}(\boldsymbol{x})$ and $C_{f.f}(\boldsymbol{x})$ denote the pressure and skin friction distributions of the high-fidelity model. The respective distributions of the low-fidelity model are denoted as $C_{p.c}(\boldsymbol{x})$ and $C_{f.c}(\boldsymbol{x})$. We will use the notation $C_{p.f}(\boldsymbol{x}) = [C_{p.f.1}(\boldsymbol{x})\, C_{p.f.2}(\boldsymbol{x}) \ldots C_{p.f.m}(\boldsymbol{x})]^T$, where $C_{p.f.j}(\boldsymbol{x})$ is the jth component of $C_{p.f}(\boldsymbol{x})$, with the components corresponding to different coordinates along the x/L axis.

At iteration i, the surrogate model $C_{p.s}{}^{(i)}$ of the pressure distribution $C_{p.f}$ is constructed as

$$C_{p.s}^{(i)}(\boldsymbol{x}) = \left[C_{p.s.1}^{(i)}(\boldsymbol{x}) \quad C_{p.s.2}^{(i)}(\boldsymbol{x}) \quad \ldots \quad C_{p.s.m}^{(i)}(\boldsymbol{x})\right]^T \tag{6.4}$$

$$C_{p.s.j}^{(i)}(\boldsymbol{x}) = A_{p.j}^{(i)} \cdot C_{p.c.j}(\boldsymbol{x}) \tag{6.5}$$

where $j = 1, 2, \ldots, m$, and

$$A_{p.j}^{(i)} = \frac{C_{p.f.j}^{(i)}\left(\boldsymbol{x}^{(i)}\right)}{C_{p.c.j}^{(i)}\left(\boldsymbol{x}^{(i)}\right)} \tag{6.6}$$

Similar definition holds for the skin friction distribution model $C_{f.s}{}^{(i)}$. One of the issues of the model (6.4)–(6.6) is that (6.6) is not defined whenever $C_{p.c.j}(\boldsymbol{x}^{(i)})$ equals

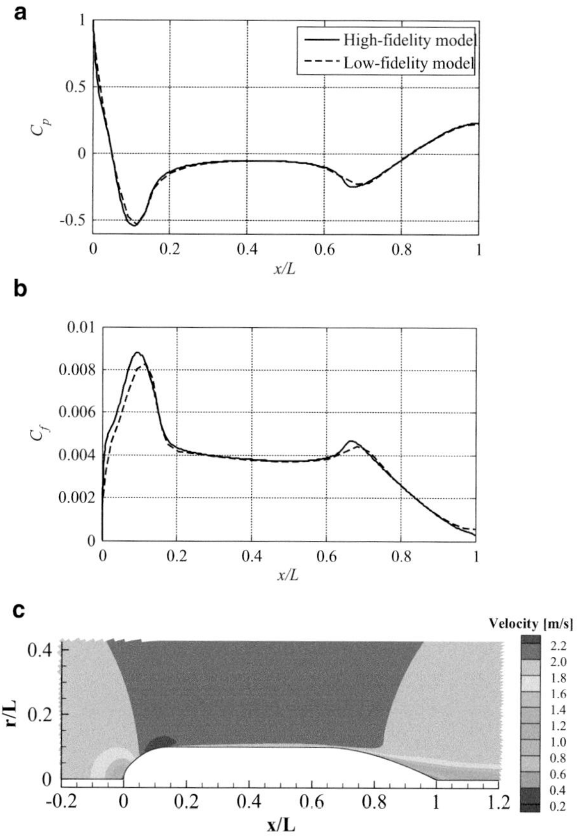

Fig. 6.4 (**a**) Comparison of the high- and low-fidelity model responses of the axisymmetric hull of diameter ratio $L/D = 5$ at a speed of 2 m/s and Reynolds number of two million, (**a**) pressure distributions, and (**b**) skin friction distributions. (**c**) Velocity contours of the flow past an axisymmetric torpedo shape obtained by the high-fidelity CFD model

zero, and that the values of $A_{p.j}{}^{(i)}$ are very large when $C_{p.c.j}(\boldsymbol{x}^{(i)})$ is close to zero. This may be a source of substantial distortion of the surrogate model response. In order to alleviate this problem, the original surrogate model response is "smoothened" in the vicinity of the regions where $A_{p.j}{}^{(i)}$ is large (which indicates the problems mentioned above). Let $j_{\max}$ be such that $|A_{p.j\max}{}^{(i)}| \gg 1$ assumes (locally) the largest value. Let Δj be the user-defined index range (typically, $\Delta j = 0.01 \cdot m$). The original values of $A_{p.j}{}^{(i)}$ are replaced, for $j = j_{\max} - \Delta j, \ldots, j_{\max} - 1,\ j_{\max},\ j_{\max} + 1, \ldots, j_{\max} + \Delta j$, by the interpolated values:

$$\overline{A}_{p.j}^{(i)} = I\Big(\{[j_{\max} - 2\Delta j \ \ldots \ j_{\max} - \Delta j - 1] \cup [j_{\max} + \Delta j - 1 \ \ldots \ j_{\max} + 2\Delta j]\}, \\ \Big\{\Big[A_{p.j_{\max}-2\Delta j}^{(i)} \ \cdots \ A_{p.j_{\max}-\Delta j-1}^{(i)}\Big] \cup \Big[A_{p.j_{\max}-2\Delta j}^{(i)} \ \cdots \ A_{p.j_{\max}-\Delta j-1}^{(i)}\Big]\Big\}, j\Big) \tag{6.7}$$

where $I(X,Y,Z)$ is a function that interpolates the function values Y defined over the domain X onto the set Z. Here, we use cubic splines. In other words, the values of

$A_{p.j}^{(i)}$ in the neighborhood of $j_{\max}$ are "restored" using the values of $A_{p.j}^{(i)}$ from the surrounding of $j = j_{\max} - \Delta j, \ldots, j_{\max} + \Delta j$.

Below, we report the results of applying the above scheme to hydrodynamic shape optimization of torpedo-type hulls, involving both the direct and inverse design approaches. Designs are obtained with the surrogate model optimized using the pattern-search algorithm (Koziel 2010c). For comparison purposes, designs obtained with direct optimization of the high-fidelity model using the pattern-search are also presented. For both the direct and the inverse design approaches, the design variable vector is $\boldsymbol{x} = [a\ x_n\ y_n\ x_t\ y_t]^T$, where a is the nose length, (x_n, y_n) and (x_t, y_t) are the coordinates of the free control points on the nose and tail Bézier curves, respectively, i.e., points 3 and 8 in Fig. 6.3. The lower and upper bounds of design variables are $\boldsymbol{l} = [0\,0\,0\,80\,0]^T$ cm and $\boldsymbol{u} = [30\,30\,10\,100\,10]^T$ cm, respectively. Other geometrical shape parameters are, for both cases, $L = 100$ cm, $d = 20$ cm, and $b = 50$ cm. The flow speed is 2 m/s and the Reynolds number is two million.

Numerical results for a direct design case are presented in Table 6.2. The hull drag coefficient is minimized by finding the appropriate shape and length of the nose and tail sections for a given hull length, diameter, and cylindrical section length. In this case, the drag coefficient is reduced by 6.3 %. This drag reduction comes from a reduction in skin friction and a lower pressure peak where the nose and tail connect with the midsection (Fig. 6.5).

These changes are due to a more streamlined nose (longer by 6 cm) and a fuller tail, when compared to the initial design. SBO requires 3 high-fidelity and 300 - low-fidelity model evaluations. The ratio of the high-fidelity model evaluation time to the corrected low-fidelity model evaluation time varies between 11 and 45, depending on whether the flow solver converges to the residual limit of 10^{-6},

Table 6.2 Numerical results for design Cases 1 and 2[a]

	Case 1 (drag minimization)			Case 2 (inverse design)		
Variable	Initial	Pattern-search	SBO	Initial	Pattern-search	SBO
a	15.0000	21.8611	20.9945	18.000	24.7407	24.7667
x_n	5.0000	5.6758	5.6676	7.0000	7.3704	6.8333
y_n	5.0000	2.7022	2.7531	8.0000	4.7407	4.5667
x_t	90.0000	98.000	96.6701	85.0000	88.1111	88.6333
y_t	5.0000	0.8214	3.0290	7.0000	5.5926	5.3000
F	–	–	–	0.0204	1.64×10^{-5}	1.93×10^{-5}
C_D	0.0915	0.0853	0.0857	0.0925	0.0894	0.0893
N_c	–	0	300	–	0	500
N_f	–	282	3	–	401	5
Total cost	–	282	13	–	401	<22

[a]Design goals: axisymmetric hull direct drag minimization (Case 1) and inverse design with a target pressure distribution (Case 2). In both cases, the flow speed is 2 m/s and the Reynolds number is 2×10^6. All the numerical values are from the high-fidelity model. N_c and N_f are the number of low- and high-fidelity model evaluations, respectively. F is the norm of the difference between the target and the design shapes

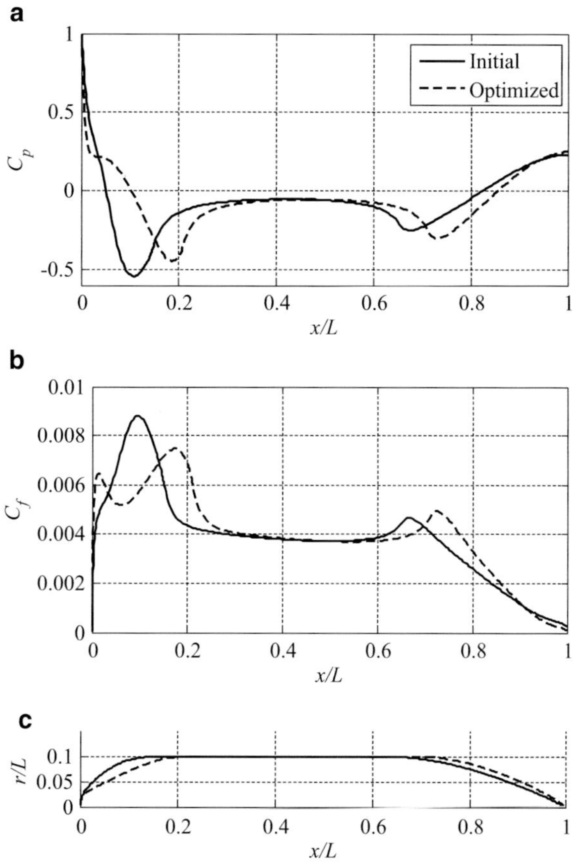

Fig. 6.5 Case 1 results for the direct hull drag minimization, showing the initial and optimized: (**a**) pressure distributions, (**b**) skin friction distributions and (**c**) hull shapes

or the maximum iteration limit of 1,000. We express the total optimization cost of the presented method in the equivalent number of high-fidelity model evaluations. For the sake of simplicity, we use a fixed value of 30 as the high- to low-fidelity model evaluation time ratio. The results show that the total optimization cost of the presented approach is around 13 equivalent high-fidelity model evaluations. The direct optimization method yields very similar design, but at a substantially higher computational cost of 282 high-fidelity model evaluations.

Inverse design of the hull shape was performed by prescribing a target pressure distribution. The objective is to minimize the norm of the difference between the pressure distribution of the hull design and the target pressure distribution. The numerical results of the inverse design are presented in Table 6.2, see also Fig. 6.6. The SBO algorithm matched the target pressure distribution (the norm of the distributions is less than 2×10^{-5}) using less than 22 equivalent high-fidelity model evaluations. The direct optimization of the high-fidelity model using the pattern-search algorithm required 401 function calls to yield a comparable matching with the target.

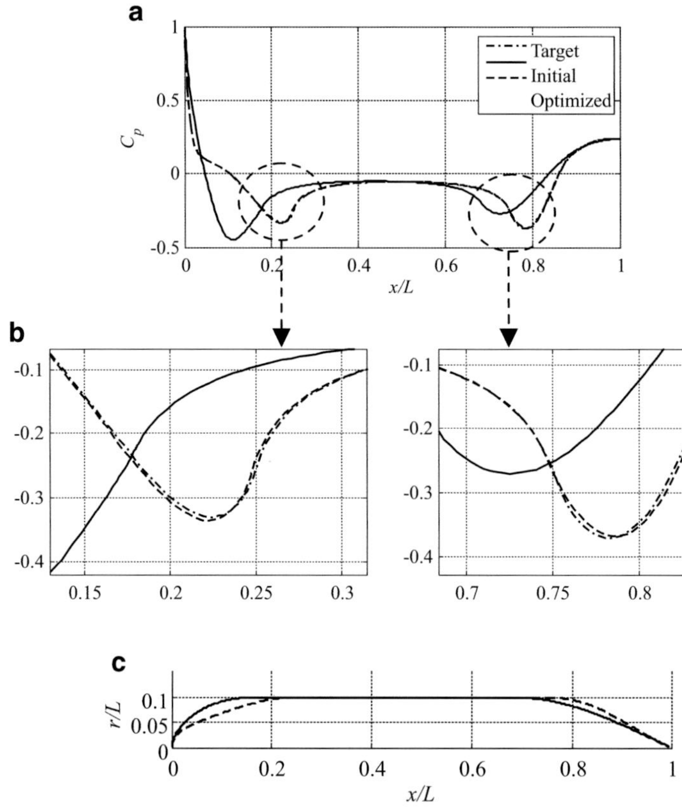

Fig. 6.6 Case 2 results of the inverse design optimization with a prescribed target pressure distribution: (**a**) target, initial, and optimized pressure distributions, (**b**) zoom on two regions of (**a**), and (**c**) the initial and optimized hull shapes

6.2 Multi-point Space Mapping

Multi-point space mapping (Koziel and Leifsson 2012b) is a generalization of the OSM concept in which the parameters of the low-fidelity model correction are determined using the high-fidelity model data at several points (designs). Consequently, the multi-point SM results in better generalization capability of the surrogate at the expense of more complex formulation of the model.

6.2.1 Surrogate Model Construction

The multi-point SM surrogate is formulated as follows (Koziel and Leifsson 2012b)

$$s^{(i)}(x) = \Lambda^{(i)} \circ c(x) + \Delta_r^{(i)} + \delta^{(i)}, \tag{6.8}$$

where $\Lambda^{(i)}$, $\Delta_r^{(i)}$, and $\delta^{(i)}$ are column vectors, whereas $\circ$ denotes component-wise multiplication. The global response correction parameters $\Lambda^{(i)}$ and $\Delta_r^{(i)}$ are obtained as

$$\left[\Lambda^{(i)}, \Delta_r^{(i)}\right] = \arg \min_{[\Lambda, \Delta_r]} \sum\nolimits_{k=0}^{i} \left|\left|f\left(x^{(k)}\right) - \Lambda \circ c\left(x^{(k)}\right) + \Delta_r\right|\right|^2 \tag{6.9}$$

i.e., the response scaling is supposed to improve the matching for all previous iteration points. The (local) additive response correction term $\delta^{(i)}$ is then defined as

$$\delta^{(i)} = f\left(x^{(i)}\right) - \left[\Lambda^{(i)} \circ c\left(x^{(i)}\right) + \Delta_r^{(i)}\right], \tag{6.10}$$

i.e., it ensures a perfect match between the surrogate and the high-fidelity model at the current design $x^{(i)}$, $s^{(i)}(x^{(i)}) = f(x^{(i)})$. It should be noted that all of the correction terms $\Lambda^{(i)}$, $\Delta_r^{(i)}$ and $\delta^{(i)}$ can be obtained analytically. Let $\Lambda^{(i)} = [\lambda_1^{(i)} \lambda_2^{(i)} \ldots \lambda_m^{(i)}]^T$, $\Delta_r^{(i)} = [\delta_1^{(i)} \delta_2^{(i)} \ldots \delta_m^{(i)}]^T$, $c(x) = [c_1(x)\ c_2(x) \ldots c_m(x)]^T$, and $f(x) = [f_1(x)\ f_2(x) \ldots f_m(x)]^T$. We have

$$\begin{bmatrix} \lambda_k^{(i)} \\ \delta_k^{(i)} \end{bmatrix} = \left(C_k^T C_k\right)^{-1} C_k^T F_k, \tag{6.11}$$

where

$$C_k = \begin{bmatrix} c_k\left(x^{(0)}\right) & \cdots & c_k\left(x^{(i)}\right) \\ 1 & \cdots & 1 \end{bmatrix}^T \tag{6.12}$$

$$F_k = \left[f_k\left(x^{(0)}\right) \quad \ldots \quad f_k\left(x^{(i)}\right)\right]^T \tag{6.13}$$

for $k = 1, \ldots, m$, which is a least-square optimal solution to the linear regression problems $c_k \lambda_k^{(i)} + \delta_k^{(i)} = f_k$, $k = 1, \ldots, m$, equivalent to (6.9). Note that the matrices $C_k^T C_k$ are non-singular for $i > 1$. For $i = 1$ only the multiplicative correction with $\Lambda^{(i)}$ components are used (calculated in a similar way).

6.2.2 Multi-point SM for Antenna Optimization

Consider a dielectric resonator antenna (DRA) shown in Fig. 6.7 (Koziel and Ogurtsov 2011c). The DRA is suspended above the ground plane in order to achieve enhanced impedance bandwidth. The dielectric resonator is placed on the Teflon slabs as shown in Fig. 6.7b–c, bringing additional degrees of freedom into the design. Additionally, we consider polycarbonate housing with relative permittivity of 2.8 and loss tangent of 0.01. The fixed dimensions of the housing are: $d_x = d_y = d_z = 1$ mm, $d_{zb} = 2$ mm, $b_x = 2$ mm, and $c_x = 6.5$ mm. The design specifications imposed on the reflection coefficient of the antenna are $|S_{11}| \leq -15$ dB for 5.1–5.9 GHz. The design variables are $\boldsymbol{x} = [a_x\, a_y\, a_z\, a_c\, u_s\, w_s\, y_s\, g_1\, b_y]^T$. The high-fidelity model $\boldsymbol{f}$ is simulated using the CST MWS transient solver (CST 2013) (~800,000 mesh cells, evaluation time 20 min). The low-fidelity model $\boldsymbol{c}$ is also evaluated in CST (~30,000 mesh cells, evaluation time 40 s). The initial design is $\boldsymbol{x}^{(0)} = [8.0\ 14.0\ 8.0\ 0.0\ 2.0\ 8.0\ 4.0\ 2.0\ 6.0]^T$ mm.

The optimization process of the DRA has been conducted twice using the SBO process (4.1) working with (1) the output SM surrogate (6.2), and (2) the multi-point SM model (6.7). The algorithm working with the OSM surrogate yields a design with $|S_{11}| \leq -15.6$ dB for 5.1–5.9 GHz in six iterations (total design cost ~17 evaluations of $\boldsymbol{f}$, cf. Table 6.3), whereas the algorithm working with multi-point response correction yields a design $\boldsymbol{x}^* = [7.7\ 13.6\ 8.8\ 0.0\ 1.75\ 10.07\ 2.8\ 1.53\ 5.9]^T$ mm in three iterations ($|S_{11}| \leq -15.8$ dB for 5.1–5.9 GHz, design cost ~9 $\times \boldsymbol{f}$), see Fig. 6.8. Reduction of the design cost is mostly due to the fact that the multi-point surrogate has better generalization capability as illustrated in Fig. 6.9.

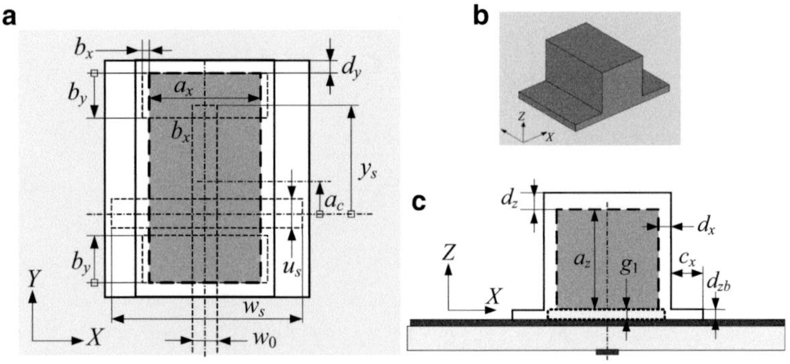

Fig. 6.7 Suspended DRA (Koziel and Ogurtsov 2011c): (**a**) 3D view of its housing, top (**b**) and front (**c**) views

Table 6.3 Suspended DRA: design cost

Surrogate model	Algorithm component	Number of model evaluations	CPU time	
			Absolute (min)	Relative to $\boldsymbol{f}$
Output space mapping (6.2)	Optimizing $\boldsymbol{s}$	$315 \times \boldsymbol{c}$	210	10.5
	Evaluating $\boldsymbol{f}$[a]	$6 \times \boldsymbol{f}$	120	6.0
	Total cost	N/A	330	16.5
Multi-point correction (6.7)	Optimizing $\boldsymbol{s}$	$162 \times \boldsymbol{c}$	108	5.4
	Evaluating $\boldsymbol{f}^*$	$3 \times \boldsymbol{f}$	60	3.0
	Total cost	N/A	168	8.4

[a]Excludes high-fidelity model evaluation at the initial design

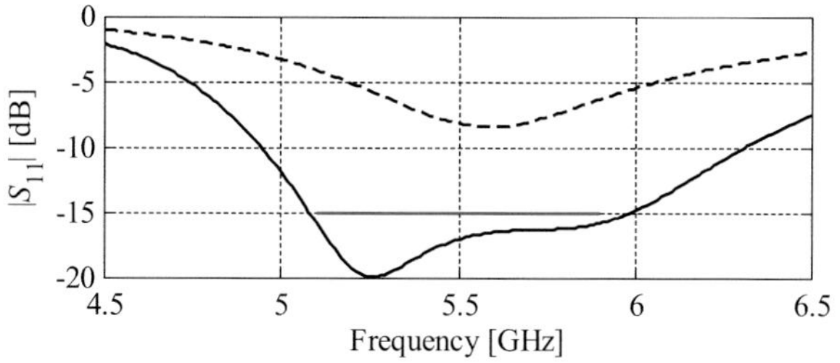

Fig. 6.8 Suspended DRA: high-fidelity model response at the initial design (*dashed line*) and at the optimized design obtained using multi-point response correction (*solid line*)

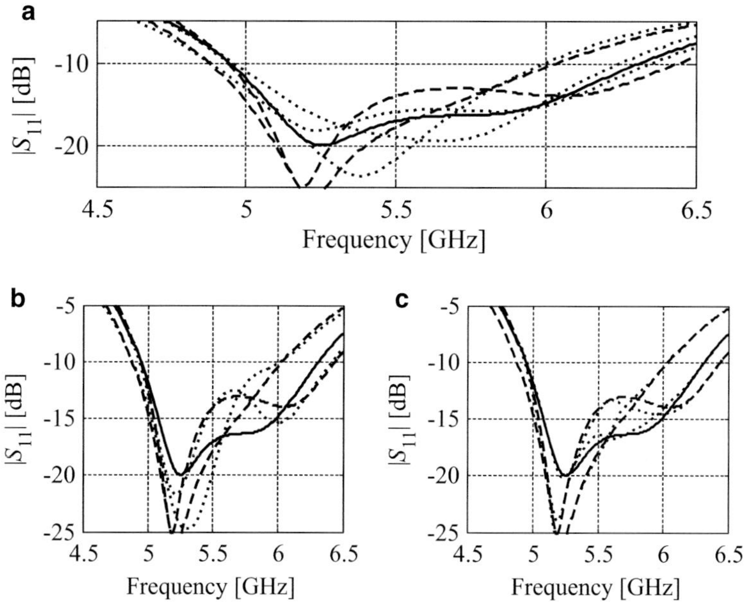

Fig. 6.9 Suspended DRA: (**a**) low- (*dotted line*) and high- (*dashed line* and *solid line*) fidelity model responses at three designs, (**b**) OSM-corrected low- (*dotted line*) and high- (*dashed line* and *solid line*) fidelity model responses at the same designs (OSM correction at the design marked *solid line*), (**c**) multi-point-corrected low- (*dotted line*) and high- (*dashed line* and *solid line*) fidelity model responses

The misalignment between $\boldsymbol{c}$ and $\boldsymbol{f}$ shown in Fig. 6.9a is perfectly removed by OSM at the design where it is established, but the alignment is not as good for other designs. On the other hand, as shown in Fig. 6.9c, the multi-point response correction improves model alignment at all designs involved in the model construction (Fig. 6.9c only shows the global part $\boldsymbol{\Lambda}^{(i)} \circ \boldsymbol{c}(\boldsymbol{x}) + \boldsymbol{\Delta}_r^{(i)}$ without $\boldsymbol{\delta}^{(i)}$ which would give a perfect alignment at $\boldsymbol{x}^{(i)}$).

6.2.3 Multi-point SM for Transonic Airfoil Optimization

In this section, we illustrate the use of multi-point OSM for aerodynamic shape optimization of transonic airfoils. The objective is to minimize the drag coefficient, C_d, given the constraints on the lift coefficient, C_l, and/or the cross-sectional area of the airfoil, A. Alternatively, the objective might be maximization of the lift coefficient with certain constraints imposed for both the drag coefficient and the cross-sectional area. The vector of design variables, $\boldsymbol{x}$, represents the airfoil geometry parameterization. Both the high- and low-fidelity models are computational fluid dynamics models evaluated using the computer code FLUENT (2015), see also Fig. 6.10. The average simulation time of the high-fidelity model is 90 min with about 400,000 mesh cells. The average time evaluation ratio between the high- and low-fidelity model is 80 min (~30,000 mesh cells used for the latter). The high-fidelity airfoil model response represents the lift coefficient $C_{l.f}$, the drag coefficient $C_{d.f}$, and the cross-sectional area A, so that we have $\boldsymbol{f}(\boldsymbol{x}) = [C_{l.f}(\boldsymbol{x}) C_{d.f}(\boldsymbol{x}) A_f(\boldsymbol{x})]^T$ (similarly for $\boldsymbol{c}(\boldsymbol{x})$). The design constraints are denoted $C(f(\boldsymbol{x})) = [c_1(f(\boldsymbol{x})) \ldots c_k(f(\boldsymbol{x}))]$. For lift maximization one may have has two nonlinear inequality constraints for drag and area, $c_1(f(\boldsymbol{x})) \le 0$ with $c_1(f(\boldsymbol{x})) = C_{d.f}(\boldsymbol{x}) - C_{d.\max}$, and $c_2(f(\boldsymbol{x})) \le 0$ with $c_2(f(\boldsymbol{x})) = -A_f(\boldsymbol{x}) + A_{\min}$.

The surrogate model is constructed using (6.8)–(6.13). Two cases are considered: (1) lift maximization (Case 1), and (2) drag minimization (Case 2). The airfoil shape is parameterized using a NACA four-digit airfoil with m (the maximum ordinate of the mean camberline as a fraction of chord c), p (the chordwise position of the maximum ordinate), and t/c (the thickness-to-chord ratio) as the design variables (Abbott and Von Doenhoff 1959). The bounds on the design variables are $0 \le m \le 0.1$, $0.2 \le p \le 0.8$, and $0.05 \le t/c \le 0.2$. Details of the test cases and optimization results are given in Table 6.4. The results are compared with the direct design optimization of the high-fidelity model using the pattern-search algorithm.

In Case 1, the initial design is NACA 2412 airfoil and it is feasible for the assumed constraints. The multi-point SM is able to obtain a better design than the

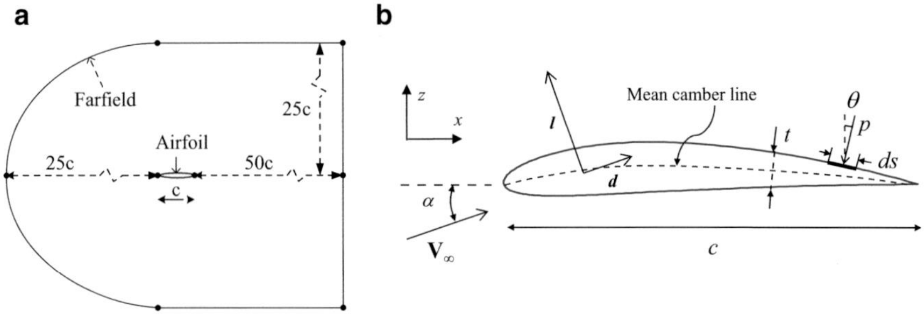

Fig. 6.10 Transonic airfoil simulation and design: (**a**) a sketch of the computational domain; (**b**) NACA 2412 airfoil section in the x-z plane with the free-stream of velocity V_∞ at an angle of attack α. Section lift is denoted by l and section drag is denoted by d. The pressure p acts normal to a surface panel element which is of length ds and at an angle θ relative to the x-axis

Table 6.4 Transonic airfoil shape optimization results (N_c and N_f are the number of low- and high-fidelity model evaluations, respectively)

	Case 1 (Lift maximization)			Case 2 (Drag minimization)		
	$M_\infty = 0.70$, $\alpha = 1°$, $C_{d,\max} = 0.006$, $A_{\min} = 0.075$			$M_\infty = 0.70$, $\alpha = 1°$, $C_{l,\min} = 0.6$, $A_{\min} = 0.075$		
Variable	Initial	Direct[a]	Multi-point SM[b]	Initial	Direct[a]	Multi-point SM[b]
m	0.02	0.02	0.0237	0.02	0.018	0.018
p	0.4	0.7467	0.6531	0.4	0.5207	0.529
t/c	0.12	0.12	0.1148	0.12	0.1141	0.1113
C_l	0.5963	0.8499	0.8909	0.5963	0.6001	0.6002
C_{dw}	0.0047	0.006	0.006	0.0047	0.0019	0.0017
A	0.0808	0.0808	0.0773	0.0808	0.0768	0.0749
N_c	–	0	260	–	0	160
N_f	–	59	5	–	110	3
Total cost[c]	–	59	<9	–	110	5

[a]Design obtained through direct optimization of the high-fidelity CFD model using the pattern-search algorithm (Koziel et al. 2010c)

[b]Design obtained using the multi-point SM algorithm; surrogate model optimization performed using pattern-search (Koziel et al. 2010c)

[c]The total optimization cost is expressed in the equivalent number of high-fidelity model evaluations

direct method. The effects of the geometry change on the pressure distribution can be observed in Fig. 6.11a: the shock strength is reduced by reducing the thickness, and the aft camber location opens up the pressure distribution behind the shock to increase the lift. The overall optimization cost corresponds to less than 9 equivalent high-fidelity model evaluations. In the drag minimization case (Case 2), the initial design is NACA 2412 and the lift constraint is slightly violated. Similar optimized designs are obtained by the direct method and the proposed method. The camber is reduced, maximum camber moved aft, and the thickness reduced. As a result, the shock is weakened, and the lift improved by opening the pressure distribution behind the shock (Fig. 6.11c). The optimization cost is 5 equivalent high-fidelity model evaluations (160 surrogate and 3 high-fidelity), whereas the direct required 110 evaluations of $\boldsymbol{f}$.

6.3 Manifold Mapping

Manifold mapping (MM) (Echeverria and Hemker 2005; Echeverría and Hemker 2008) is an interesting response correction technique that is capable of comprehensive exploitation of available high-fidelity model data. Here, we discuss the basic version of MM described in Echeverria and Hemker (2005).

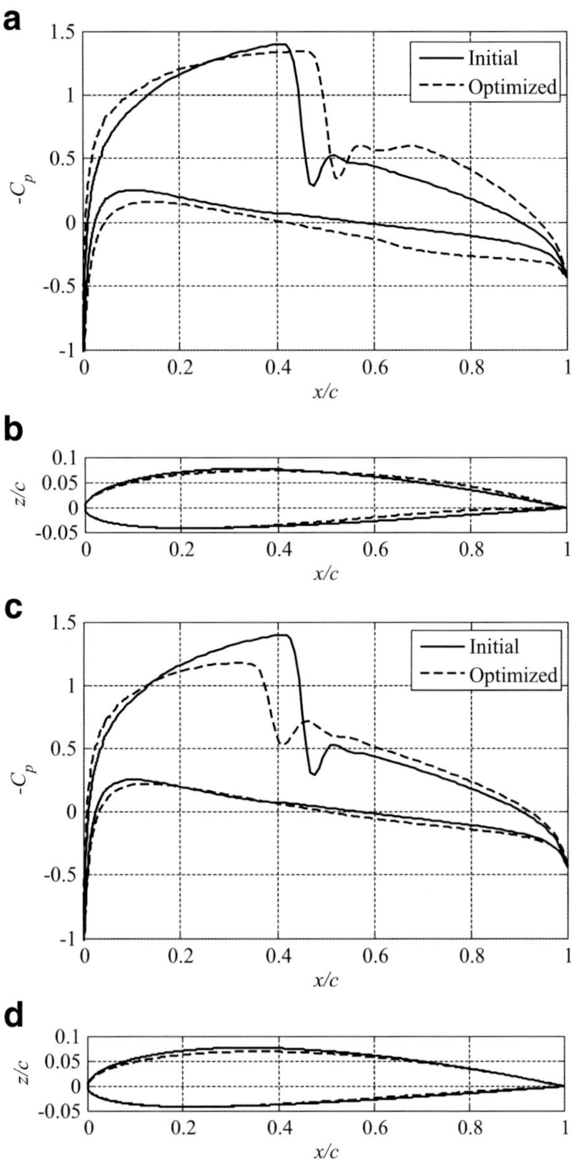

Fig. 6.11 (**a**) Pressure distribution of initial (*solid*) and optimized (*dashed*) airfoils of Case 1, (**b**) initial (*solid*) and optimized (*dashed*) airfoil shapes of Case 1, (**c**) pressure distribution of initial (*solid*) and optimized (*dashed*) airfoils of Case 2, (**d**) initial (*solid*) and optimized (*dashed*) airfoil shapes of Case 2

6.3.1 Surrogate Model Construction

The MM surrogate model is defined as (Echeverria and Hemker 2005)

$$\boldsymbol{s}^{(i)}(\boldsymbol{x}) = \boldsymbol{f}\left(\boldsymbol{x}^{(i)}\right) + \boldsymbol{S}^{(i)}\left(\boldsymbol{c}(\boldsymbol{x}) - \boldsymbol{c}\left(\boldsymbol{x}^{(i)}\right)\right) \tag{6.14}$$

with $\boldsymbol{S}^{(i)}$, being the $m \times m$ correction matrix, defined as

$$S^{(i)} = \Delta F \cdot \Delta C^{\dagger}, \tag{6.15}$$

where

$$\Delta F = \left[f\left(x^{(i)}\right) - f\left(x^{(i-1)}\right) \quad \ldots \quad f\left(x^{(i)}\right) - f\left(x^{(\max\{i-n,0\})}\right) \right], \tag{6.16}$$

$$\Delta C = \left[c\left(x^{(i)}\right) - c\left(x^{(i-1)}\right) \quad \ldots \quad c\left(x^{(i)}\right) - c\left(x^{(\max\{i-n,0\})}\right) \right]. \tag{6.17}$$

Here, the pseudoinverse, denoted by †, is defined as

$$\Delta C^{\dagger} = V_{\Delta C} \Sigma_{\Delta C}^{\dagger} U_{\Delta C}^{T}, \tag{6.18}$$

where $U_{\Delta C}$, $\sum_{\Delta C}$, and $V_{\Delta C}$ are the factors in the singular value decomposition of the matrix ΔC. The matrix $\Sigma_{\Delta C}{}^{\dagger}$ is the result of inverting the nonzero entries in $\sum_{\Delta C}$, leaving the zeroes invariant. Figure 6.12 shows the effect of applying the mapping (6.14) to the low-fidelity model. Upon convergence, the linear correction S^* (being the limit of $S^{(i)}$ with $i \to \infty$) maps the point $c(x^*)$ to $f(x^*)$, and the tangent plane for $c(x)$ at $c(x^*)$ to the tangent plane for $f(x)$ at $f(x^*)$ (Echeverría and Hemker 2008).

It should be noted that although MM does not explicitly use sensitivity information, the surrogate and the high-fidelity model Jacobians become more and more similar to each other towards the end of the MM optimization process (i.e., when $\|x^{(i)} - x^{(i-1)}\| \to 0$) so that the surrogate (approximately) satisfies both zero- and first-order consistency conditions (Alexandrov and Lewis 2001) with f. This allows for a more precise identification of the high-fidelity model optimum. On the other hand, the correction matrix $S^{(i)}$ can be defined using exact Jacobians of the low- and high-fidelity models if available.

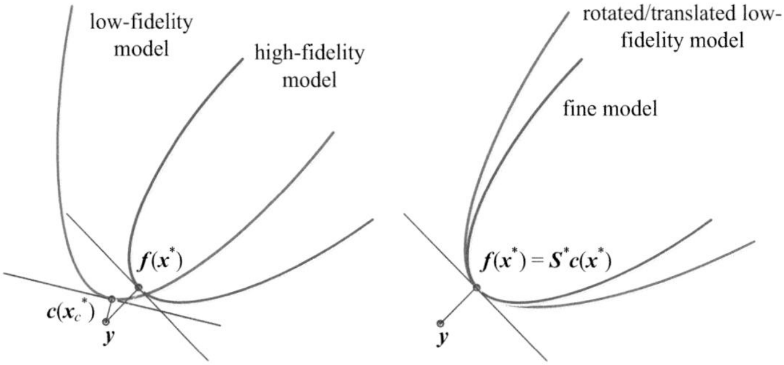

Fig. 6.12 The concept of the MM model alignment for a least-squares optimization problem: x_c^* is the low-fidelity model minimizer, and y is the vector of design specifications. The *straight lines* denote the tangent planes for f and c at their optimal designs, respectively. By the linear correction S^*, the point $c(x^*)$ is mapped to $f(x^*)$, and the tangent plane for $c(x)$ at $c(x^*)$ to the tangent plane for $f(x)$ at $f(x^*)$ (Koziel et al. 2011)

6.3.2 Manifold Mapping Optimization of UWB Monopole Antenna

Consider the UWB monopole shown in Fig. 6.13. Design variables are $\boldsymbol{x} = [h_0\, w_0\, a_0\, s_0\, h_1\, w_1\, l_{gnd}\, w_s]^T$. Other parameters are fixed: $l_s = 25$, $w_m = 1.25$, $h_p = 0.75$ (all in mm). The microstrip input of the monopole is fed through an edge mount SMA connector (SMA 2013). Simulation time of the low-fidelity model $\boldsymbol{c}$ (~150,000 mesh cells) is 2 min, and that of the high-fidelity model $\boldsymbol{f}$ (~1,200,000 mesh cells) is 45 min (both at the initial design). Both models are evaluated using the transient solver of CST Microwave Studio (CST 2013). The design specifications for antenna reflection are $|S_{11}| \leq -10$ dB for 3.1–10.6 GHz.

The initial design is $\boldsymbol{x}^{(0)} = [18\ 12\ 2\ 0\ 5\ 1\ 15\ 40]^T$ mm. Because the low-fidelity model is relatively expensive, the MM algorithm is using the underlying kriging interpolation model $\boldsymbol{c}_{kr}$ created in the vicinity of the approximate optimum of $\boldsymbol{c}$ (obtained at the cost of 100 $\boldsymbol{c}$ evaluations) using 100 $\boldsymbol{c}$ samples. Optimization performed using the MM algorithm yields the final design $\boldsymbol{x}^* = [19.13\ 20.13\ 1.95\ 1.33\ 1.79\ 6.32\ 15.03\ 36.36]^T$ mm ($|S_{11}| < -15$ dB in the frequency band of interest). The total design cost is about 21 high-fidelity model evaluations (Table 6.5). Figure 6.14a shows reflection responses of the high- and low-fidelity models at the initial design as well as the high-fidelity model response at the final design. The convergence plot for the MM algorithm is shown in Fig. 6.14b.

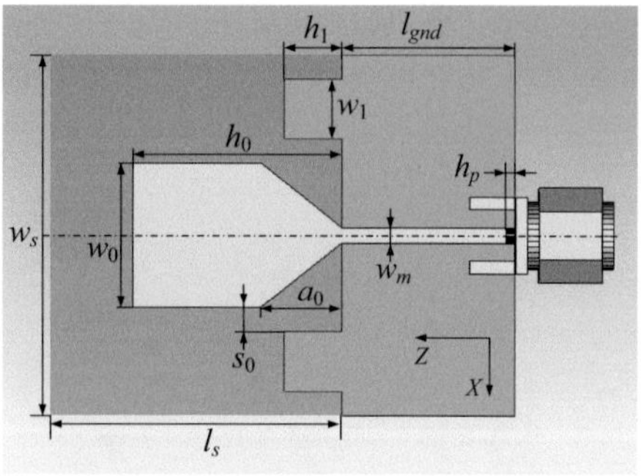

Fig. 6.13 UWB monopole: top view, substrate shown transparent. Magnetic-symmetry wall is shown with the *dash-dot line*

Table 6.5 UWB monopole: design cost

		CPU time	
Algorithm component	Number of model evaluations	Absolute (min)	Relative to $\boldsymbol{f}$
Evaluation of $\boldsymbol{c}$	$200 \times \boldsymbol{c}$	400	8.8
Evaluation of $\boldsymbol{f}$	$12 \times \boldsymbol{f}$	540	12.0
Total cost	N/A	940	20.8

6.3.3 Manifold Mapping Optimization of Microstrip Filter

As an illustration example, consider the miniature dual-mode bandpass filter (Lin et al. 2007) shown in Fig. 6.15a. The design parameters are $\boldsymbol{x} = [L\ s\ p\ g]^T$; $W = 1$ mm, $W_c = 0.5$ mm. The high-fidelity model is simulated in FEKO (FEKO 2008). The coarse model is the circuit model implemented in Agilent ADS (Agilent 2011) (Fig. 6.15b). The design specifications imposed on the transmission coefficient of the filter are $|S_{21}| \geq -1$ dB for 2.35 GHz $\leq \omega \leq$ 2.45 GHz, $|S_{21}| \leq -20$ dB for 1.6 GHz $\leq \omega \leq$ 2.2 GHz and for 2.6 GHz $\leq \omega \leq$ 3.2 GHz. The low-fidelity model was initially corrected by tuning the substrate height and dielectric constants corresponding to the microstrip models of the equivalent circuit. The MM algorithm yields an optimized design $\boldsymbol{x}^* = [13.143\ 0.792\ 1.466\ 0.1285]^T$ with the

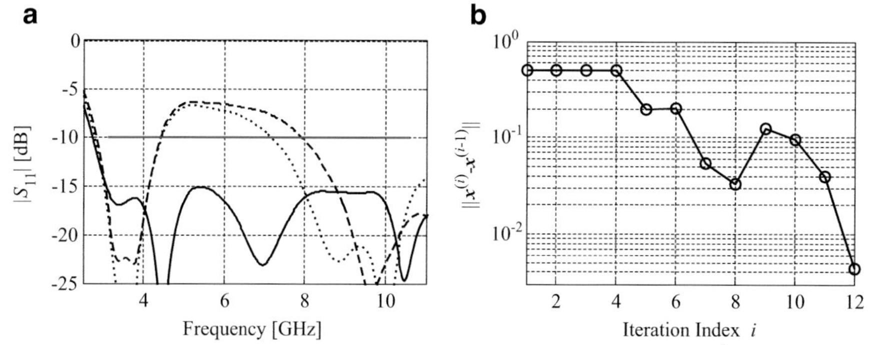

Fig. 6.14 UWB monopole: (**a**) high- (*dashed line*) and low-fidelity (*dotted line*) model responses at the initial design, as well as high-fidelity model (*solid line*) at the final design $\boldsymbol{x}^*$; (**b**) convergence plot of the MM algorithm

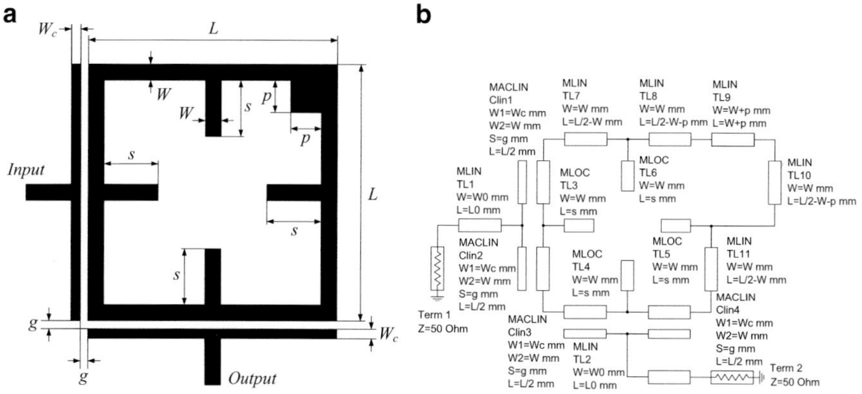

Fig. 6.15 Miniature dual-mode bandpass filter: (**a**) geometry (Lin et al. 2007); (**b**) low-fidelity model (Agilent ADS)

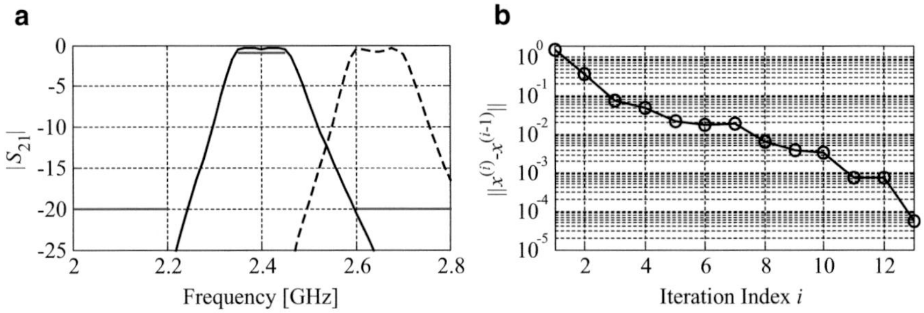

Fig. 6.16 Miniature dual-mode bandpass filter: (**a**) initial (*dashed line*) and optimized (*solid line*) $|S_{21}|$ versus frequency; optimized design obtained using MM algorithm; (**b**) convergence plot

corresponding specification error of −0.45 dB (Fig. 6.16a). The optimization cost is 13 fine model evaluations (termination condition: $||\boldsymbol{x}^{(i)} - \boldsymbol{x}^{(i-1)}|| < 10^{-4}$). The convergence plot is shown in Fig. 6.16b.

6.4 Multi-point Response Correction

The last technique discussed in this chapter is multi-point response correction (Koziel 2010d), which is yet another way of generalizing output SM in order to exploit as much of available high-fidelity model data as possible. The surrogate model is formulated as

$$\mathbf{s}^{(i)}(\mathbf{x}) = \mathbf{c}(\mathbf{x}) + \boldsymbol{\Delta}_g^{(i)}(\mathbf{x}) \tag{6.19}$$

with the design-variable-dependent correction term $\boldsymbol{\Delta}_g^{(i)}$ defined to have the following features:

- $\boldsymbol{\Delta}_g^{(i)}(\boldsymbol{x}^{(i)}) = \boldsymbol{f}(\boldsymbol{x}^{(i)}) - \boldsymbol{c}(\boldsymbol{x}^{(i)})$, i.e., $\boldsymbol{s}^{(i)}(\boldsymbol{x}^{(i)}) = \boldsymbol{f}(\boldsymbol{x}^{(i)})$
- $\boldsymbol{s}^{(i)}(\boldsymbol{x}^{(k)}) = \boldsymbol{f}(\boldsymbol{x}^{(k)})$, i.e., $\boldsymbol{\Delta}_g^{(i)}(\boldsymbol{x}^{(k)}) = \boldsymbol{f}(\boldsymbol{x}^{(k)}) - \boldsymbol{c}(\boldsymbol{x}^{(k)})$ for as many $k < i$ as possible
- $\boldsymbol{\Delta}_g^{(i)}$ is a linear function of $\boldsymbol{x}$ (so that it does not degrade the ability of $\boldsymbol{c}$ to model higher order nonlinearities of $\boldsymbol{f}$)

Let $S^{(i)}$ be a set of all affinely independent subsets of $X^{(i)} = \{\boldsymbol{x}^{(1)}, \boldsymbol{x}^{(2)}, \ldots, \boldsymbol{x}^{(i)}\}$ containing $\boldsymbol{x}^{(i)}$. Let $s^{(i)} \in S^{(i)}$ be such that: (1) $s^{(i)}$ is maximal, i.e., $|s^{(i)}| = n_i = \max\{|s| : s \in S^{(i)}\}$, and (2) $s^{(i)}$ is the smallest in the sense of its diameter, i.e., $d(s^{(i)}) = \min\{d(s) : s \in S^{(i)} \wedge |s| = n_i\}$, where $d(s) \equiv \max\{||\boldsymbol{x} - \boldsymbol{y}|| : \boldsymbol{x}, \boldsymbol{y} \in s\}$.

We have $|s^{(i)}| \le \min(i, n+1)$. Let $\boldsymbol{x}^k$, $k = 1, 2, \ldots, n_i$, denote the elements of $s^{(i)}$ (here, $\boldsymbol{x}^1 = \boldsymbol{x}^{(i)}$). Define $\boldsymbol{r}_k = \boldsymbol{f}(\boldsymbol{x}^k) - \boldsymbol{c}(\boldsymbol{x}^k)$, $k = 1, 2, \ldots, n_i$. Let us also define the $m \times n_i$ matrix $\mathbf{R}^{(i)} = [r_1 r_2 \ldots r_{n_i}]$. Let $M^{(i)} \subseteq R^n$ be a linear manifold generated by $s^{(i)}$, i.e.,

$$M^{(i)} = \left\{ x \in R^n \ : \ x = \sum\nolimits_{k=1}^{n_i} \alpha_k x^k, \ \sum\nolimits_{k=1}^{n_i} \alpha_k = 1 \right\}. \tag{6.20}$$

$M^{(i)}$ is well defined because $s^{(i)}$ is affinely independent. Let $B^{(i)} = \{\mathbf{b}_1, \ldots, \mathbf{b}_{n_i-1}\}$, where $\boldsymbol{b}_k = \boldsymbol{x}^{k+1} - \boldsymbol{x}^1$, $k = 1, 2, \ldots, n_i - 1$. Let $Y^{(i)} \subseteq R^n$ be a linear subspace generated by $B^{(i)}$, i.e.,

$$Y^{(i)} = \left\{ x \in R^n \ : \ x = \sum\nolimits_{k=1}^{n_i-1} \beta_k \mathbf{b}_k, \ \beta_k \in R, \ k = 1, 2, \ldots, n_i - 1 \right\} \tag{6.21}$$

Clearly, $Y^{(i)} = M^{(i)} - \boldsymbol{x}^{(i)}$. For any $\boldsymbol{x} \in R^n$ define $P_i(\boldsymbol{x})$ as the orthogonal projection of $\boldsymbol{x} - \boldsymbol{x}^{(i)}$ onto $Y^{(i)}$. For any $\boldsymbol{y} \in M^{(i)}$ define $\boldsymbol{\lambda}_i(\boldsymbol{y}) = [\lambda_1 \lambda_2 \ldots \lambda_{n_i}]^T$ so that $\boldsymbol{y} = \lambda_1 \boldsymbol{x}^1 + \lambda_2 \boldsymbol{x}^2 + \ldots + \lambda_{n_i} \boldsymbol{x}^{n_i}$. $\boldsymbol{\lambda}_i$ is uniquely determined and $\lambda_1 + \lambda_2 + \ldots + \lambda_{n_i} = 1$.

We can now define $\boldsymbol{\Delta}_g^{(i)}(\boldsymbol{x})$, $\boldsymbol{x} \in R^n$, as

$$\boldsymbol{\Delta}_g^{(i)}(\boldsymbol{x}) = \begin{cases} \boldsymbol{f}\left(\boldsymbol{x}^{(i)}\right) - \boldsymbol{c}\left(\boldsymbol{x}^{(i)}\right) & \text{for} \quad i = 1 \\ \boldsymbol{R}^{(i)} \cdot \boldsymbol{\lambda}_\mathbf{i}\left(P_i\left(\boldsymbol{x} - \boldsymbol{x}^{(i)}\right) + \boldsymbol{x}^{(i)}\right) & \text{for} \quad i > 1 \end{cases} \tag{6.22}$$

In (6.22), we distinguished two cases because some of the above objects (e.g., $Y^{(i)}$) are not defined for $i = 1$. Figure 6.17 illustrates the concepts used to define $\boldsymbol{\Delta}_g^{(i)}(\cdot)$. The vector $\boldsymbol{\lambda}_i(\boldsymbol{y})$ can be determined by solving a linear system

$$\boldsymbol{T}^{(i)} \boldsymbol{\lambda}_i = \boldsymbol{y}^{(i)}, \tag{6.23}$$

where $\boldsymbol{T}^{(i)}$ is $(n+1) \times n_i$ matrix defined as

$$\boldsymbol{T}^{(i)} = \begin{bmatrix} & \boldsymbol{B}^{(i)} & 0 \\ [1 & 1 \ \ldots \ 1]_{1 \times (n_i - 1)} & 1 \end{bmatrix} \tag{6.24}$$

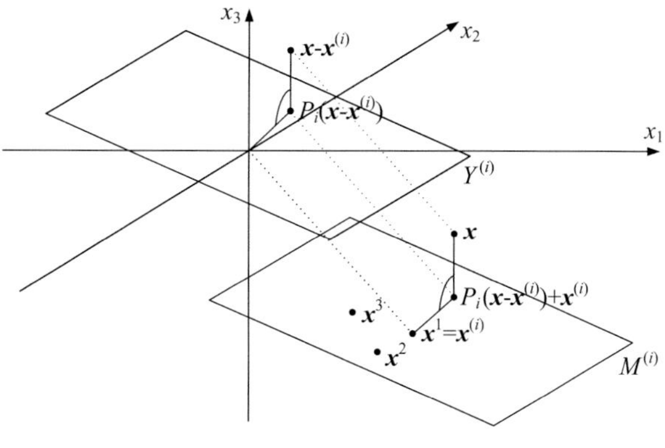

Fig. 6.17 Illustration of the objects used to define the multipoint response correction model $\boldsymbol{\Delta}_g^{(i)}(\boldsymbol{x})$ for $n = 3$ (three design variables) (Koziel 2011a, b)

with $\boldsymbol{B}^{(i)}$ being the $n\times(n_i-1)$ matrix such that its kth column is a vector $\boldsymbol{b}_k$, and $\boldsymbol{y}^{(i)}$ is $(n+1)\times 1$ vector

$$\boldsymbol{y}^{(i)} = \begin{bmatrix} P_i\left(\boldsymbol{x}-\boldsymbol{x}^{(i)}\right)+\boldsymbol{x}^{(i)} \\ 1 \end{bmatrix}. \tag{6.25}$$

As vectors $\boldsymbol{b}_k$, $k=1, 2, \ldots, n_i-1$, are linearly independent, solution to (6.23) can be found as

$$\boldsymbol{\lambda}_i = \left(\boldsymbol{T}^{(i)T}\boldsymbol{T}^{(i)}\right)^{-1}\boldsymbol{T}^{(i)T}\boldsymbol{y}^{(i)}. \tag{6.26}$$

It can be shown that $\boldsymbol{\Delta}_g^{(i)}$ satisfies the properties listed at the beginning of the section (Koziel 2010d). Also, $\boldsymbol{\Delta}_g^{(i)}$ contains some information about the Jacobian of $\boldsymbol{f}(\boldsymbol{x}^{(i)})-\boldsymbol{c}(\boldsymbol{x}^{(i)})$. In particular, the Jacobian of $\boldsymbol{\Delta}_g^{(i)}$ is the approximation of the Jacobian reduced to the subspace $Y^{(i)}$: it can be shown, that if $d(s^{(i)})\rightarrow 0$, then $\boldsymbol{J}[\boldsymbol{\Delta}_g^{(i)}(\boldsymbol{x}^{(i)})]\rightarrow \boldsymbol{J}[(\boldsymbol{f}(\boldsymbol{x}^{(i)})-\boldsymbol{c}(\boldsymbol{x}^{(i)})]$ on $Y^{(i)}$. In particular, if $|s^{(i)}|=n+1$, then the Jacobian of $\boldsymbol{\Delta}_g^{(i)}$ becomes a good approximation of the Jacobian of $\boldsymbol{f}(\boldsymbol{x}^{(i)})-\boldsymbol{c}(\boldsymbol{x}^{(i)})$ on R^n, provided that $d(s^{(i)})$ is sufficiently small. See Koziel (2011a) for the discussion of some practical issues related to the conditioning of the problem (6.23). To improve the algorithm performance, a trust-region approach (Koziel 2011a) can be used as a convergence safeguard.

The multi-point response correction is illustrated using the second-order capacitively coupled dual-behavior resonator (CCDBR) microstrip filter (Manchec et al. 2006) shown in Fig. 6.18a. The design parameters are $\boldsymbol{x}=[L_1\, L_2\, L_3\, S]^T$. The high-fidelity model $\boldsymbol{f}$ is simulated in FEKO (FEKO 2008). The low-fidelity model $\boldsymbol{c}$ is the circuit model implemented in Agilent ADS (Agilent 2011), see Fig. 6.18b. The design specifications for the filter are $|S_{21}|\leq -20$ dB for 2.0 GHz $\leq\omega\leq$ 3.2 GHz, $|S_{21}|\geq -3$ dB for 3.8 GHz $\leq\omega\leq$ 4.2 GHz, and $|S_{21}|\leq$ -20 dB for 4.8 GHz $\leq\omega\leq$ 6.0 GHz. The initial design is the low-fidelity model optimal solution, $\boldsymbol{x}^{(0)}=[2.415\; 6.093\; 1.167\; 0.082]^T$ mm.

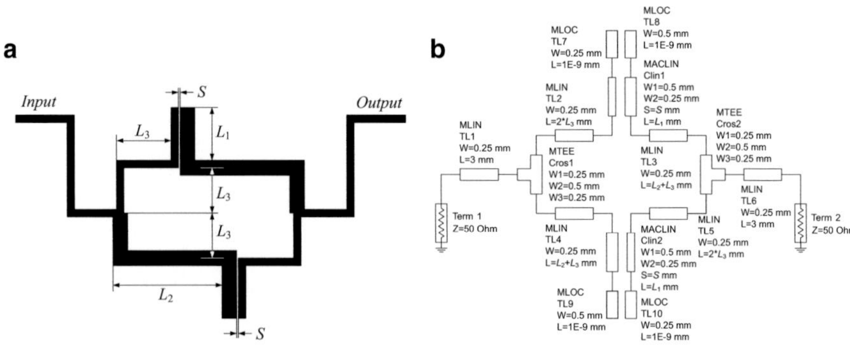

Fig. 6.18 CCDBR filter: (**a**) geometry (Manchec et al. 2006), (**b**) coarse model (Agilent ADS)

Table 6.6 CCDBR filter: optimization results

Surrogate model	SM algorithm	Specification error		Number of high-fidelity model evaluations
		Final (dB)	Best found (dB)	
$c(\boldsymbol{x}+\boldsymbol{q})$	SM_{STD}	−1.1	−1.9	20
	SM_{TR}	−2.3	−2.3	18
	SM_{MRC}	−2.1	−2.1	13
$\mathbf{c}_I(\mathbf{x})$	SM_{STD}	−1.2	−2.1	21[a]
	SM_{TR}	−2.0	−2.0	15
	SM_{MRC}	−2.1	−2.1	13
$c_F(\boldsymbol{x}+\boldsymbol{q})$	SM_{STD}	−0.9	−1.8	13
	SM_{TR}	−1.8	−1.8	11
	SM_{MRC}	−2.6	−2.6	14
$\mathbf{c}_{FI}(\mathbf{x})$	SM_{STD}	−0.5	−0.5	10
	SM_{TR}	−1.6	−1.6	16
	SM_{MRC}	−2.2	−2.2	17
$c_I(\boldsymbol{x}+\boldsymbol{q})$	SM_{STD}	+0.8	−0.4	8
	SM_{TR}	−1.6	−1.6	16
	SM_{MRC}	−2.5	−2.5	15

[a]Convergence not obtained; algorithm terminated after 20 iterations

The performance of the multi-point response correction is compared to the performance of the SM algorithm using the standard OSM (SM_{STD}) (cf. (6.2)), and SM enhanced by trust region approach (SM_{TR}). Because performance of SM algorithms generally depends on the selection of the underlying surrogate model (Koziel et al. 2008b), our test problem is solved using various combinations of SM transformations. This allows us to verify the robustness of a given algorithm implementation with respect to the surrogate model choice.

The optimization results are shown in Table 6.6. We consider five different types of SM surrogate models: (1) input SM model $\boldsymbol{c}(\boldsymbol{x}+\boldsymbol{q})$ (Bandler et al. 2004a), (2) implicit SM model $\boldsymbol{c}_I(\boldsymbol{x})$ in which $\boldsymbol{c}_I$ is the low-fidelity model with the substrate height and dielectric constants used as preassigned parameters (Bandler et al. 2004b), (3) the combination of input and frequency SM, $\boldsymbol{c}_F(\boldsymbol{x}+\boldsymbol{q})$ where $\boldsymbol{c}_F$ is the low-fidelity model evaluated at frequencies different from the original sweep according to the linear mapping $\omega \rightarrow f_1 + f_2\omega$ (f_1 and f_2 are obtained using the usual parameter extraction process), (4) the combination of implicit and frequency SM, and (5) the combination of input and implicit SM.

The standard SM algorithm converges for most surrogate model types considered. In all cases, the SM_{MRC} algorithm consistently yields better design (specification error better than −2 dB in all cases) than both SM_{STD} and SM_{TR}. This indicates that the multi-point correction may improve the SM algorithm ability to locate the high-fidelity model optimum when compared to the output SM. Figure 6.19a shows the initial fine model response and the optimized fine model response obtained using SM_{MRC} algorithm with the $\boldsymbol{c}_F(\boldsymbol{x}+\boldsymbol{q})$ model. Figure 6.19b, c show the convergence plot and specification error versus iteration index

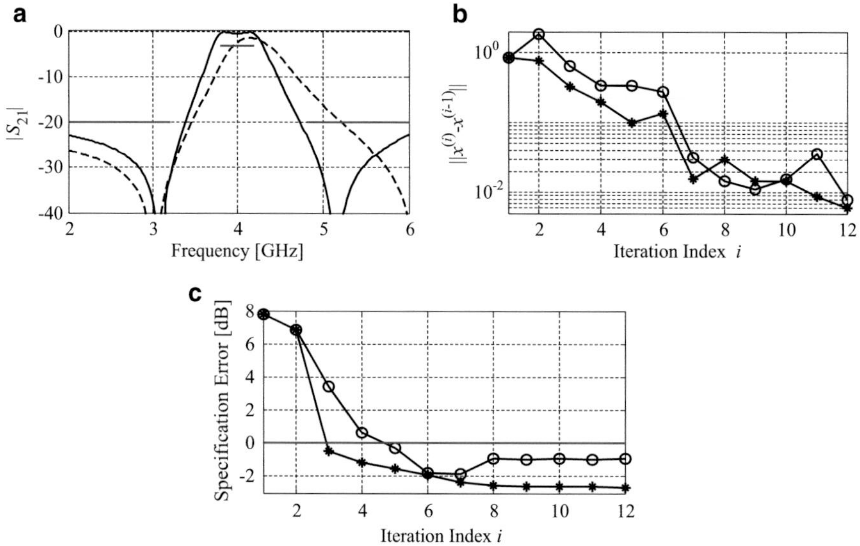

Fig. 6.19 CCDBR filter: (**a**) Initial (*dashed line*) and optimized (*solid line*) $|S_{21}|$ versus frequency; optimization using SM_{MRC} algorithm with the $\boldsymbol{c}_F(\boldsymbol{x}+\boldsymbol{q})$ model; (**b**) Convergence plots for SM_{STD} (*open circle*) and SM_{MRC} (*asterisk*), both using surrogate model $\boldsymbol{c}_F(\boldsymbol{x}+\boldsymbol{q})$, versus iteration index; (**c**) Specification error versus plots for SM_{STD} (*open circle*) and SM_{MRC} (*asterisk*), both using surrogate model $\boldsymbol{c}_F(\boldsymbol{x}+\boldsymbol{q})$, versus iteration index

for SM_{MRC} and SM_{STD}, both using the $\boldsymbol{c}_F(\boldsymbol{x}+\boldsymbol{q})$ model. We can observe that although the convergence pattern looks similar for both algorithms, the SM_{MRC} algorithm improves the design in each iteration, which is not the case for SM_{STD}.

6.5 Summary and Discussion

This chapter aimed at discussing selected parametric response correction techniques for surrogate-based optimization of expensive computational models in various engineering disciplines. Parametric correction techniques are simple to implement because the surrogate model parameters are obtained from available low- and high-fidelity model data using explicit formulas or by solving simple linear regression problems. As demonstrated using a number of real-world tasks, even simple one-point correction (additive or multiplicative) may be sufficient for certain problems where the low- and high-fidelity models are well correlated. More complex methods such as multi-point SM or manifold mapping allow exploitation of all available high-fidelity model data usually resulting in better generalization capability of the surrogate model. Appropriate selection of the correction technique may require sufficient engineering insight into the problem at hand. At the same time, response correction methods can (and, in some cases, should) be used together with other techniques for low-fidelity model enhancement such as input or implicit space mapping.

Chapter 7
Nonparametric Response Correction Techniques

The response correction techniques described in Chap. 6 utilized explicit formulation where the relationship between the high-fidelity model and the surrogate is quantified by a number of parameters that need to be extracted in order to identify the model. In this chapter, we focus on nonparametric methods, where the relationship between the low- and high-fidelity models is identified, or better said, directly extracted from the model responses; however, it is not explicitly given by any formula. In particular, in the optimization context, the surrogate model prediction may be obtained by tracking the response changes of the low-fidelity model and applying these changes to the known high-fidelity model response at a certain reference design. The formulation of nonparametric techniques is generally more complex and involves more restrictive assumptions regarding their applicability. However, these methods are normally characterized by a better generalization capability than the parametric techniques (cf. Chap. 6). The particular techniques described in this chapter are adaptive response correction, adaptive response prediction, and shape-preserving response prediction. We provide their formulations and illustrate their performance using design problems that involve airfoil shapes, and microwave devices, as well as antenna structures.

7.1 Adaptive Response Correction

The adaptive response correction (ARC) technique was introduced in microwave engineering (Koziel et al. 2009). In ARC, the surrogate model is implemented by translating the changes of the low-fidelity model response due to the adjustment of the design variables, into the corresponding changes of the high-fidelity model response without evaluating the latter. Unlike space mapping (Bandler et al. 1994), the ARC technique does not require any parameter extraction. Therefore, it is particularly suited to work with relatively expensive low-fidelity models. Moreover, ARC exploits the knowledge embedded in the low-fidelity model to the fuller

S. Koziel, L. Leifsson, *Simulation-Driven Design by Knowledge-Based Response Correction Techniques*, DOI 10.1007/978-3-319-30115-0_7

extent than many parametric techniques. Also, as opposed to other related techniques (e.g., SPRP, cf. Sect. 7.3, or feature-based optimization, cf. Chap. 9) it does not have limitations regarding response shape similarities of models of different fidelity. The ARC optimization algorithm uses the generic SBO iterative process (cf. (4.1)). Below, we describe in detail the ARC formulation and provide several design examples illustrating the use of the technique and its performance.

7.1.1 ARC Formulation

ARC works for vector-valued models in which the components of the response vector are function of a certain free parameters such as frequency, time, or some geometrical parameter such as a chord line location, etc. For such responses, one may consider a "shape" that can be stretched, scaled, or otherwise distorted with respect to this free parameters. The ARC surrogate $\boldsymbol{s}_{ARC}{}^{(i)}$ is defined as (Koziel et al. 2009)

$$\boldsymbol{s}_{ARC}^{(i)}(\boldsymbol{x}) = \boldsymbol{c}(\boldsymbol{x}) + \boldsymbol{d}_{ARC}(\boldsymbol{x}, \boldsymbol{x}^{(i)}), \tag{7.1}$$

where $\boldsymbol{d}_{ARC}(\boldsymbol{x}, \boldsymbol{x}^{(i)})$ is the response correction term dependent on the design variables $\boldsymbol{x}$. We want to maintain a perfect match between $\boldsymbol{f}$ and the surrogate at $\boldsymbol{x}^{(i)}$, i.e., $\boldsymbol{d}_{ARC}(\boldsymbol{x}, \boldsymbol{x}^{(i)})$ should satisfy

$$\boldsymbol{d}_{ARC}(\boldsymbol{x}^{(i)}, \boldsymbol{x}^{(i)}) = \boldsymbol{d}^{(i)} = \boldsymbol{f}(\boldsymbol{x}^{(i)}) - \boldsymbol{c}(\boldsymbol{x}^{(i)}) \tag{7.2}$$

with $\boldsymbol{d}^{(i)}$ being a basic response correction term (Cheng et al. 2010). The idea behind ARC is to account for the difference between $\boldsymbol{c}(\boldsymbol{x})$ and $\boldsymbol{c}(\boldsymbol{x}^{(i)})$ and to modify the correction term $\boldsymbol{d}^{(i)}$ so that this modification reflects changes of $\boldsymbol{c}$ during the surrogate optimization: if $\boldsymbol{c}$ shifts or changes its shape with respect to frequency (used throughout this section as the aforementioned free parameter), then ARC should track these changes.

Figure 7.1 shows $\boldsymbol{f}$ and $\boldsymbol{c}$ at two different designs and the corresponding additive correction terms $\boldsymbol{d}$ from (7.2). The relation between these terms is similar to the relation between the low-fidelity model responses so that tracking changes of $\boldsymbol{f}$ helps to determine necessary changes to $\boldsymbol{d}^{(i)}$. For the purpose of the ARC formulation, we can consider the explicit dependence of the model responses on frequency ω, so that $\boldsymbol{f}(\boldsymbol{x},\omega)$ is the value of $\boldsymbol{c}(\boldsymbol{x})$ at ω. The core of ARC is a function $F^{(i)}: X \times \Omega \rightarrow [\omega_{\min}, \omega_{\max}]$, where X stands for the low- and high-fidelity model domain, Ω is a frequency band of interest, and $[\omega_{\min}, \omega_{\max}] \supseteq \Omega$ is its possible expansion. $F^{(i)}$ is established at iteration i so that the difference between $\boldsymbol{c}(\boldsymbol{x},\omega)$, and $\boldsymbol{c}(\boldsymbol{x}^{(i)}, F^{(i)}(\boldsymbol{x},\omega))$ (the F-scaled $\boldsymbol{c}(\boldsymbol{x},\omega)$) is minimized in the L-square sense, i.e.,

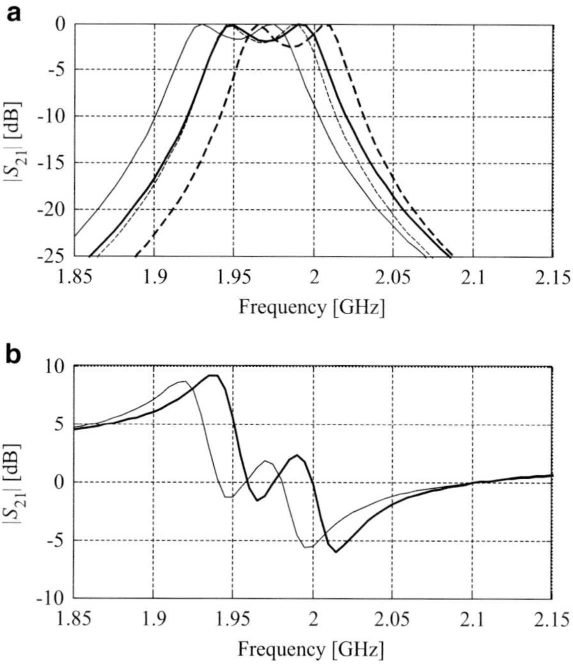

Fig. 7.1 Illustration of the additive correction terms: (**a**) high- and low-fidelity model responses, ***f*** (*solid line*) and ***c*** (*dashed line*), respectively, of a certain design, and ***f*** (*thick solid line*) and ***c*** (*dashed line*) at a different design; and (**b**) the additive correction terms corresponding to responses of (**a**)

$||\boldsymbol{c}(\boldsymbol{x}^{(i)},F^{(i)}(\boldsymbol{x},\omega)) - \boldsymbol{c}(\boldsymbol{x},\omega)||$ is minimized. Thus, $F^{(i)}$ is supposed to be defined so that the mapped frequency reflects the change of the $\boldsymbol{c}$ response at $\boldsymbol{x}$ with respect to $\boldsymbol{x}^{(i)}$.

In general, Ω may be a proper subset of $[\omega_{\min},\omega_{\max}]$ so that if the response of the model is not a constant function of frequency around $\min(\Omega)$ and $\max(\Omega)$, one may need to allow the mapped frequencies exceed the interval Ω in order to make a good alignment between $\boldsymbol{c}(\boldsymbol{x},\omega)$ and $\boldsymbol{c}(\boldsymbol{x}^{(i)},F^{(i)}(\boldsymbol{x},\omega))$ possible (Koziel et al. 2009).

The ARC correction term is defined as follows (here, the dependence of $\boldsymbol{d}_{ARC}(\boldsymbol{x}, \boldsymbol{x}^{(i)})$ on ω is shown explicitly):

$$\boldsymbol{d}_{ARC}(\boldsymbol{x},\boldsymbol{x}^{(i)};\omega) = \boldsymbol{f}\left(\boldsymbol{x}^{(i)},F^{(i)}(\boldsymbol{x},\omega)\right) - \boldsymbol{c}\left(\boldsymbol{x}^{(i)},F^{(i)}(\boldsymbol{x},\omega)\right). \tag{7.3}$$

Realization of the mapping $F^{(i)}$ is problem dependent. Here, we describe two approaches depending on the model responses.

Mapping 1: For problems in which the low-fidelity model response is rather smooth as a function of frequency (i.e., without sharp local minima and maxima) and such that the range of the response (as a function of frequency) does not change much with $\boldsymbol{x}$ the following mapping may be used. Let the mapping $F^{(i)}$ be implemented using a polynomial of the form

$$F^{(i)}(\boldsymbol{x},\omega) = [\lambda^{(i)}(\boldsymbol{x})]^T \boldsymbol{v}(\omega), \tag{7.4}$$

where $\boldsymbol{\lambda}^{(i)}(\boldsymbol{x}) = [\lambda_1{}^{(i)}(\boldsymbol{x})\ \lambda_2{}^{(i)}(\boldsymbol{x}) \ldots \lambda_p{}^{(i)}(\boldsymbol{x})]^T$ are scaling coefficients determined to minimize $\|\boldsymbol{c}(\boldsymbol{x}^{(i)},\boldsymbol{\lambda}^{(i)}(\boldsymbol{x})\boldsymbol{v}(\Omega)) - \boldsymbol{c}(\boldsymbol{x},\omega)\|$, while $\boldsymbol{v}(\omega) = [v_1(\omega)\ v_2(\omega) \ldots v_p(\omega)]^T$ are basis functions. The above minimization problem is solved subject to the following constraints:

1. $d([\boldsymbol{\lambda}^{(i)}(\boldsymbol{x})]^T\boldsymbol{v}(\omega))/d\omega > 0$
2. $[\boldsymbol{\lambda}^{(i)}(\boldsymbol{x})]^T\boldsymbol{v}(\min(\Omega)) \geq \omega_{\min}$, and
3. $[\boldsymbol{\lambda}^{(i)}(\boldsymbol{x})]^T\boldsymbol{v}(\max(\Omega)) \leq \omega_{\max}$, i.e., we need to ensure that $F^{(i)}(\boldsymbol{x},\omega)$ is a monotonic function of ω within its range in $[\omega_{\min}, \omega_{\max}]$

A third-order polynomial scaling is often used, i.e., $p = 4$ and $\boldsymbol{v}(\omega) = [1\ \omega\ \omega^2\ \omega^3]^T$. Furthermore, $\omega_{\min}$ is, typically, set slightly smaller than ω_1 (e.g., by 10–20 %), and $\omega_{\max}$ slightly larger than ω_m.

Mapping 2: For problems in which the coarse model response contains clearly defined characteristic frequencies, e.g., sharp minima corresponding to zeros of the transfer function, the following mapping of $F^{(i)}$ can be used. Let $\Omega_1 = \{\omega_{1.1}, \omega_{1.2}, \ldots, \omega_{1.N}\}$ and $\Omega_2 = \{\omega_{2.1}, \omega_{2.2}, \ldots, \omega_{2.N}\}$ be the sets of characteristic frequencies of $\overline{\boldsymbol{c}}^{(i)}(\boldsymbol{x}^{(i)})$ and $\overline{\boldsymbol{c}}^{(i)}(\boldsymbol{x})$, respectively. The mapped frequency $F^{(i)}(\boldsymbol{x},\omega)$ is defined as an interpolation function such that $F^{(i)}(\boldsymbol{x},\omega_{2.j}) = \omega_{1.j}$, $j = 1, 2, \ldots, N$. Here, F can be implemented using cubic splines. If there is a different number of characteristic frequencies for $\overline{\boldsymbol{c}}^{(i)}(\boldsymbol{x})$ and $\overline{\boldsymbol{c}}^{(i)}(\boldsymbol{x}^{(i)})$ both Ω_1 and Ω_2 are adjusted using an auxiliary algorithm in order to maintain the correspondence between the two sets. In order to have the mapped frequency $F^{(i)}(\boldsymbol{x},\omega)$ well defined on $[\omega_1, \omega_m]$, we typically set $\omega_{1.1} = \omega_{2.1} = \omega_1$ and $\omega_{1.N} = \omega_{2.N} = \omega_m$. However, if there is a difference between $\overline{\boldsymbol{c}}^{(i)}(\boldsymbol{x})$ and $\overline{\boldsymbol{c}}^{(i)}(\boldsymbol{x}^{(i)})$ at ω_1, then it is better to set $\omega_{1.1}$ to $\omega = \arg\min_{\omega} \left|I\left(\overline{\boldsymbol{c}}^{(i)}(\boldsymbol{x}^{(i)}), \Omega, \omega\right) - \overline{\boldsymbol{c}}^{(i)}(\boldsymbol{x}, \omega_1)\right|$, in order to obtain a better overall match between $I\left(\overline{\boldsymbol{c}}^{(i)}(\boldsymbol{x}^{(i)}), \Omega, F^{(i)}(\boldsymbol{x}, \Omega)\right)$ and $\overline{\boldsymbol{c}}^{(i)}(\boldsymbol{x})$. A similar adjustment can be done for $\omega_{1.N}$.

Figure 7.2 illustrates the operation of the ARC technique. Compared to the conventional, additive response correction defined using the constant vector $\boldsymbol{d}^{(i)} = \boldsymbol{f}(\boldsymbol{x}^{(i)}) - \boldsymbol{c}(\boldsymbol{x}^{(i)})$ (cf. (7.2)), the adaptive response correction is capable to better utilize the knowledge about the structure under consideration contained in the low-fidelity model $\boldsymbol{c}$, which leads to a smaller number of design iterations (and, thus, lower computational cost) necessary to find the optimized design. This was demonstrated by Koziel et al. (2009) for microwave filters.

In the next three sections, we illustrate the ARC technique using design problems involving microwave devices (filters and antennas), and airfoil shapes.

7.1.2 Wideband Bandstop Filter Design with ARC

Consider the following example: the wideband bandstop microstrip filter (Hsieh and Wang 2005) shown in Fig. 7.3a. The design parameters are $\boldsymbol{x} = [L_r\, W_r\, L_c\, W_c\, G_c]^T$. The fine model $\boldsymbol{f}$ is simulated in FEKO (2008). The low-fidelity

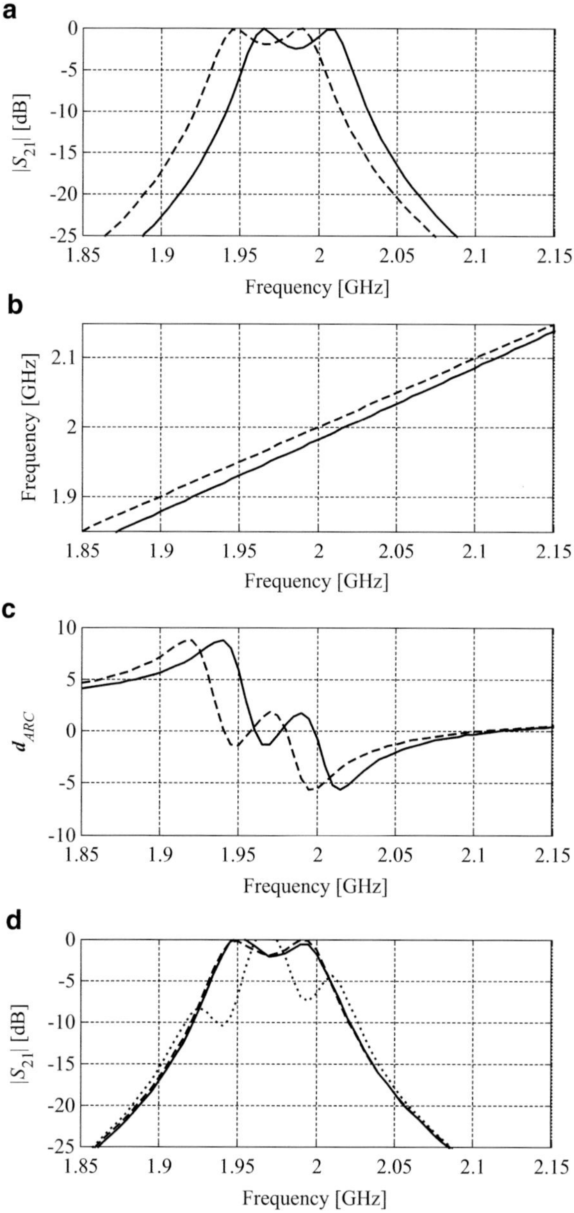

Fig. 7.2 (**a**) Low-fidelity model responses $\boldsymbol{c}(\boldsymbol{x}^{(i)})$ (*dashed line*) and $\boldsymbol{c}(\boldsymbol{x})$ (*solid line*); (**b**) Mapping function $F^{(i)}(\boldsymbol{x})$ (*solid line*), and the identity function (which equals $F^{(i)}(\boldsymbol{x}^{(i)})$) (*dashed line*); (**c**) ARC correction terms $\boldsymbol{d}_{ARC}(\boldsymbol{x}^{(i)}, \boldsymbol{x}^{(i)})$ (*dashed line*) and $\boldsymbol{d}_{ARC}(\boldsymbol{x}, \boldsymbol{x}^{(i)})$ (*solid line*); (**d**) Predicted $\boldsymbol{f}$ response at $\boldsymbol{x}^{(i)}$ obtained with ARC, i.e., $\boldsymbol{s}_{ARC}{}^{(i)}(\boldsymbol{x}) = \boldsymbol{c}(\boldsymbol{x}) + \boldsymbol{d}_{ARC}(\boldsymbol{x}, \boldsymbol{x}^{(i)})$ (*solid line*), actual $\boldsymbol{f}$ response $\boldsymbol{f}(\boldsymbol{x})$ (*dashed line*), and predicted high-fidelity model response $\boldsymbol{s}^{(i)}(\boldsymbol{x}) = \boldsymbol{c}(\boldsymbol{x}) + \boldsymbol{d}^{(i)}$ (*dotted line*)

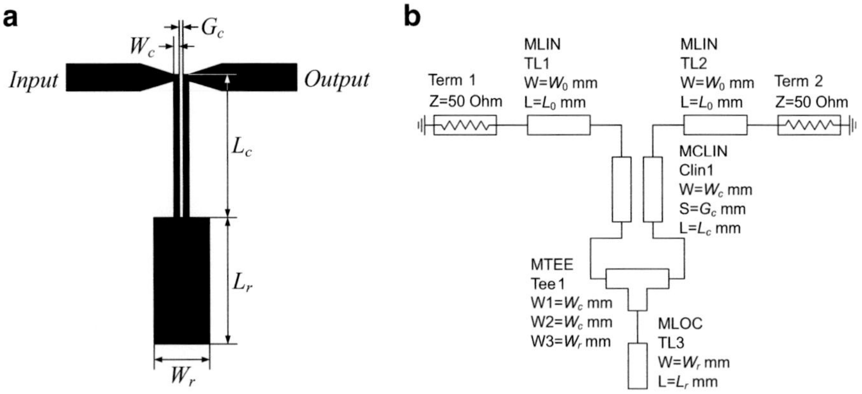

Fig. 7.3 A wideband bandstop microstrip filter (Hsieh and Wang 2005): (**a**) geometry (Koziel et al. 2009), (**b**) coarse model (Agilent ADS 2008)

Table 7.1 Optimization results for the wideband bandstop microstrip filter (Koziel et al. 2009)

Method	Final specification error (dB)	Number of fine model evaluations[a]
Standard output space mapping	−1.1	8
Adaptive response correction	−2.1	3

[a]Excludes the fine model evaluation at the starting point

model $\boldsymbol{c}$ is the circuit model implemented in Agilent ADS (2008) (Fig. 7.3b). The design specifications on the transmission coefficient $|S_{21}|$ are

- $|S_{21}| \geq -3$ dB for 1.0 GHz $\leq \omega \leq$ 2.0 GHz
- $|S_{21}| \leq -20$ dB for 3.0 GHz $\leq \omega \leq$ 9.0 GHz, and
- $|S_{21}| \geq -3$ dB for 10.0 GHz $\leq \omega \leq$ 11.0 GHz

We optimize the filter using the standard SBO approach (cf. Chap. 5) and the ARC (cf. (7.1)) with the mapping function $F^{(i)}$ realized through the second approach described in Sect. 7.1.1 as the filter response contains three clearly defined characteristic points.

Table 7.1 shows the optimization results as well as a comparison with standard output space mapping (Bandler et al. 2003; Koziel et al. 2006a). Note that the quality of solution found with the adaptive response correction is better than the one obtained with the standard method. Also, the number of fine model evaluations required to find the solution is substantially smaller for the adaptive correction technique. Figure 7.4 shows the response of the initial surrogate model $\boldsymbol{s}_{ARC}{}^{(0)}$ at the starting point and at its optimal solution $\boldsymbol{x}^{(1)}$. Figure 7.5 shows the correction terms $\boldsymbol{\Delta}_r(\boldsymbol{x}^{(0)}, \boldsymbol{x}^{(0)}) = \boldsymbol{d}^{(0)}$ and $\boldsymbol{\Delta}_r(\boldsymbol{x}^{(1)}, \boldsymbol{x}^{(0)})$. Figure 7.6 shows the plot of the scaling function $F^{(0)}(\boldsymbol{x}^{(1)},\omega)$. The fine model response at the design found by the space mapping algorithm with the adaptive response correction is shown in Fig. 7.7.

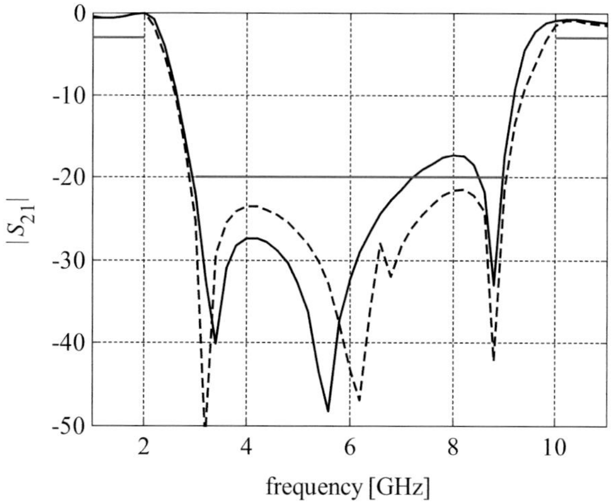

Fig. 7.4 Wideband bandstop filter (Koziel et al. 2009): surrogate model response $\boldsymbol{s}_{ARC}{}^{(0)}(\boldsymbol{x}^{(0)})$ (*solid line*), and optimized surrogate model response $\boldsymbol{s}_{ARC}{}^{(0)}(\boldsymbol{x}^{(1)})$ (*dashed line*). Plots obtained for the adaptive response correction method

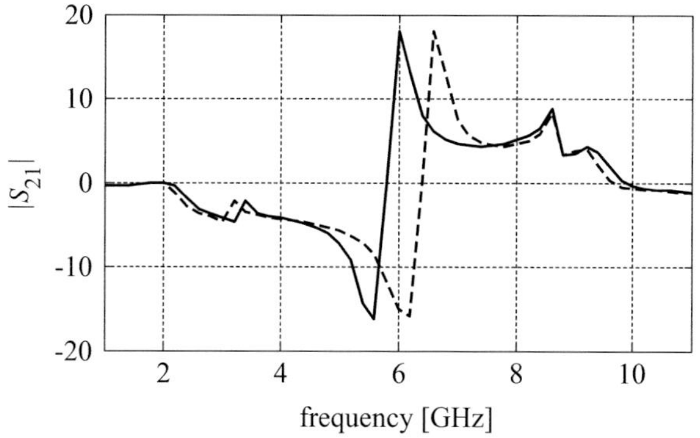

Fig. 7.5 Wideband bandstop filter (Koziel et al. 2009): correction terms $\boldsymbol{d}_{ARC}(\boldsymbol{x}^{(0)}, \boldsymbol{x}^{(0)})$ (*solid line*), and $\boldsymbol{d}_{ARC}(\boldsymbol{x}^{(1)}, \boldsymbol{x}^{(0)})$ (*dashed line*). Plots obtained for the adaptive response correction method

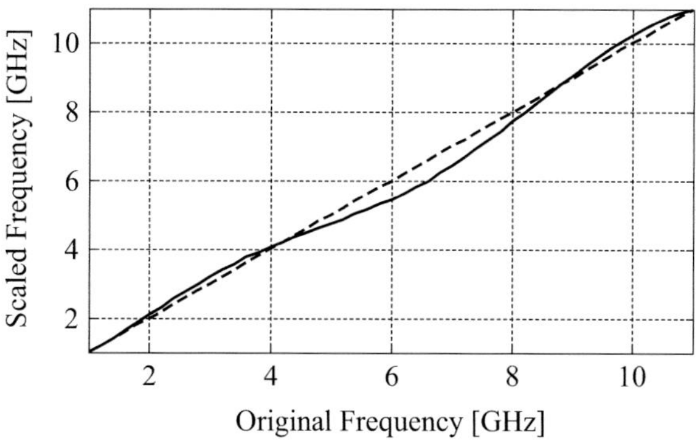

Fig. 7.6 Wideband bandstop filter (Koziel et al. 2009): the scaling function $F^{(0)}(\boldsymbol{x}^{(1)},\omega)$ (*solid line*). Plot obtained for the adaptive response correction method. The identity function plot is shown as *dashed line*

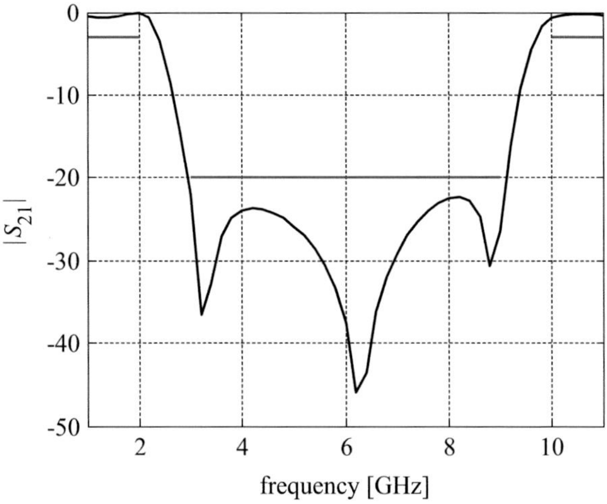

Fig. 7.7 Wideband bandstop filter (Koziel et al. 2009): fine model response at the design found by the space mapping optimization algorithm with the adaptive response correction

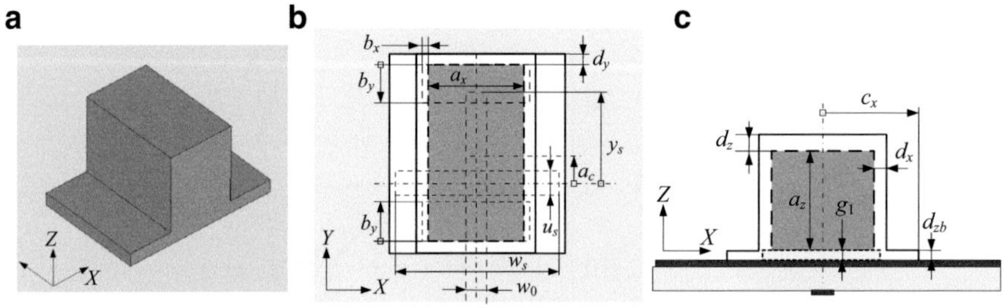

Fig. 7.8 DRA (Koziel and Ogurtsov 2011d): (**a**) 3D view, (**b**) top, (**c**) front

7.1.3 Dielectric Resonator Antenna (DRA) Design with ARC

Consider the DRA in a polycarbonate housing (Koziel and Ogurtsov 2011d) shown in Fig. 7.8a comprising a rectangular dielectric resonator (DR) residing on two Teflon slabs above a metal ground. The DR is energized with a 50 Ω microstrip through a ground plane slot. The DR core permittivity and loss tangent are 10 and 1e−4 at 6.5 GHz respectively. The substrate is a 0.51 mm thick R04003C layer. The microstrip signal trace width, w_0, is 1.15 mm. The substrate and the ground plane are modeled with infinite lateral extends.

The design variables are $\boldsymbol{x} = [a_x\, a_y\, a_c\, u_s\, w_s\, y_s\, g_1]^T$, where a_x and a_y are the lateral dimensions of the DR; a_c is the Y-offset of the DR relatively the slot; u_s and w_s are the ground plane slot dimensions; y_s is the length of the open ended stub; and g_1 is the thickness of the Teflon slabs. Other dimensions, shown in Fig. 7.8b, c, are fixed/dependent as follows: $a_z = a_x$, $b_x = 2$, $b_y = 6$, $c_x = 7.5$, $d_x = d_y = d_z = 1$, and $d_{zb} = g_1 + 1$, all in mm.

Table 7.2 Dielectric resonator antenna: optimization cost (Koziel and Ogurtsov 2011d)

Algorithm	Algorithm component	Number of model evaluations	CPU time	
			Absolute (min)	Relative to $\boldsymbol{f}$
ARC	Evaluation of $\boldsymbol{c}$[a]	$240 \times \boldsymbol{c}$	180	9.0
	Evaluation of $\boldsymbol{f}$	$4 \times \boldsymbol{f}$	80	4.0
	Total cost	N/A	260	13.0
Output space mapping	Evaluation of $\boldsymbol{c}$[a]	$405 \times \boldsymbol{c}$	303	15.2
	Evaluation of $\boldsymbol{f}$	$7 \times \boldsymbol{f}$	140	7.0
	Total cost	N/A	443	22.2
Pattern search	Evaluation of $\boldsymbol{f}$	$151 \times \boldsymbol{f}$	3000	151.0

[a]Total number of the low-fidelity model evaluations necessary to optimize the ARC surrogate

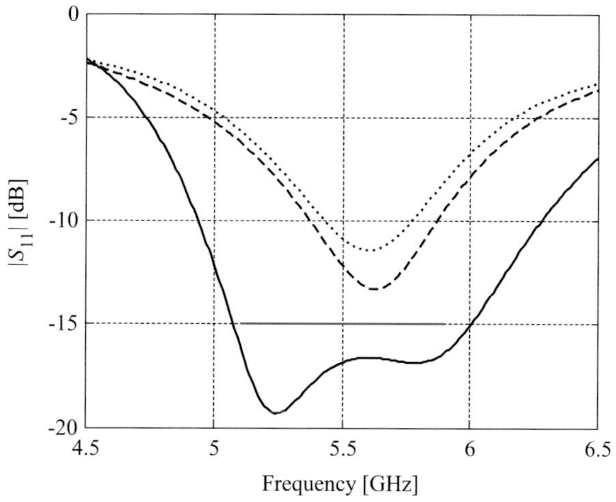

Fig. 7.9 DRA reflection responses (Koziel and Ogurtsov 2011d): the low-fidelity model $\boldsymbol{c}$ at $\boldsymbol{x}^{(0)}$ (*dotted line*) and the high-fidelity model $\boldsymbol{f}$ at $\boldsymbol{x}^{(0)}$ (dashed line) and at the final design $\boldsymbol{x}^{(4)}$ (*solid line*)

The high-fidelity model $\boldsymbol{f}$ is simulated using the CST MWS (2013) transient solver (763,840 mesh cells at the initial design, evaluation time of 20 min). The design objective is to obtain $|S_{11}| \leq -15$ dB from 5.1 to 5.9 GHz. Design constraints imposed on the radiation characteristics are the following: (1) realized gain is to be not less than 3 dB for the zero zenith angle, and (2) realized gain of back radiation is to be less than −5 dB. Both gain constraints are to be imposed over the impedance bandwidth of interest. The initial design is $\boldsymbol{x}^{(0)} = [8.0\ 14.0\ 0.0\ 2.0\ 10.0\ 4.0\ 2.0]^T$ mm. In this case, we exploit a low-fidelity model $\boldsymbol{c}$ evaluated in CST (30,720 mesh cells at $\boldsymbol{x}^{(0)}$, evaluation time of 45 s).

The DRA was optimized using the ARC surrogate model (cf. (7.1)) with mapping 1 described in Sect. 7.1.1. The optimized design is $\boldsymbol{x}^{(4)} = [8.42\ 13.51\ 1.40\ 1.51\ 10.33\ 2.73\ 1.61]^T$ mm. The design cost corresponds to 13 evaluations of $\boldsymbol{f}$ (Table 7.2). The final design reflection response with $|S_{11}| < -16$ dB for 5.1–5.9 GHz is shown in Fig. 7.9 and the radiation response is shown in Fig. 7.10. Direct optimization of $\boldsymbol{f}$ using pattern search (Kolda et al. 2003) yields a design of the

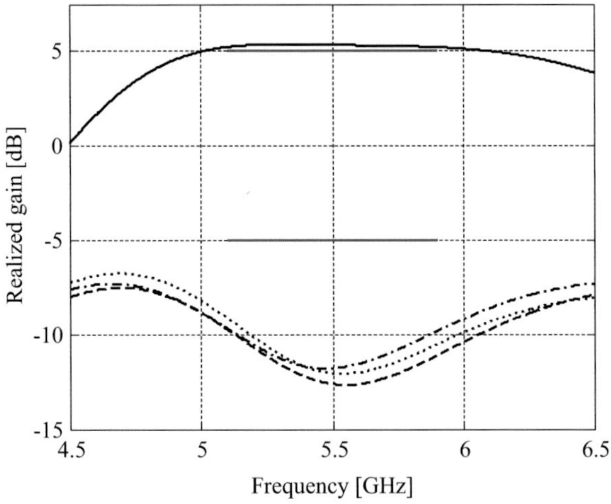

Fig. 7.10 Radiation response of the DRA, $\boldsymbol{R}_f$ at the final design (Koziel and Ogurtsov 2011d): for zenith angle of 0° (*solid line*); and back radiation, for zenith angles of 135° (positive Y-direction, *dot line*), 180° (*dash line*), and 135° (negative Y-direction *dash-dot line*). Design constraints are shown with the *horizontal lines*

same quality but at a much higher cost of 151 f evaluations. Optimization using output space mapping with the correction term $\boldsymbol{d}^{(i)}$ (cf. (7.3)) yields a comparable design with $|S_{11}| < -16.4$ dB for 5.1–5.9 GHz, but at higher computational cost equivalent to about 22 evaluations of the high-fidelity model with the cost budget listed in Table 7.2.

It should be emphasized that the ARC algorithm only needs four iterations to converge. Further reduction of the computational cost could be obtained by using more efficient algorithm to optimize the ARC surrogate (note that surrogate optimization contributes to about 2/3 of the total design cost). In this case, the surrogate is optimized using pattern search.

7.1.4 Airfoil Shape Optimization with ARC

Aerodynamic shape optimization involves finding an optimal streamlined surface, such as an airfoil or a wing, in a fluid flow. Typically, the objective is to minimize the drag (the component of force acting on the surface which is parallel to the flow stream direction) while maintaining a given lift (the component of the force acting on the surface which is perpendicular to the flow stream direction).

The forces are normally presented in a nondimensional form. The nondimensional drag coefficient for a two-dimensional airfoil is defined to be $C_d \equiv d/(q_\infty c)$, where d is the drag force, c is the chord (the linear distance from the leading edge of the airfoil to the trailing edge), and $q_\infty \equiv 1/2 \rho_\infty {V_\infty}^2$ is the free stream dynamic pressure where ρ_∞ is the free stream density and V_∞ is the free stream velocity. Similarly, the nondimensional lift coefficient is defined as $C_l \equiv l/(q_\infty c)$, where l is the lift force.

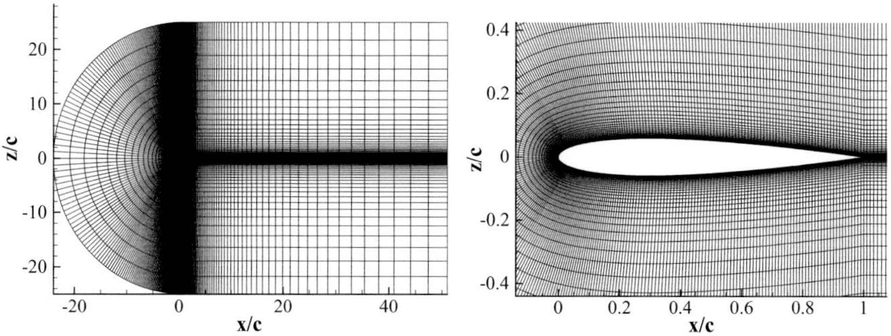

Fig. 7.11 An example computational mesh with a structured C-topology for the NACA 0012 airfoil showing the whole computational domain (*left*), and a view of the mesh close to the surface (*right*)

The evaluation of the aerodynamic characteristics, i.e., the forces in this case, requires a computational analysis of the flow field. A high-fidelity computational analysis, termed computational fluid dynamics (CFD), involves solving a set of partial differential equations describing the physics, as well as several state equation models. These equations are solved on a computational grid, like the one shown in Fig. 7.11, and the computational time, on a parallel high performance computing cluster, can be on the order of few minutes up to several hours or even days.

The outcome of the CFD simulation yields the flow variables, such as velocity (in all directions), pressure, and density, in the entire flow field. Figure 7.12a shows the contours of the Mach number, defined as the local velocity over the local speed of sound, in transonic flow past an airfoil. One can observe a strong shock wave on the upper surface where the flow accelerates past Mach 1, the speed of sound, and then suddenly decelerates below Mach 1. Figure 7.12b shows the variation of the nondimensional pressure, defined as $C_p \equiv (p - p_\infty)/\mathrm{q}_\infty$ where p is a local pressure and p_∞ is the free-stream pressure, on the airfoil surface with the chordwise location (x/c). Another important variable is the skin friction distribution along the surface which is defined as $C_f \equiv \tau/\mathrm{q}_\infty$, where τ is the local shear stress acting on the surface due to friction.

The figures of merit, the lift and drag coefficients in this case, are typically calculated using the pressure and skin friction distributions. In particular, the forces acting on the surface can be uniquely determined from those distributions by integrating them around the entire profile. Assuming the inviscid flow, and thereby neglecting viscous effects (giving $C_f = 0$), the lift coefficient C_l and the drag coefficient, called the wave drag coefficient for high-speed inviscid flow, C_{dw}, are calculated as: $C_l = -C_x \sin\alpha + C_z \cos\alpha$, and $C_{dw} = C_x \cos\alpha + C_z \sin\alpha$, where $C_x = \oint C_p \sin\theta \; ds$, and $C_z = -\oint C_p \cos\theta \; ds$ with ds as the length of a surface panel element along the airfoil surface and θ is the angle the panel makes relative to the x-axis (horizontal).

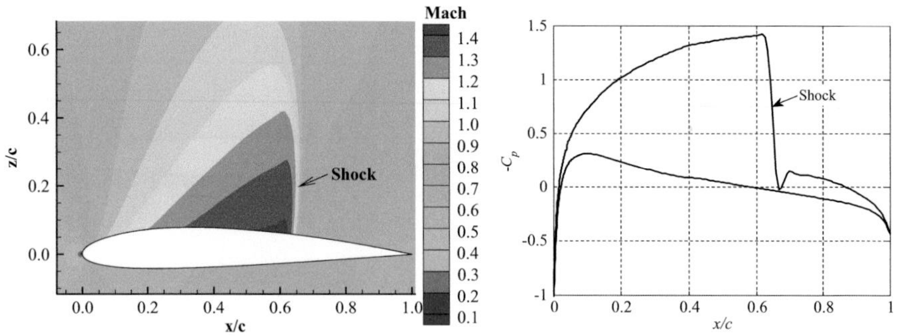

Fig. 7.12 Transonic flow past the NACA 2412 airfoil at a Mach number of $M_\infty = 0.75$, and a lift coefficient of $C_l = 0.67$. The *left figure* shows a Mach contour plot of the flow field, and the *right figure* shows the pressure coefficient (C_p) distribution on the airfoil surface

In the case of airfoil shape optimization, the ARC surrogate is constructed for the pressure distribution, an intermediate result of the fluid flow analysis, which then yields the figures of merit, i.e., the lift and drag coefficients. We will now demonstrate how this is derived and utilized based on the description of the ARC method in Sect. 7.1.1.

For the airfoil pressure distribution, the ARC surrogate model at the design iteration i, $C_{p.ARC}{}^{(i)}$, is defined as (Koziel and Leifsson 2012c)

$$C_{p.ARC}^{(i)}(\boldsymbol{x}) = C_{p.c}(\boldsymbol{x}) + \Delta C_{p.ARC}(\boldsymbol{x}, \boldsymbol{x}^{(i)}), \tag{7.5}$$

where $\Delta C_{p.ARC}(\boldsymbol{x}, \boldsymbol{x}^{(i)})$ is the response correction term dependent on the design variable vector $\boldsymbol{x}$. A perfect match between the high-fidelity model and the surrogate is maintained at $\boldsymbol{x}^{(i)}$, therefore, we must have

$$\Delta C_{p.ARC}(\boldsymbol{x}^{(i)}, \boldsymbol{x}^{(i)}) = \Delta C_p^{(i)} = C_{p.f}(\boldsymbol{x}^{(i)}) - C_{p.c}(\boldsymbol{x}^{(i)}). \tag{7.6}$$

As explained earlier, the idea behind the ARC is to account for the difference between the low-fidelity model responses at $\boldsymbol{x}$ and at $\boldsymbol{x}^{(i)}$ and to modify the correction term accordingly with the initial correction $\Delta C_p{}^{(i)}$ taken as a reference. In particular, we want to modify the initial correction term $\Delta C_p{}^{(i)}$ so that this modification reflects changes to $C_{p.c}$ during the process of surrogate model optimization. Moreover, if the response of $C_{p.c}$ shifts or changes its shape with respect to the coordinate value x/c, the response correction term should track these changes.

As an illustration, consider Fig. 7.13a that shows the high- and low-fidelity model responses at the two different designs. It is seen that the general relation between the low- and high-fidelity model responses is preserved (e.g., the pressure shock shifts in a similar way for both models). Figure 7.13b shows the additive correction terms ΔC_p corresponding to the responses shown in Fig. 7.13a. As we can observe, the relation between these terms as functions of x/c is similar to the

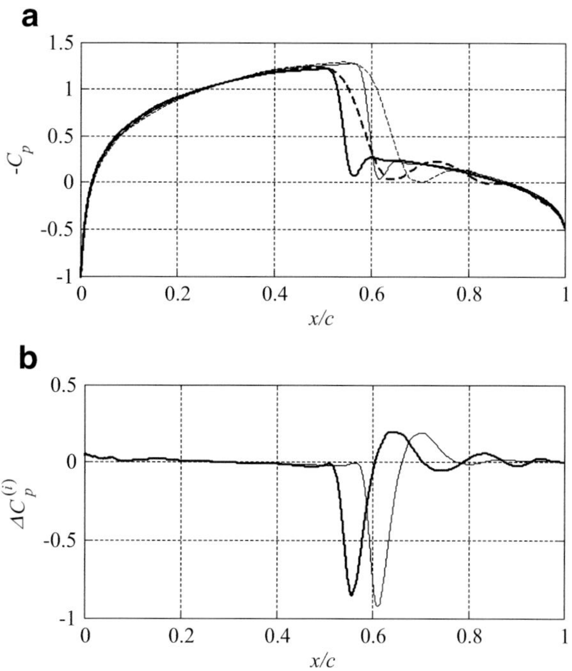

Fig. 7.13 An illustration of pressure distribution shape change tracking by ARC (Koziel and Leifsson 2012c): (**a**) High- (*thin solid line*) and low-fidelity model (*thin dashed line*) responses at certain design, as well as high- (*thick solid line*) and low-fidelity model (*thick dashed line*) responses at a different design; (**b**) the additive correction terms corresponding to responses shown in (**a**)

relation between the low-fidelity model responses so that proper tracking of the low-fidelity model changes help determine the necessary changes to the correction terms.

We will now apply the general formulation given in Sect. 7.1.1 to the airfoil pressure distribution ARC. Let $w = x/c$. We can consider the pressure distribution as a function of two variables, $\boldsymbol{x}$ and w, so that $C_{p.c}(\boldsymbol{x},w)$ is the value of $C_{p.c}(\boldsymbol{x})$ at the coordinate $w = x/c$.

The mapping $F^{(i)}: X \times \Omega \rightarrow \Omega$, where X stands for the low- and high-fidelity model domains and $\Omega = [0,1]$ (the range of normalized coordinate $w = x/c$) is established at iteration i in such a way that the difference between $C_{p.c}(\boldsymbol{x},w)$, i.e., the low-fidelity model response at $\boldsymbol{x}$, and $C_{p.c}(\boldsymbol{x}^{(i)},F^{(i)}(\boldsymbol{x},w))$, i.e., $C_{p.c}(\boldsymbol{x}^{(i)})$ evaluated at $F^{(i)}(\boldsymbol{x},w)$, is minimized in a given sense, e.g., the following norm $\int_w[(C_{p.c}(\boldsymbol{x}^{(i)},F^{(i)}(\boldsymbol{x},w)) - C_{p.c}(\boldsymbol{x},w)]^2 dw$ to realize an L-square minimization.

The function $C_{p.c}(\boldsymbol{x}^{(i)},F^{(i)}(\boldsymbol{x},w))$ is nothing but $C_{p.c}(\boldsymbol{x}^{(i)},w)$ scaled with respect to the normalized coordinate in order to be as similar to $C_{p.c}(\boldsymbol{x},w)$ as possible in a given sense. In other words, the mapping $F^{(i)}$ is supposed to be defined in such a way that mapped coordinate $F^{(i)}(\boldsymbol{x},w)$ reflects the shape change of the low-fidelity model response at $\boldsymbol{x}$ with respect to its original shape at $\boldsymbol{x}^{(i)}$. For the sake of consistency with condition (7.6), $F^{(i)}(\boldsymbol{x}^{(i)},w)$ should be equal w, i.e., the mapped coordinate should be the same as the original coordinate at $\boldsymbol{x}^{(i)}$.

Having the mapping $F^{(i)}$, we can define the response correction term as follows (here, the dependence of $\Delta C_{p.ARC}(\boldsymbol{x}, \boldsymbol{x}^{(i)})$ on w is shown explicitly):

$$\Delta C_{p.ARC}(\boldsymbol{x}, \boldsymbol{x}^{(i)}; w) = C_{p.f}\left(\boldsymbol{x}^{(i)}, F^{(i)}(\boldsymbol{x}, w)\right) - C_{p.c}\left(\boldsymbol{x}^{(i)}, F^{(i)}(\boldsymbol{x}, w)\right). \tag{7.7}$$

Mapping 1 can be implemented for this particular case. Using (7.4), the mapping function is defined as

$$F^{(i)}(\boldsymbol{x}, w) = [\lambda^{(i)}(\boldsymbol{x})]^T \boldsymbol{v}(w), \tag{7.8}$$

where $\boldsymbol{\lambda}^{(i)}(x) = [\lambda_1{}^{(i)}(x)\ \lambda_2{}^{(i)}(x) \ldots \lambda_p{}^{(i)}(x)]^T$ are scaling coefficients determined as

$$\lambda^{(i)}(\boldsymbol{x}) = \arg\min_{\lambda} \int_0^1 [C_{p.c}(\boldsymbol{x}^{(i)}, \lambda^T v(w)) - C_{p.c}(\boldsymbol{x}, w)]^2 dw, \tag{7.9}$$

while $\boldsymbol{v}(\omega) = [v_1(\omega)\ v_2(\omega) \ldots v_p(\omega)]^T$ are basis functions. The optimization problem (7.8), which is readily solved, has the following constraints: (1) $d([\boldsymbol{\lambda}^{(i)}(\boldsymbol{x})]^T \boldsymbol{v}(w))/dw > 0$, i.e., scaling function F is monotonic, (2) $[\boldsymbol{\lambda}^{(i)}(\boldsymbol{x})]^T \boldsymbol{v}(0) = 0$, and (3) $[\boldsymbol{\lambda}^{(i)}(\boldsymbol{x})]^T \boldsymbol{v}(1) = 1$, i.e., $F^{(i)}$ maps 0 into 0 and 1 into 1. Here, we use a third-order polynomial scaling, i.e., we have $p = 4$ and $\boldsymbol{v}(w) = [1\ w\ w^2\ w^3]^T$.

Figure 7.14 illustrates the operation of the ARC technique for the airfoil pressure distribution. The responses of the low-fidelity model at two designs, $\boldsymbol{x}^{(i)}$ and $\boldsymbol{x}$ are shown in Fig. 7.14a. Figure 7.14b shows the mapping function $F^{(i)}$ obtained using (7.8) and (7.9). In Fig. 7.14c, we can observe the ARC correction term $\Delta C_{p.ARC}$ at both $\boldsymbol{x}^{(i)}$, i.e., $\Delta C_{p.ARC}(\boldsymbol{x}^{(i)}, \boldsymbol{x}^{(i)})$, and at $\boldsymbol{x}$, i.e., $\Delta C_{p.ARC}(\boldsymbol{x}, \boldsymbol{x}^{(i)})$. Recall that $\Delta C_{p.ARC}(\boldsymbol{x}^{(i)}, \boldsymbol{x}^{(i)}) = \Delta C_p{}^{(i)}$. Finally, Fig. 7.14d shows the predicted high-fidelity model response at $\boldsymbol{x}^{(i)}$ obtained using the ARC, i.e., $C_{p.ARC}{}^{(i)}(\boldsymbol{x}) = C_{p.c}(\boldsymbol{x}) + \Delta C_{p.ARC}(\boldsymbol{x}, \boldsymbol{x}^{(i)})$, which is in a good agreement with the actual high-fidelity model response $C_{p.f}(\boldsymbol{x})$.

An example airfoil shape optimization with the ARC surrogate involving lift maximization is given. Details of the test cases and optimization results are given in Table 7.3. The lift maximization objective function is defined as

$$H\left(C_p(\boldsymbol{x})\right) = -C_{l.s}\left(C_p(\boldsymbol{x})\right) + \beta\left[\Delta C_{dw.s}\left(C_p(\boldsymbol{x})\right)\right]^2, \tag{7.10}$$

where $\Delta C_{dw.s} = 0$ if $C_{dw.s} \le C_{dw.s.max}$ and $\Delta C_{dw.s} = C_{dw.s} - C_{dw.s.max}$ otherwise. The numerical experiments we use a penalty factor of $\beta = 1{,}000$. Here, the pressure distribution for the surrogate model is $C_p = C_{p.s}$, and for the high-fidelity model $C_p = C_{p.f}$. Also, $C_{l.s}$ and $C_{dw.s}$ denote the lift and wave drag coefficients (both being functions of the pressure distribution).

The airfoil shape is parameterized using the NACA four-digit airfoil with m (the maximum ordinate of the mean camberline as a fraction of chord), p (the chordwise position of the maximum ordinate), and t/c (the thickness-to-chord ratio) as the design variables (see Abbott and von Doenhoff (1959) for details). The design variable bounds are $0 \le m \le 0.1$, $0.2 \le p \le 0.8$, and $0.05 \le t/c \le 0.2$.

The airfoil performance is obtained through CFD models which are implemented using the ICEM CFD (2012) grid generator and the FLUENT

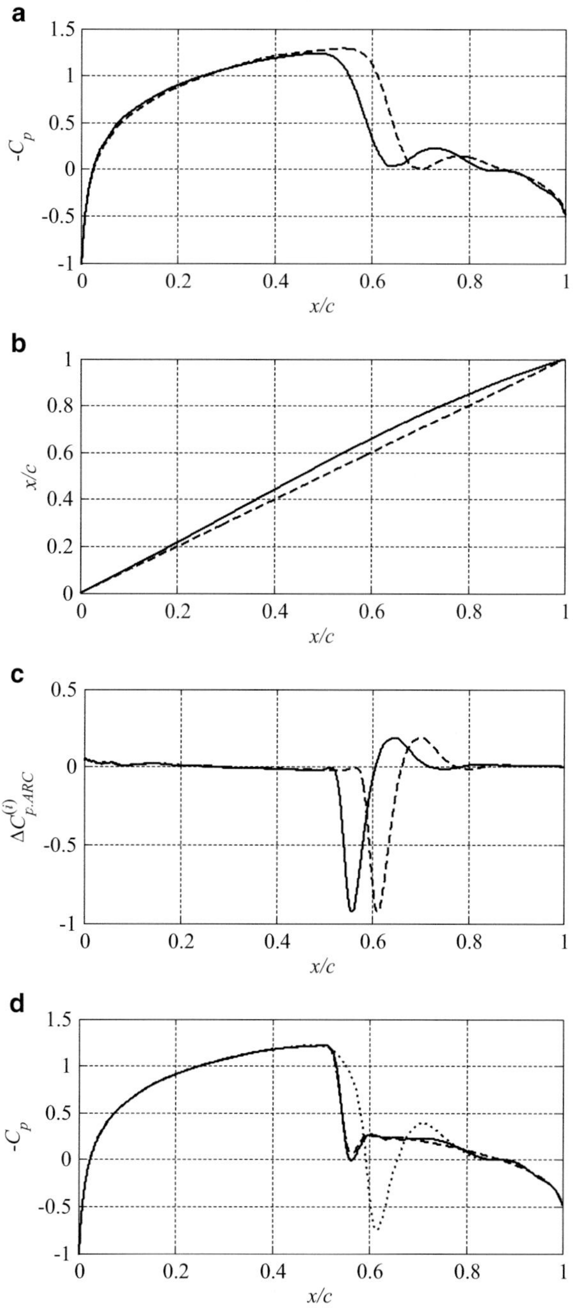

Fig. 7.14 An illustration of the ARC technique for the pressure distribution of the upper surface of an airfoil in transonic flow (Koziel and Leifsson 2012c): (**a**) Low-fidelity model response at two designs, $\boldsymbol{x}^{(i)}$ and $\boldsymbol{x}$, $C_{p.c}(\boldsymbol{x}^{(i)})$ (*dashed line*), and $C_{p.c}(\boldsymbol{x})$ (*solid line*); (**b**) Mapping function $F^{(i)}(\boldsymbol{x})$ (*solid line*) obtained using (7.8) and (7.9); the identity function (which equals $F^{(i)}(\boldsymbol{x}^{(i)})$) is marked using *dashed line*; (**c**) ARC correction terms $\Delta C_{p.ARC}(\boldsymbol{x}^{(i)}, \boldsymbol{x}^{(i)})$ (*dashed line*) and $\Delta C_{p.ARC}(\boldsymbol{x}, \boldsymbol{x}^{(i)})$ (*solid line*); (**d**) Predicted high-fidelity model response at $\boldsymbol{x}^{(i)}$ obtained using ARC, i.e., $C_{p.ARC}{}^{(i)}(\boldsymbol{x}) = C_{p.c}(\boldsymbol{x}) + \Delta C_{p.ARC}(\boldsymbol{x}, \boldsymbol{x}^{(i)})$ (*solid line*), actual high-fidelity model response $C_{p.f}(\boldsymbol{x})$ (*dashed line*), and predicted high fidelity model response $C_{p.s}{}^{(i)}(\boldsymbol{x}) = C_{p.c}(\boldsymbol{x}) + \Delta C_p{}^{(i)}$ (*dotted line*)

Table 7.3 Optimization results for the airfoil lift maximization (Koziel and Leifsson 2012c)

	Lift maximization			
	$M_\infty = 0.75$, $\alpha = 0°$, $C_{d,\max} = 0.0050$, $A_{\min} = 0.075$			
Variable	Initial	Direct[a]	Kriging[b]	ARC[c]
m	0.02	0.0140	0.0131	0.0140
p	0.40	0.7704	0.8000	0.7925
t/c	0.12	0.1150	0.1147	0.1120
C_l	0.4745	0.5572	0.5607	0.5874
C_{dw}	0.0115	0.0050	0.0050	0.0050
A	0.0808	0.0774	0.0772	0.0754
N_c	–	0	0	210
N_f	–	96	25	7
Total cost[d]	–	96	25	<10

N_c and N_f are the number of low- and high-fidelity model evaluations, respectively. All the numerical values are from the high-fidelity model

[a]Design obtained through direct optimization of the high-fidelity CFD model using the pattern-search algorithm (Koziel 2010c)

[b]Design obtained through surrogate-based optimization using kriging with updates (Forrester and Keane 2009)

[c]Design obtained using the ARC algorithm; surrogate model optimization performed using pattern-search (Koziel 2010c)

[d]The total optimization cost is expressed in the equivalent number of high-fidelity model evaluations. The ratio of the high-fidelity model evaluation time to the corrected low-fidelity model evaluation time varies between 43.8 and 110 depending on the design. For the sake of simplicity we use a fixed value of 80 here

(2012) flow solver. The high-fidelity CFD model $\boldsymbol{f}$ is a two-dimensional steady-state Euler analysis with roughly 400,000 mesh cells and an overall simulation time around 67 min. The low-fidelity CFD model $\boldsymbol{c}$ is the same as the high-fidelity one, but with a coarser mesh (roughly 30,000 cells) and relaxed convergence criteria (100 flow solver iterations). The low-fidelity model is roughly 80 times faster than the high-fidelity one.

The ARC approach is compared to the following benchmark optimization techniques: (1) direct design optimization of the high-fidelity CFD model using a pattern-search algorithm (Koziel 2010c), and (2) surrogate-based optimization using adaptively updated kriging interpolation surrogate (Forrester and Keane 2009). The latter approach is implemented as follows. The initial kriging model (Forrester and Keane 2009; Journel, and Huijbregts 1981) is constructed using 20 fine model samples distributed in the design space using Latin Hypercube Sampling (Forrester and Keane 2009). The fine model is evaluated at the optimum of the kriging model and this data is used as an infill point to update the surrogate. Surrogate model optimization is constrained using the trust region framework. The termination condition is the same as for the ARC algorithm, i.e., convergence in argument ($\|\boldsymbol{x}^{(i+1)} - \boldsymbol{x}^{(i)}\| < 10^{-4}$) or reduction of the trust region radius (Conn et al. 2000) below a user-defined threshold, here, $\lambda < 10^{-4}$.

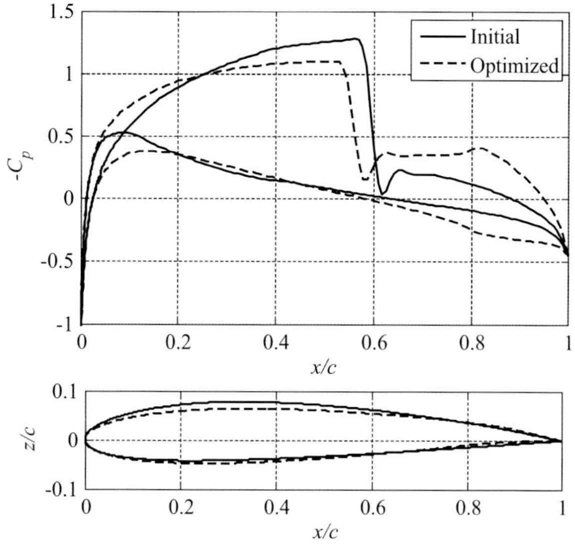

Fig. 7.15 Initial and optimized airfoil shapes (*below*) and the corresponding pressure distributions (*above*) (Koziel and Leifsson 2012c)

The initial number of samples (i.e., 20) was selected based on the initial experiments so that it suits the scale of the problem (number of design variables and the design space size). In particular, the kriging-based algorithm starting from 10 initial samples normally fails to find a satisfactory design, in particular, the solution found typically violates constraints and/or the obtained lift/drag coefficients (depending on the test case) are much worse than was obtained by ARC.

As shown in Table 7.3, the initial airfoil design is the NACA 2412 airfoil and the drag constraint is violated. The direct method and the ARC method obtain comparable optimized designs by reducing the camber (m), placing the location of the maximum camber relatively aft (p), and reducing the thickness (t/c). The effects on the pressure distribution can be observed in Fig. 7.15: the shock strength is reduced as a result of reduced camber and thickness, and the pressure distribution opens up behind the shock due to the aft camber to increase lift. The proposed method required 210 surrogate model evaluations and 7 high-fidelity model evaluations, yielding an equivalent number of high-fidelity model evaluations of less than 10. Direct high-fidelity model optimization required 96 high-fidelity model evaluations. The SBO approach with the function-approximation surrogate model, kriging interpolation, required 25 high-fidelity model evaluations.

The optimization history is given in Fig. 7.16. In particular, one can observe a convergence plot (Fig. 7.16a), as well as the evolution of the objective function (Fig. 7.16b), the lift coefficient (Fig. 7.16c) and the drag coefficient (Fig. 7.16d). It follows that the algorithm exhibits a good convergence pattern and that the mechanisms introduced in the algorithm (in particular the trust region approach and the penalty function) enforce the drag limitation to be satisfied while increasing the lift coefficient as much as possible.

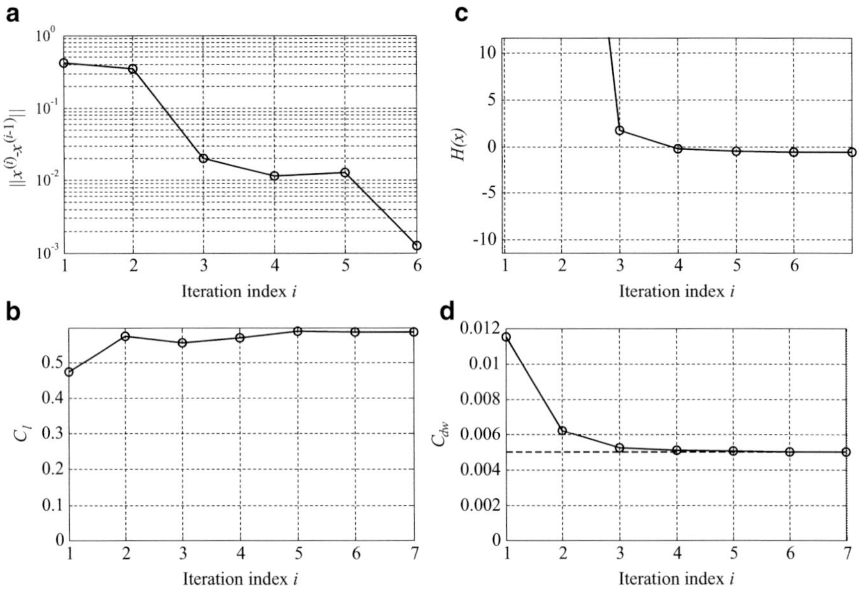

Fig. 7.16 Optimization history (Koziel and Leifsson 2012c): (**a**) convergence plot; (**b**) evolution of the objective function; (**c**) evolution of the lift coefficient; and (**d**) evolution of the drag coefficient (drag constraint marked using a *dashed horizontal line*). The graphs show all high-fidelity function evaluations performed in the optimization

7.2 Adaptive Response Prediction

The mapping $F^{(i)}$ (cf. (7.4)) describes the "horizontal" change of the low-fidelity model response while moving from $\boldsymbol{x}^{(i)}$ to $\boldsymbol{x}$. This type of mapping is the essence of the ARC technique. The adaptive response prediction (ARP) technique (Koziel and Leifsson 2012a) enhances ARC by using an additional correction stage, where we account for the "vertical" difference between $C_{p.c}(\boldsymbol{x}^{(i)},F^{(i)}(\boldsymbol{x},w))$ (i.e., $C_{p.c}(\boldsymbol{x}^{(i)},w)$ after applying "horizontal" scaling) and $C_{p.c}(\boldsymbol{x},w)$. This is particularly important when the model responses at $\boldsymbol{x}^{(i)}$ and at $\boldsymbol{x}$ exhibit substantial "vertical" difference, such as the pressure shock in Fig. 7.12a. In fact, the origins of the ARP come from the area of aerodynamic shape optimization. We will now describe the ARP formulation and give an application of it to airfoil design.

7.2.1 ARP Formulation

The vertical changes can be described by a mapping $G^{(i)}: \Omega \times R \rightarrow \Omega \times R$, where $\Omega = [0,1]$ is the range of the normalized coordinate $w = x/c$, and R is a set of real numbers. $G^{(i)}$ is established at iteration i in such a way that the difference between $C_{p.c}(\boldsymbol{x},w)$, i.e., the low-fidelity model response at $\boldsymbol{x}$, and $G^{(i)}(w,C_{p.c}(\boldsymbol{x}^{(i)},F^{(i)}(\boldsymbol{x},w)))$, i.e., $C_{p.c}(\boldsymbol{x}^{(i)}, F^{(i)}(\boldsymbol{x},w))$ corrected by $G^{(i)}(w,\cdot)$, is minimized in a given sense, e.g., the norm $\int_w[(G^{(i)}(w,C_{p.c}(\boldsymbol{x}^{(i)},F^{(i)}(\boldsymbol{x},w))) - C_{p.c}(\boldsymbol{x},w)]^2dw$ to realize an L-square minimization.

The function $G^{(i)}(w,C_{p.c}(\boldsymbol{x}^{(i)},F^{(i)}(\boldsymbol{x},w)))$ realizes a coordinate-dependent vertical correction of the F-scaled low-fidelity model response. For the sake of consistency, we have to ensure that $G^{(i)}(w,C_{p.c}(\boldsymbol{x}^{(i)},F^{(i)}(\boldsymbol{x}^{(i)},w)))$, which is the same as $G^{(i)}(w,C_{p.c}(\boldsymbol{x}^{(i)},w))$, equals $C_{p.c}(\boldsymbol{x}^{(i)},w)$.

While $G^{(i)}$ can be implemented in various ways, we describe one possible realization where $G^{(i)}$ is based on low-order polynomials. The function $G^{(i)}(w,\cdot)$ is defined as (Koziel and Leifsson 2012a)

$$G^{(i)}(w, y(w)) = [\alpha^{(i)}]^T\boldsymbol{s}(w) + [\beta^{(i)}]^T\boldsymbol{r}(w)\cdot y(w), \tag{7.11}$$

where $\boldsymbol{\alpha}^{(i)} = [\alpha_1{}^{(i)}\alpha_2{}^{(i)}\ldots\alpha_{p.\alpha}{}^{(i)}]^T$ and $\boldsymbol{\beta}^{(i)} = [\beta_1{}^{(i)}\beta_2{}^{(i)}\ldots\beta_{p.\beta}{}^{(i)}]^T$ are coefficients to be determined; $p.\alpha$ and $p.\beta$ are user-defined model orders; $\boldsymbol{s}(w) = [s_1(w)\ s_2(w)\ldots s_{p.\alpha}(w)]^T$ and $\boldsymbol{r}(w) = [r_1(w)\ r_2(w)\ldots r_{p.\beta}(w)]^T$ are basis functions. For simplicity of notation, we use $y(w)$, which is $y(w) = C_{p.c}(\boldsymbol{x}^{(i)},F^{(i)}(\boldsymbol{x}^{(i)},w))$. Here, we aim at identifying the coefficients of the function $G^{(i)}$ so that $G^{(i)}(w,C_{p.c}(\boldsymbol{x}^{(i)},F^{(i)}(\boldsymbol{x},w)))$ matches $C_{p.c}(\boldsymbol{x},w)$ as well as possible in an L-square sense. This, in conjunction with the model $G^{(i)}$ defined by (7.11), forms a linear regression problem which is very convenient as the globally optimal coefficient $\boldsymbol{\alpha}^{(i)}$ and $\boldsymbol{\beta}^{(i)}$ can be found analytically by solving the following system:

$$\boldsymbol{A}\begin{bmatrix}\alpha^{(i)}\\ \beta^{(i)}\end{bmatrix} = \boldsymbol{Y}, \tag{7.12}$$

where

$$\boldsymbol{A} = \begin{bmatrix} s_1(w_1) & \cdots & s_{p.\alpha}(w_1) & r_1(w_1)y(w_1) & \cdots & r_{p.\beta}(w_1)y(w_1) \\ s_1(w_2) & \cdots & s_{p.\alpha}(w_2) & r_1(w_2)y(w_2) & \cdots & r_{p.\beta}(w_2)y(w_2) \\ \vdots & \ddots & \vdots & \vdots & \ddots & \vdots \\ s_1(w_K) & \cdots & s_{p.\alpha}(w_K) & r_1(w_K)y(w_K) & \cdots & r_{p.\beta}(w_K)y(w_K) \end{bmatrix} \tag{7.13}$$

and

$$\boldsymbol{Y} = [y(w_1) \quad y(w_2) \quad \cdots \quad y(w_K)]^T. \tag{7.14}$$

Here, w_1 to w_K are K discrete values of the coordinate x/c used to identify the coefficients of the mapping $G^{(i)}$. The L-square optimum solution for the regression problem (7.12) is given by

$$\begin{bmatrix} \alpha^{(i)} \\ \beta^{(i)} \end{bmatrix} = (A^T A) A^T Y. \tag{7.15}$$

The basis functions can be, in particular, monomials of the form $s(w) = [1\ w\ w^2 \ldots w^{p.\alpha}]^T$ and $r(w) = [1\ w\ w^2 \ldots w^{p.\beta}]^T$.

The mappings $F^{(i)}$ and $G^{(i)}$ describe the change of the low-fidelity model response while moving from the reference design $x^{(i)}$ to x. As mentioned before, the low- and high-fidelity models are physically related. The adaptive response prediction technique assumes that, because of this relation the high-fidelity model response at x can be predicted by applying $F^{(i)}$ and $G^{(i)}$ to the high-fidelity model response at $x^{(i)}$, $C_{p.f}(x^{(i)},w)$. More specifically, the ARP surrogate model with the reference design $x^{(i)}$ is defined as follows (Koziel and Leifsson 2012a)

$$C^{(i)}_{s.ARP}(x, w) = G^{(i)}\left(w, C_{p.f}\left(x^{(i)}, F^{(i)}(x, w)\right)\right). \tag{7.16}$$

Figure 7.17 illustrates the operation of the ARP technique. The responses of the low-fidelity model at two designs, $x^{(i)}$ and x are shown in Fig. 7.17a. Figure 7.17b shows the mapping function $F^{(i)}$ obtained using (7.8) and (7.9). Figure 7.17c shows the high-fidelity model prediction obtained using ARP. We can observe that the ARP prediction is in good agreement with the actual high-fidelity model response at x, particularly having in mind that only a single evaluation of the high-fidelity model was used to create the surrogate model and that the response at x is substantially different from that at the reference design $x^{(i)}$. For the sake of comparison, Fig. 7.17c also shows the prediction obtained using ARC which is of much worse quality for this particular case. This demonstrates that the vertical correction introduced by the ARP technique may be important to maintain a good generalization capability of the surrogate model.

7.2.2 Airfoil Shape Optimization with ARP

The application of ARP is demonstrated using the same problem described in Sect. 7.1.4. The optimization results are shown in Table 7.4 (with the results form Sect. 7.1.4 repeated for convenience of comparison). We see that the ARP technique arrives at nearly the same design as the ARC one (the only difference is in the location of the maximum camber, which slightly more forward for the ARP design). The ARP design slightly violates, by 4 %, the drag constraint. However, the design cost is significantly lower with the ARP technique, requiring 1 less high-fidelity simulation and 30 low-fidelity ones than the ARC technique.

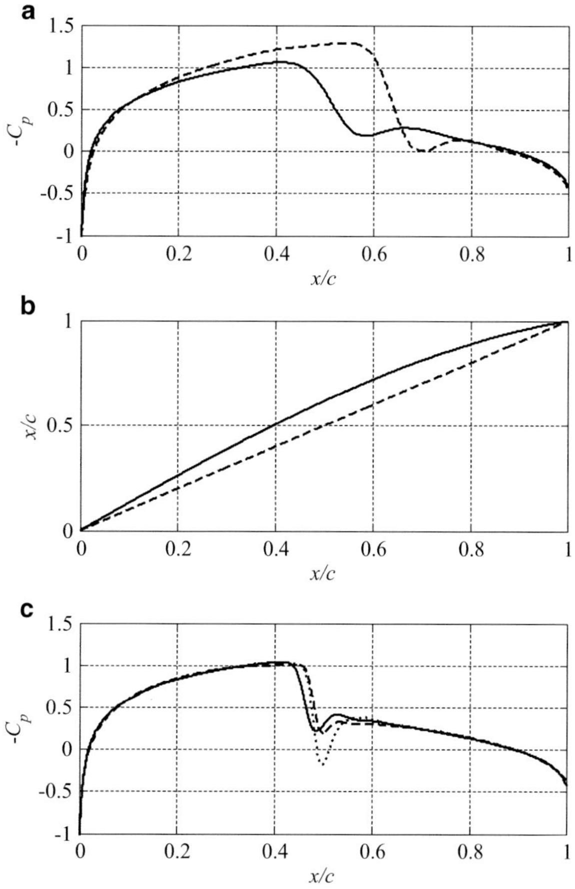

Fig. 7.17 An illustration of the ARP technique for the pressure distribution of the upper surface of an airfoil in transonic flow (Koziel and Leifsson 2012a): (**a**) Low-fidelity model response at two designs, $\boldsymbol{x}^{(i)}$ and $\boldsymbol{x}$, $C_{p.c}(\boldsymbol{x}^{(i)})$ (*dashed line*), and $C_{p.c}(\boldsymbol{x})$ (*solid line*); (**b**) mapping function $F^{(i)}(\boldsymbol{x})$ (*solid line*) obtained using (7.8) and (7.9); the identity function (which equals $F^{(i)}(\boldsymbol{x}^{(i)})$) is marked using a *dashed line*; (c) predicted high-fidelity model response at $\boldsymbol{x}$ obtained using ARC (*dotted line*), predicted high-fidelity model response obtained using ARP, i.e., $C_{p.ARP}{}^{(i)}(\boldsymbol{x})$ (*dashed line*), as well as the actual high-fidelity model response $C_{p.f}(\boldsymbol{x})$ (*solid line*)

7.3 Shape-Preserving Response Prediction

Shape-preserving response prediction (SPRP) (Koziel 2010a, e) was initially introduced in microwave engineering to reduce the cost of optimizing electromagnetic (EM)-simulated structures such as filters. In SPRP, the surrogate model is constructed assuming that the change of the high-fidelity model response due to the adjustment of the design variables can be predicted using the actual response changes of the auxiliary low-fidelity (or coarse) model $\boldsymbol{c}: X_c \rightarrow R^m$, $X_c \subseteq R^n$, that describes the same structure as the high-fidelity model; $\boldsymbol{c}$ is less accurate but much faster to evaluate than $\boldsymbol{f}$.

The choice of the low-fidelity model very much depends on the engineering discipline. In microwave engineering, the coarse model might be an equivalent circuit of the considered microwave structure, describing the structure using circuit theory tools rather than through solution of the Maxwell equations. It is critically important for SPRP that the coarse model is physically based, which ensures that

Table 7.4 Optimization results for the airfoil lift maximization (Koziel and Leifsson 2012a)

	Lift maximization				
	$M_\infty = 0.75$, $\alpha = 0°$, $C_{d,\max} = 0.0050$, $A_{\min} = 0.075$				
Variable	Initial	Direct[a]	Kriging[b]	ARC[c]	ARP[d]
m	0.02	0.0140	0.0131	0.0140	0.0140
p	0.40	0.7704	0.8000	0.7925	0.7883
t/c	0.12	0.1150	0.1147	0.1120	0.1120
C_l	0.4745	0.5572	0.5607	0.5874	0.5804
C_{dw}	0.0115	0.0050	0.0050	0.0050	0.0048
A	0.0808	0.0774	0.0772	0.0754	0.0754
N_c	–	0	0	210	180
N_f	–	96	25	7	6
Total cost[e]	–	96	25	<10	<9

N_c and N_f are the number of low- and high-fidelity model evaluations, respectively. All the numerical values are from the high-fidelity model

[a]Design obtained through direct optimization of the high-fidelity CFD model using the pattern-search algorithm (Koziel 2010c)

[b]Design obtained through surrogate-based optimization using kriging with updates (Forrester and Keane 2009)

[c]Design obtained using the ARC algorithm; surrogate model optimization performed using pattern-search (Koziel 2010c)

[d]Design obtained using the ARP algorithm; surrogate model optimization performed using pattern-search (Koziel 2010c)

[e]The total optimization cost is expressed in the equivalent number of high-fidelity model evaluations. The ratio of the high-fidelity model evaluation time to the corrected low-fidelity model evaluation time varies between 43.8 and 110 depending on the design. For the sake of simplicity we use a fixed value of 80 here

the effect of the design parameter variations on the model response is similar for both the fine and coarse models. The change of the coarse model response is described by the translation vectors corresponding to certain (finite) number of characteristic points of the model's response. These translation vectors are subsequently used to predict the change of the fine model response with the actual response of $\boldsymbol{f}$ at the current iteration point, $\boldsymbol{f}(\boldsymbol{x}^{(i)})$, treated as a reference.

Here, the concept of SPRP is explained using the specific case of a microwave filter. Figure 7.18a shows the example of the coarse model response $|S_{21}|$ in the frequency range 8–18 GHz, at the design $\boldsymbol{x}^{(i)}$, as well as the coarse model response at some other design $\boldsymbol{x}$. The responses come from the double folded stub bandstop filter (Koziel 2010e). Circles denote five characteristic points of $\boldsymbol{c}(\boldsymbol{x}^{(i)})$, here, selected to represent $|S_{21}| = -3$ dB, $|S_{21}| = -20$ dB, and the local $|S_{21}|$ maximum (at about 13 GHz). Squares denote corresponding characteristic points for $\boldsymbol{c}(\boldsymbol{x})$, while small line segments represent the translation vectors that determine the "shift" of the characteristic points of $\boldsymbol{c}$ when changing the design variables from $\boldsymbol{x}^{(i)}$ to $\boldsymbol{x}$. Because the coarse model is physics-based, the fine model response at the given design, here, $\boldsymbol{x}$, can be predicted using the same translation vectors applied to the corresponding characteristic points of the fine model response at $\boldsymbol{x}^{(i)}$, $\boldsymbol{f}(\boldsymbol{x}^{(i)})$.

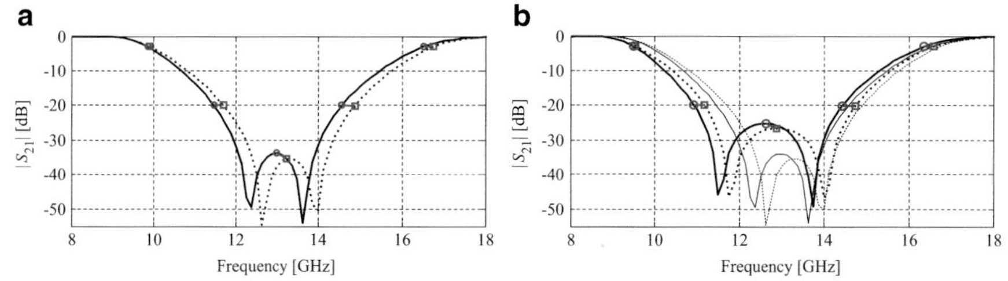

Fig. 7.18 The SPRP concept (Koziel 2010a, e): (**a**) Example coarse model response at the design $\boldsymbol{x}^{(i)}$, $\boldsymbol{c}(\boldsymbol{x}^{(i)})$ (*solid line*), the coarse model response at $\boldsymbol{x}$, $\boldsymbol{c}(\boldsymbol{x})$ (*dotted line*), characteristic points of $\boldsymbol{c}(\boldsymbol{x}^{(i)})$ (*open circle*) and $\boldsymbol{c}(\boldsymbol{x})$ (*open square*), and the translation vectors (*short lines*); (**b**) Fine model response at $\boldsymbol{x}^{(i)}$, $\boldsymbol{f}(\boldsymbol{x}^{(i)})$ (*solid line*) and the predicted fine model response at $\boldsymbol{x}$ (*dotted line*) obtained using SPRP based on characteristic points of Fig. 7.18a; characteristic points of $\boldsymbol{f}(\boldsymbol{x}^{(i)})$ (*open circle*) and the translation vectors (*short lines*) were used to find the characteristic points (*open square*) of the predicted fine model response; coarse model responses $\boldsymbol{c}(\boldsymbol{x}^{(i)})$ and $\boldsymbol{c}(\boldsymbol{x})$ are plotted using *thin solid* and *dotted line*, respectively

This is illustrated in Fig. 7.18b where the predicted fine model response at $\boldsymbol{x}$ is shown, as well as the actual response, $\boldsymbol{f}(\boldsymbol{x})$, with a good agreement between both curves.

7.3.1 SPRP Formulation

A rigorous formulation of SPRP (Koziel 2010a, e) follows. Let $\boldsymbol{f}(\boldsymbol{x}) = [f(\boldsymbol{x},\omega_1) \ldots f(\boldsymbol{x},\omega_m)]^T$ and $\boldsymbol{c}(\boldsymbol{x}) = [c(\boldsymbol{x},\omega_1) \ldots c(\boldsymbol{x},\omega_m)]^T$, where ω_j, $j = 1, \ldots, m$, is the frequency sweep (it can be assumed without loss of generality that the model responses are parameterized by frequency). Let $p_j^f = [\omega_j^f r_j^f]^T$, $p_j^{c0} = [\omega_j^{c0} r_j^{c0}]^T$, and $p_j^c = [\omega_j^c r_j^c]^T$, $j = 1, \ldots, K$, denote the sets of characteristic points of $\boldsymbol{f}(\boldsymbol{x}^{(i)})$, $\boldsymbol{c}(\boldsymbol{x}^{(i)})$ and $c(\boldsymbol{x})$, respectively. Here, ω and r denote the frequency and magnitude components of the respective point. The translation vectors of the coarse model response are defined as $t_j = [\omega_j^t r_j^t]^T$, $j = 1, \ldots, K$, where $\omega_j^t = \omega_j^c - \omega_j^{c0}$ and $r_j^t = r_j^c - r_j^{c0}$.

The shape-preserving response prediction surrogate model is defined as follows

$$\boldsymbol{s}^{(i)}(\boldsymbol{x}) = [s^{(i)}(\boldsymbol{x},\omega_1) \quad \ldots \quad s^{(i)}(\boldsymbol{x},\omega_m)]^T, \tag{7.17}$$

where

$$\boldsymbol{s}^{(i)}(\boldsymbol{x},\omega_j) = \bar{\boldsymbol{f}}\left(\boldsymbol{x}^{(i)}, F(\omega_j, \{-\omega_k^t\}_{k=1}^K)\right) + R(\omega_j, \{r_k^t\}_{k=1}^K) \tag{7.18}$$

for $j = 1, \ldots, m$. $\bar{\boldsymbol{f}}(\boldsymbol{x},\omega)$ is an interpolation of $\{f(\boldsymbol{x},\omega_1), \ldots, f(\boldsymbol{x},\omega_m)\}$ onto the frequency interval $[\omega_1,\omega_m]$. The scaling function F interpolates the data pairs

$\{\omega_1,\omega_1\}$, $\{\omega_1^f,\omega_1^f-\omega_1^t\},\ldots,\{\omega_K^f,\omega_K^f-\omega_K^t\}$, $\{\omega_m,\omega_m\}$, onto the frequency interval $[\omega_1,\omega_m]$. The function R does a similar interpolation for data pairs $\{\omega_1,r_1\}$, $\{\omega_1^f,r_1^f-r_1^t\},\ldots,\{\omega_K^f,r_K^f-r_K^t\}$, $\{\omega_m,r_m\}$; here $r_1=c(\boldsymbol{x},\omega_1)-c(\boldsymbol{x}^r,\omega_1)$ and $r_m=c(\boldsymbol{x},\omega_m)-c(\boldsymbol{x}^r,\omega_m)$. In other words, the function F translates the frequency components of the characteristic points of $\boldsymbol{f}(\boldsymbol{x}^{(i)})$ to the frequencies at which they should be located according to the translation vectors t_j, while the function R adds the necessary magnitude component. The interpolation onto $[\omega_1,\omega_m]$ is necessary because the original frequency sweep is a discrete set. Formally, both the translation vectors t_j and their components should have an additional index (i) indicating that they are determined at iteration i of the optimization algorithm; however, this is omitted for the sake of simplicity.

As follows from its formulation, SPRP is developed assuming that the frequency components of the translation vectors are zero at the edges of the frequency spectrum (i.e., at ω_1 and ω_m). This limitation can be easily overcome either by extending the frequency range of the coarse model and applying extrapolation (cf. Koziel 2010e). Also, it is assumed that the overall shape of both the fine and coarse model response is similar. This means, in particular, that the characteristic points of responses of both the coarse model $\boldsymbol{c}$ and the fine model $\boldsymbol{f}$ are in one-to-one correspondence. If this assumption is not satisfied, the surrogate model (7.17), (7.18) cannot be evaluated because the translation vectors t_i are not well defined. Generalizations of SPRP that allow alleviating this difficulty in some cases can be found in Koziel (2010e).

7.3.2 *Optimization with SPRP: Dual-Band Bandpass Filter*

Consider the dual-band bandpass filter (Guan et al. 2008) (Fig. 7.19a). The design parameters are $\boldsymbol{x}=[L_1\,L_2\,S_1\,S_2\,S_3\,d\;g\;W]^T$ mm. The fine model is simulated in Sonnet ***em*** (2010). The design specifications are $|S_{21}|\geq-3$ dB for $0.85\text{ GHz}\leq\omega\leq0.95$ GHz and $1.75\text{ GHz}\leq\omega\leq1.85$ GHz, and $|S_{21}|\leq-20$ dB for $0.5\text{ GHz}\leq\omega\leq0.7$ GHz, $1.1\text{ GHz}\leq\omega\leq1.6$ GHz, and $2.0\text{ GHz}\leq\omega\leq2.2$ GHz.

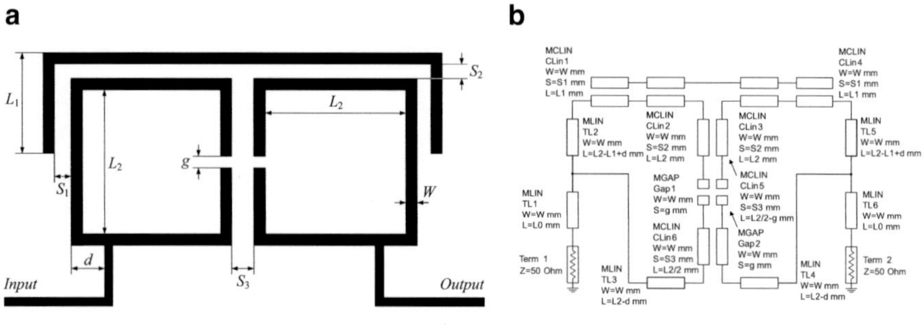

Fig. 7.19 Dual-band bandpass filter: (**a**) geometry (Guan et al. 2008), (**b**) coarse model (Agilent ADS 2011)

Table 7.5 SPRP optimization results for dual-band bandpass filter (Koziel 2010a)

Algorithm	Final specification error (dB)	Number of fine model evaluations[a]
SPRP	−2.0[b]	3
SPRP + ISM[c]	−1.9[d]	2

[a]Excludes the fine model evaluation at the starting point
[b]Design specifications satisfied after the first iteration (spec. error −1.2 dB)
[c]The surrogate model is of the form $s^{(i)}(\boldsymbol{x}) = c(\boldsymbol{x} + \boldsymbol{c}^{(i)})$; $\boldsymbol{c}^{(i)}$ is found using parameter extraction
[d]Design specifications satisfied after the first iteration (spec. error −1.0 dB)

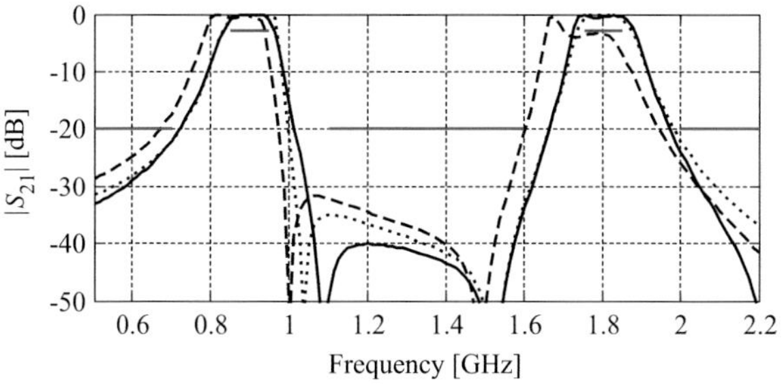

Fig. 7.20 Dual-band bandpass filter (Koziel 2010a): fine model (*dashed line*) and coarse model (*dotted line*) response at $\boldsymbol{x}^{(0)}$, and the optimized fine model response (*solid line*) at the design obtained using shape-preserving response prediction. *Thick horizontal lines* denote design specifications

The coarse model is implemented in Agilent ADS (2011) (Fig. 7.19b). The initial design is $\boldsymbol{x}^{(0)} = [16.14\ 17.28\ 1.16\ 0.38\ 1.18\ 0.98\ 0.98\ 0.20]^T$ mm (the optimal solution of $\boldsymbol{c}$).

The following characteristic points are selected to set up functions F and R: four points for which $|S_{21}| = -20$ dB, four points with $|S_{21}| = -5$ dB, as well as six additional points located between −5 dB points. This setup allows for capturing the location of the passbands as well as the transmission curve slopes. For the purpose of optimization, the coarse model was enhanced by tuning the dielectric constants and the substrate heights of the microstrip models corresponding to the design variables L_1, L_2, d, and g (original values of ε_r and H were 10.2 mm and 0.635 mm, respectively) (Koziel 2010a). The filter was optimized using two versions of SPRP, a regular one and SPRP enhanced by input SM (cf. Table 7.5). Figure 7.20 shows the initial fine model response as well as the fine model response at the design obtained using the shape-preserving response prediction method.

7.3.3 *Optimization with SPRP: Wideband Microstrip Antenna*

This section illustrates the use of SPRP for the design of antenna structures. As an example, consider the antenna shown in Fig. 7.21 (Chen 2008), where $\boldsymbol{x} =$

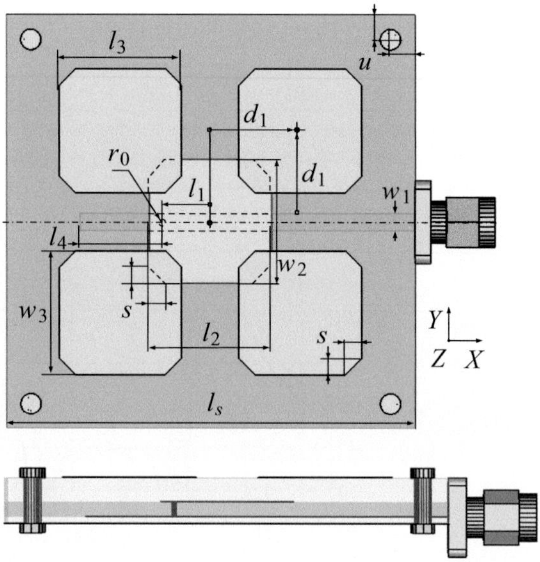

Fig. 7.21 Wideband microstrip antenna (Chen 2008): top and side views. The *dash-dot line* in the top view shows the magnetic symmetry wall (XOY)

$[l_1\, l_2\, l_3\, l_4\, w_2\, w_3\, d_1\, s]^T$ are the design variables. Multilayer substrate is $l_s \times l_s$ ($l_s = 30$ mm). The antenna stack (bottom-to-top) comprises: metal ground, 0.813 mm thick RO4003, microstrip trace ($w_1 = 1.1$ mm), 1.905 mm thick RO3006, and a trace-to-patch via ($r_0 = 0.25$ mm), driven patch, 3.048 mm thick RO4003, and four patches at the top. The antenna stack is fixed with four M1.6 bolts at the corners ($u = 3$ mm). Metallization is with thick 50 μm copper. Feeding is through an edge mount 50 Ω SMA connector with the 10 mm × 10 mm × 2 mm flange. The design objective is $|S_{11}| \leq -10$ dB for 3.1–4.8 GHz. Realized gain not less than 5 dB for the zero zenith angle is an optimization constraint over the frequency band. The initial design is $\boldsymbol{x}^{init} = [-4\ 15\ 15\ 2\ 15\ 15\ 20\ 2]^T$ mm.

Both the high-fidelity model f (2,334,312 mesh cells at the initial design, 160 min of the evaluation time) and the low-fidelity model $\boldsymbol{c}$ (122,713 mesh cells, 3 min of the evaluation time) are simulated using the CST MWS transient solver (CST Microwave Studio 2011). Here, the first step is to find the rough optimum of c, $\boldsymbol{x}^{(0)} = [-4.91\ 15.15\ 15.07\ 2.56\ 14.21\ 14.23\ 21.07\ 2.67]^T$ mm. The computational cost of this step is 82 evaluations of c (which corresponds to about 1.5 evaluations of the high-fidelity model). Figure 7.22a shows the responses of $\boldsymbol{f}$ at $\boldsymbol{x}^{init}$ and $\boldsymbol{x}^{(0)}$, as well as the response of $\boldsymbol{c}$ at $\boldsymbol{x}^{(0)}$. The final design $\boldsymbol{x}^{(4)} = [-5.21\ 15.38\ 15.57\ 2.58\ 14.41\ 13.73\ 21.07\ 2.067]^T$ mm ($|S_{11}| \leq -11$ dB for 3.1 GHz to 4.8 GHz, Fig. 7.22b) is obtained after four iterations of the SPRP-based optimization. The SPRP characteristic points were selected at the −10 dB level with additional 20 points allocated in between them. This allows for capturing the behavior and changes of that part of the response that determine −10-dB bandwidth of the antenna. The gain of the final design is shown in Fig. 7.22c which illustrates that the maximum of

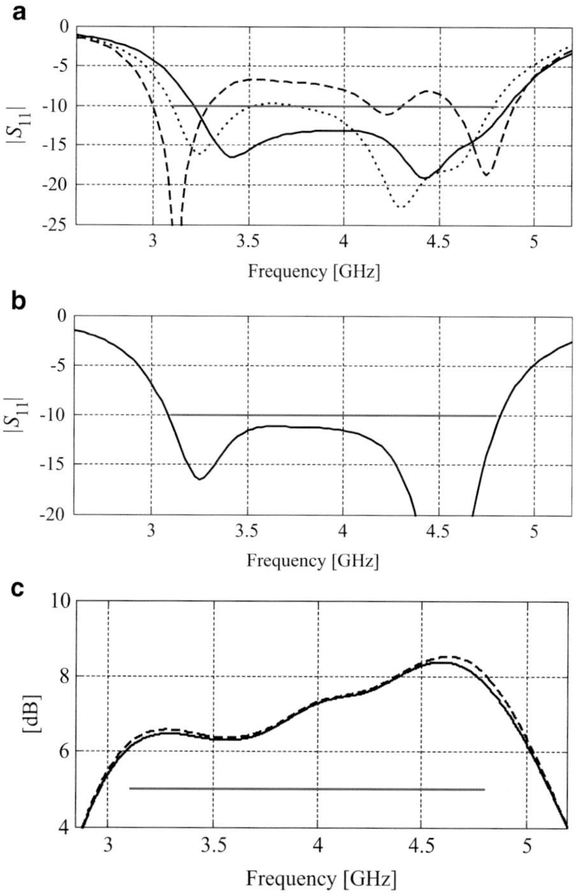

Fig. 7.22 Wideband microstrip antenna (Koziel 2010a): (**a**) high-fidelity model response (*dashed line*) at the initial design $\boldsymbol{x}^{init}$, and high- (*solid line*) and low-fidelity (*dotted line*) model responses at the approximate low-fidelity model optimum $\boldsymbol{x}^{(0)}$; (**b**) high-fidelity model $|S_{11}|$ at the final design; (**c**) realized gain at the final design for the zero zenith angle (*solid line*, XOZ co-pol.) and realized peak gain (*dash line*). Design constraint is shown with the *horizontal line* at the 5 dB level

Table 7.6 Wideband microstrip antenna: optimization cost (Koziel 2010a)

Algorithm component	Number of model evaluations	Evaluation time	
		Absolute (h)	Relative to $\boldsymbol{f}$
Evaluation of $\boldsymbol{c}^a$	$289 \times \boldsymbol{c}$	14.4	5.4
Evaluation of $\boldsymbol{f}^b$	$5 \times \boldsymbol{f}$	13.3	5.0
Total optimization time	N/A	27.7	10.4

[a]Includes initial optimization of $\boldsymbol{c}$ and optimization of SPRP surrogate
[b]Excludes evaluation of $\boldsymbol{f}$ at the initial design

radiation points along the zero zenith angle closely over the bandwidth of interest. The total design cost corresponds to about ten evaluations of the high-fidelity model (Table 7.6).

7.3.4 Optimization with SPRP: Airfoil Design

This section demonstrates the use of the SPRP technique for the aerodynamic design of airfoil sections at transonic flow conditions (Koziel and Leifsson 2013c). The airfoil shapes are parameterized with three parameters of the NACA four-digit method described in Sect. 7.1.4 with the design variable vector being $\boldsymbol{x} = [m\ p\ t/c]^T$. Moreover, the same CFD models are used here as described in Sect. 7.1.4.

In aerodynamic shape optimization, the SPRP technique is applied to the pressure distribution ($C_p(\boldsymbol{x})$) on the airfoil surface (Koziel and Leifsson 2013c). Figure 7.23 shows the pressure distributions of two different designs obtained by the low-fidelity model. Shown are the characteristic points (red circles) and the translation vectors (blue lines) at important areas of the distributions. One of the critical parts is the location and the strength of the pressure shock. The application of the translation vectors to the high-fidelity model distributions is shown in Fig. 7.24.

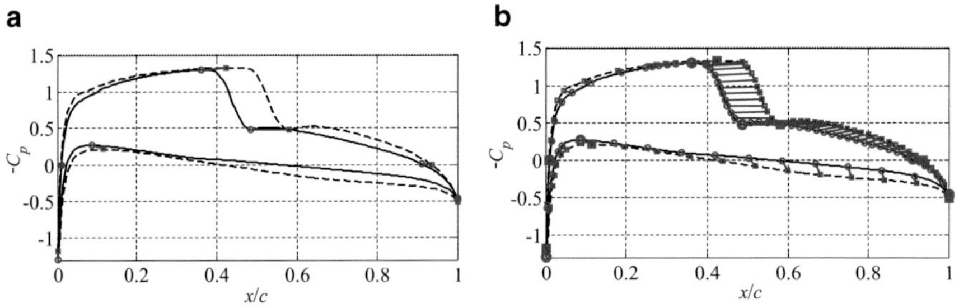

Fig. 7.23 An illustration of the SPRP technique applied to the pressure distributions obtained by the low-fidelity CFD models of two designs (Koziel and Leifsson 2013c): (**a**) initial characteristic points and translation vectors, (**b**) additional points

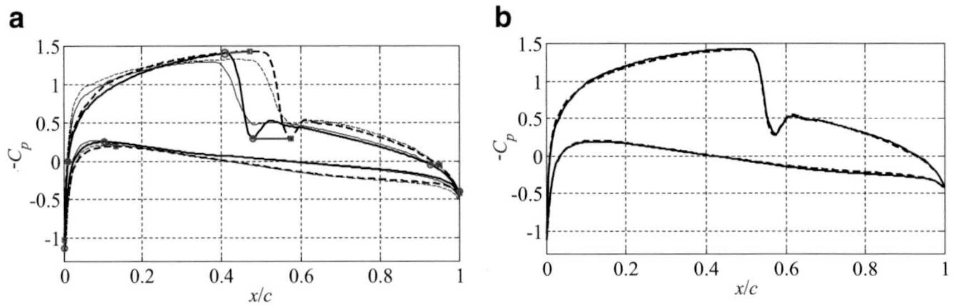

Fig. 7.24 Application of SPRP to the high-fidelity CFD model responses (*thick lines*) with (Koziel and Leifsson 2013c): (**a**) initial characteristic points and translation vectors (coarse model distributions are shown with *thin lines*), and (**b**) comparison of the actual and the predicted (*dash*) high-fidelity response

The design objective is to maximize the section lift coefficient ($C_l(\boldsymbol{x})$) subject to constraints on the section drag coefficient ($C_{dw}(\boldsymbol{x})$) and the nondimensional cross-sectional area ($A(\boldsymbol{x})$). The problem is formulated as minimization of the high-fidelity model $\boldsymbol{f}(\boldsymbol{x}) = -C_l(\boldsymbol{x})$ subject to $g_1(\boldsymbol{x}) = C_{dw}(\boldsymbol{x}) - C_{dw.max} \leq 0$, and $g_2(\boldsymbol{x}) = A_{min} - A(\boldsymbol{x}) \leq 0$, where $C_{dw.max} = 0.0041$ is the maximum drag and $A_{min} = 0.065$ the minimum cross section. The free-stream Mach number is $M_\infty = 0.75$ and the angle of attack $\alpha = 1°$. The variable bounds are $0 \leq m \leq 0.1$, $0.2 \leq p \leq 0.8$, and $0.05 \leq t \leq 0.20$. The initial design is $\boldsymbol{x}^{init} = [0.03\ 0.2\ 0.1]^T$.

Due to an unavoidable misalignment between the pressure distributions of the high-fidelity model and its SPRP surrogate, it is not convenient to handle the drag constraint directly, because the design that is feasible for the surrogate model may not be feasible for the high-fidelity model. Therefore, the objective function is defined with a penalty function as in (7.10). The cross-sectional area constraint is handled directly.

To meet the design goals, the optimizer does three fundamental shape changes:

1. The maximum ordinate of the mean camber line (m) is reduced.
2. The location of the maximum ordinate of the mean camber line (p) is moved aft, thus increasing the trailing-edge camber.
3. The thickness-to-chord ratio (t/c) is reduced.

Shape changes (1) and (3) reduce the shock strength and, thus, reduce the drag coefficient. The associated change in the pressure distribution reduces the lift coefficient. However, shape change (2) improves (or recovers a part of) the lift by opening up the pressure distribution behind the shock. These effects can be seen in the pressure distribution plot in Fig. 7.25, and the Mach contour plots in Fig. 7.26.

The optimization problem is solved by the direct optimization of the high-fidelity model using the pattern search algorithm, as well as by the SPRP algorithm. The results are presented in Table 7.7. It can be seen that both approaches are able to meet the design goals and produce similar optimized airfoil shapes. The direct approach requires 120 high-fidelity model evaluations (N_f). The SPRP algorithm requires 330 low-fidelity model evaluations (N_c) and 11 high-fidelity ones, yielding a total cost of less than 18 equivalent high-fidelity model evaluations.

7.4 Summary and Discussion

The most important difference between the nonparametric response correction methods discussed here and the parametric techniques presented in Chap. 6 is that the former methods are not described by any explicit formulas with parameters that need to be extracted in order to identify the surrogate model. The nonparametric methods attempt to exploit the knowledge embedded in the low-fidelity model to the fuller extent than what is done in parametric surrogates. Below, we provide a brief qualitative comparison of both groups of methods, as well as formulate some guidelines and recommendations for the readers interested in applying these

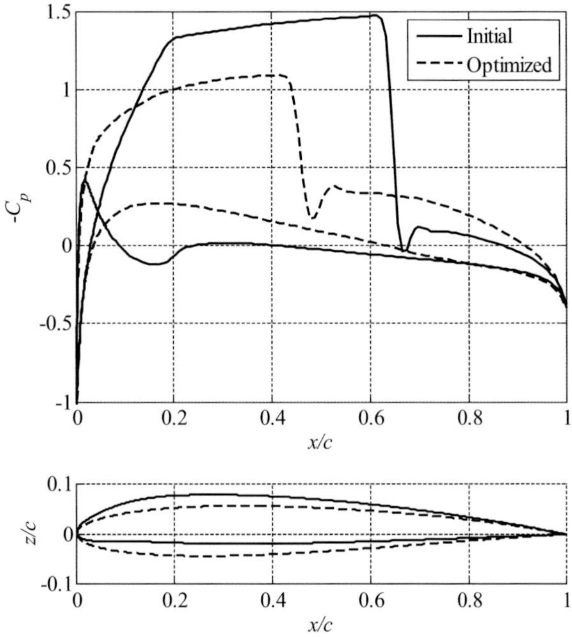

Fig. 7.25 Airfoil optimization results (Koziel and Leifsson 2013c): initial and optimized airfoils pressure distributions and shapes

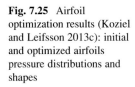

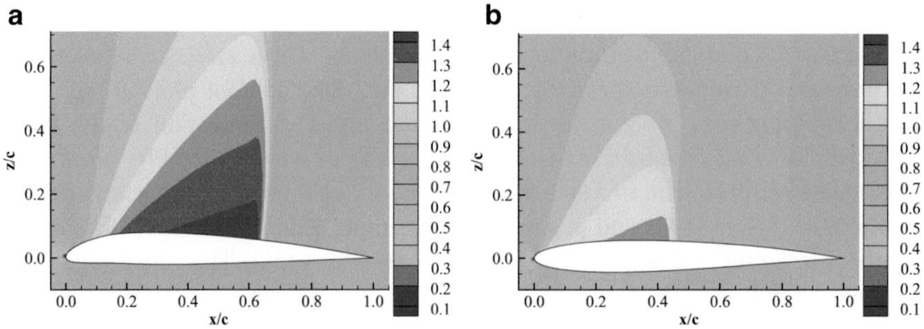

Fig. 7.26 Airfoil optimization results (Koziel and Leifsson 2013c): Mach contours at (**a**) the initial design, (**b**) the optimized design

methods in their research and/or design work. The main factors are complexity of implementation, computational efficiency, robustness, as well as range of applications.

In terms of implementation, the parametric response correction techniques are very straightforward. Techniques such as output space mapping, manifold mapping or multi-point response correction construct the surrogate model by using analytical formulas, therefore, they are easy to implement. Nonparametric methods may be more involved because the surrogate models are constructed by considering some auxiliary quantities such as the scaling functions (adaptive response correction) or

Table 7.7 Results for the airfoil design optimization (Koziel and Leifsson 2013c)

Variable	Initial	Direct (pattern search)	SPRP
m	0.0300	0.0080	0.0090
p	0.2000	0.6859	0.6732
t/c	0.1000	0.1044	0.1010
C_l	0.8035	0.4641	0.4872
C_{dw}	0.0410	0.0041	0.0040
A	0.0675	0.0703	0.0680
N_c	N/A	0	330
N_f	N/A	120	11
Total cost	N/A	120	<18

characteristic points and translation vectors (SPRP), which generally depend on the response shape. Also, in SPRP, the user is responsible for defining the characteristic points on case-to-case basis.

In terms of computational complexity, nonparametric methods prove to be more efficient. The reason is that ARC, ARP, and SPRP exploit the knowledge embedded in the low-fidelity model to a larger extent than the parametric methods which only do a local model alignment. As indicated by the examples, ARC and SPRP are capable of yielding a satisfactory design after few iterations, whereas output SM or MM typically require more iterations. On the other hand, the parametric techniques tend to be more robust, particularly when embedded in the trust-region framework which improves their convergence properties. Parametric techniques are also more generic than ARC, SPRP and AADS. For the latter, some considerations are necessary in order to select a proper realization of the scaling function (ARC), the definition of the characteristic points (SPRP), or the analysis of the model responses to properly modify design specifications (AADS). Also, the SPRP technique assumes one-to-one correspondence between characteristic points of the coarse and fine models at all considered design, which may be difficult to satisfy for certain problems.

Based on the above remarks, parametric methods, particularly output SM, are recommended for less experienced users, and when the underlying coarse model is relatively fast (e.g., equivalent circuit). More experienced users are encouraged to try either ARC or SPRP, particularly if the available low-fidelity model is relatively expensive (e.g., obtained from computer simulations). Also, these methods are expected to outperform parametric methods if certain a priori information is available, e.g., regarding the response shapes, which can be utilized to identify the characteristic points, etc. In either case, the quality of the low-fidelity model is one of the key factors, and therefore its preconditioning using suitably selected space mapping transformation may be recommended.

Chapter 8
Expedited Simulation-Driven Optimization Using Adaptively Adjusted Design Specifications

The focus of this chapter is the adaptively adjusted design specifications (AADS) technique, which, in some sense, is qualitatively different from the SBO methods discussed so far in this book. The key component of the majority of the SBO techniques is a surrogate model, constructed so as to represent well (in a given sense that might be problem dependent) the high-fidelity model being the subject of the optimization process. In the case of physics-based models, the surrogate is created through a suitable correction of an underlying low-fidelity model. A number of correction techniques have been discussed in detail in the preceding chapters of this book. The perspective offered by AADS is different: instead of correcting the low-fidelity model it modifies the design requirements for the problem at hand. The modifications are implemented to account for the discrepancies between the low- and the high-fidelity models. Given that the models are sufficiently well correlated, optimization of the (intact) low-fidelity model leads to design improvement of the high-fidelity one. There is no need for any model correction, which is an important advantage of AADS, therefore making it simple to implement. On the other hand, AADS can only be used for certain types of problems, including those for which the design specifications can be expressed in a minimax form. In this chapter, we formulate the AADS methodology and illustrate its use for solving various design problems related to computational electromagnetics, such as design of microwave filters, antennas, and high-frequency transition structures.

8.1 Adaptively Adjusted Design Specifications: Concept and Formulation

A majority of the physics-based surrogate-assisted optimization techniques—including those presented in this book so far—aim at correcting the underlying low-fidelity model so that it becomes, at least locally, an accurate representation of

S. Koziel, L. Leifsson, *Simulation-Driven Design by Knowledge-Based Response Correction Techniques*, DOI 10.1007/978-3-319-30115-0_8

the high-fidelity model. An alternative way of exploiting low-fidelity models in simulation-driven optimization is to modify the design specifications in such a way that the updated specifications reflect the discrepancy between the models. While this approach may not be universally applicable, it is extremely simple to implement because no changes of the low-fidelity model are necessary.

The adaptively adjusted design specifications (AADS) technique introduced in Koziel (2010b) consists of the following two steps:

1. Modify the design specifications of the original problem at hand in order to take into account the differences between the responses of the high-fidelity model $\boldsymbol{f}$ and the low-fidelity model $\boldsymbol{c}$ at their characteristic points.
2. Obtain a new design by optimizing the low-fidelity model with respect to the modified specifications.

AADS has been developed to handle minimax-type of specifications (Koziel 2010b) so that the characteristic points of the responses should correspond to the relevant design specification levels. They may also include local maxima/minima of the respective responses at which the specifications may not be satisfied. The key concepts of selecting and handling the characteristic points are explained in Fig. 8.1. Figure 8.1a shows the high- and low-fidelity model responses at the optimal design of $\boldsymbol{c}$, corresponding to the bandstop filter example considered in Koziel (2010f); the minimax design specifications are indicated using horizontal lines (here: $|S_{21}| \leq -30$ dB for the frequencies 12–14 GHz, and $|S_{21}| \geq -3$ dB for the frequencies 8–9 GHz and 17–18 GHz). Figure 8.1b shows the characteristic points of $\boldsymbol{f}$ and $\boldsymbol{c}$ for the bandstop filter example. The points correspond to -3 and -30 dB levels as well to the local maxima of the responses. As one can observe in Fig. 8.1b the selection of points is rather straightforward.

As mentioned above, in the first step of the AADS optimization procedure, the design specifications are modified (or mapped) so that the level of satisfying/violating the modified specifications by the low-fidelity model response corresponds to the satisfaction/violation levels of the original specifications by the high-fidelity model response. In the example of Fig. 8.1, for each edge of the specification line, the edge frequency is shifted by the difference of the frequencies of the corresponding characteristic points, e.g., the left edge of the specification line of -30 dB is moved to the right by about 0.7 GHz, which is equal to the length of the line connecting the corresponding characteristic points in Fig. 8.1b. Similarly, the specification levels are shifted by the difference between the local maxima/minima values for the respective points, e.g., the -30 dB level is shifted down by about 8.5 dB because of the difference of the local maxima of the corresponding characteristic points of $\boldsymbol{f}$ and $\boldsymbol{c}$. Modified design specifications are shown in Fig. 8.1c.

The low-fidelity model is subsequently optimized with respect to the modified specifications and the new design obtained this way is treated as an approximated solution to the original design problem (i.e., optimization of the fine model with respect to the original specifications). Steps 1 and 2 (listed above) can be repeated if necessary. If the correlation between the low- and high-fidelity models is good, a

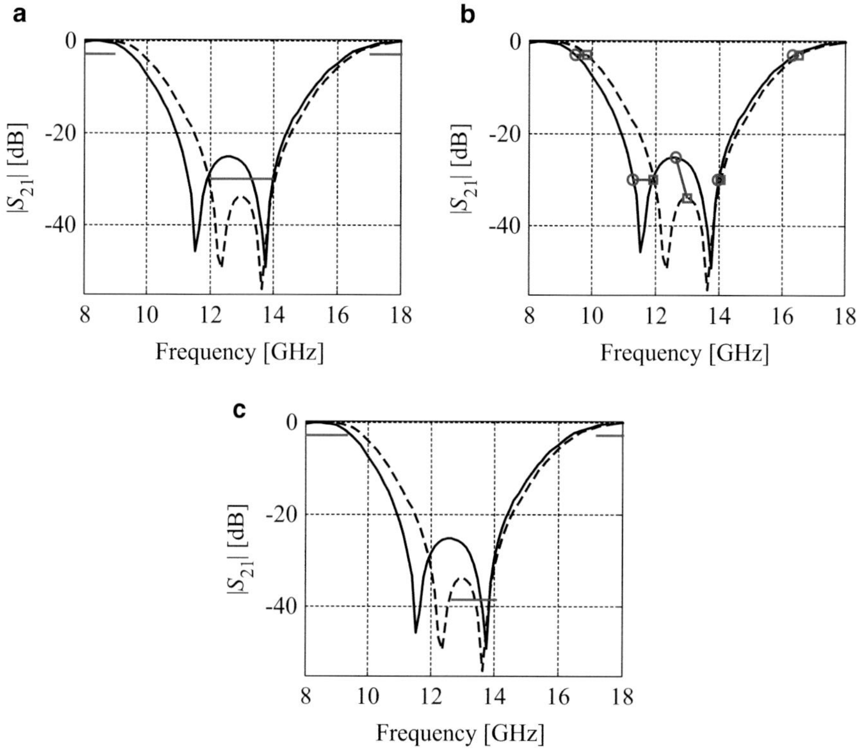

Fig. 8.1 Bandstop filter example (responses of ***f*** and ***c*** are marked with *solid* and *dashed line*, respectively): (**a**) fine and coarse model responses at the initial design (optimum of ***c***) as well as the original design specifications, (**b**) characteristic points of the responses corresponding to the specification levels (here, −3 and −30 dB) and to the local response maxima, (**c**) fine and coarse model responses at the initial design and the modified design specifications

substantial design improvement is typically observed after the first iteration; however, additional iterations may bring further enhancements (Koziel 2010b).

It should be mentioned that AADS is considerably simpler than many other SBO techniques presented before. In particular, there is no need to alter the low-fidelity model in any way as the interaction between the models is only through the design specifications. The lack of extractable parameters is its additional advantage compared to some other approached (e.g., space mapping) because the computational overhead related to the parameter extraction, while negligible for very fast coarse models, may substantially increase the overall design cost if the coarse model is relatively expensive (e.g., implemented through coarse-discretization simulation).

Figure 8.2 illustrates an iteration of the adaptively adjusted design specifications technique applied to microstrip-to-SIW transition design (Ogurtsov and Koziel 2011). It can be observed that optimizing the low-fidelity model with respect to the modified specifications results in improving the high-fidelity model design with

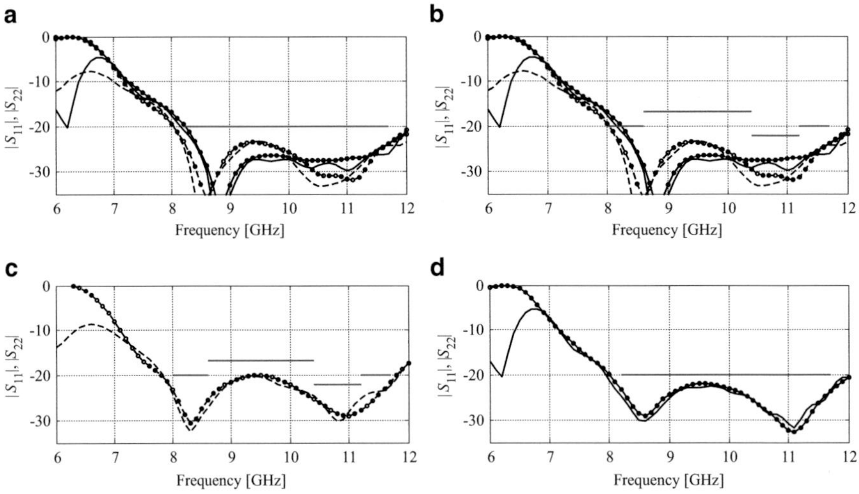

Fig. 8.2 Illustration of the adaptively adjusted design specification technique applied to optimize microstrip-to-SIW transitions. High- and low-fidelity model response denoted as *solid* and *dashed lines*, respectively. $|S_{22}|$ distinguished from $|S_{11}|$ using *circles*. Design specifications denoted by *thick horizontal lines*. (**a**) High- and low-fidelity model responses at the beginning of the iteration as well as original design specifications; (**b**) high- and low-fidelity model responses and modified design specifications that reflect the differences between the responses; (**c**) low-fidelity model optimized with respect to the modified specifications; (**d**) high-fidelity model at the low-fidelity model optimum shown versus original specifications

respect to the original specifications. Because the discrepancy between the high- and low-fidelity models may change somehow from one design to another, a few iterations may be necessary to find an optimal high-fidelity model design.

8.2 AADS for Design Optimization of Microwave Filters

In this section, we present an application of AADS for design optimization of microwave filters. As indicated in Sect. 8.2.2, in the case of considerable misalignment between the models it might be advantageous to carry out low-fidelity model preconditioning using, e.g., space mapping (Koziel 2010b).

8.2.1 Bandpass Microstrip Filter

The first example is the bandpass microstrip filter with an open stub inverter (Lee et al. 2000) as shown in Fig. 8.3a. The design parameters are $\boldsymbol{x} = [L_1\, L_2\, L_3\, S_1\, S_2\, W_1]^T$. The fine model is simulated using the full-wave electromagnetic solver FEKO (2008). The design specifications are $|S_{21}| \leq -20$ dB for 1.5 GHz $\leq \omega \leq$ 1.8 GHz,

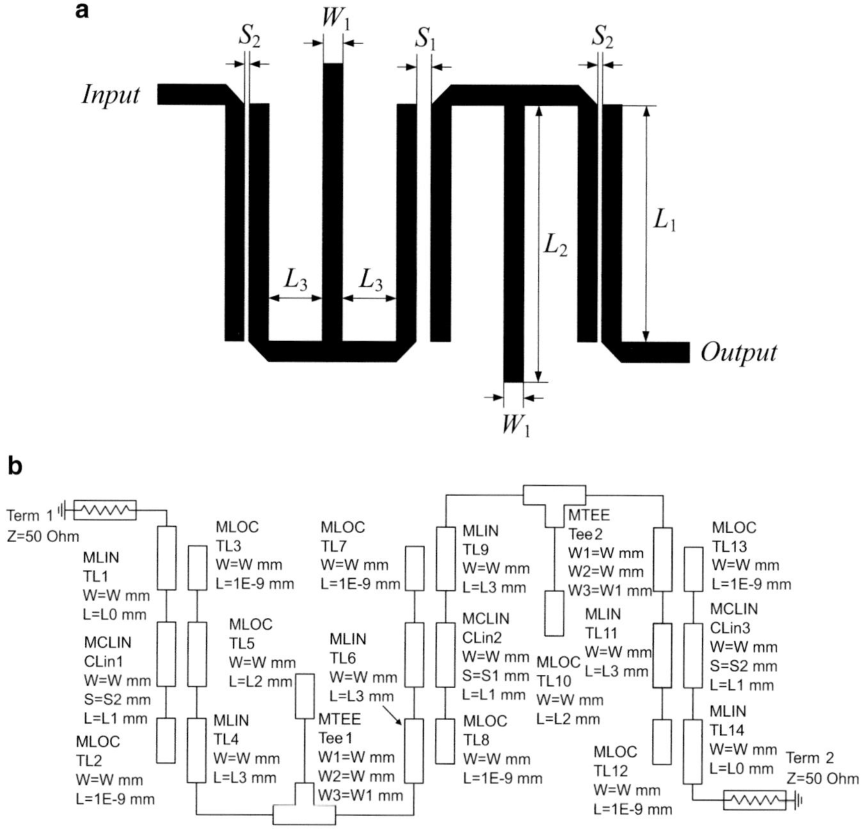

Fig. 8.3 Bandpass filter with open stub inverter: (**a**) geometry (Lee et al. 2000); (**b**) low-fidelity model (Agilent ADS)

$|S_{21}| \geq -3$ dB for 1.95 GHz $\leq \omega \leq$ 2.05 GHz and $|S_{21}| \leq -20$ dB for 2.2 GHz $\leq \omega \leq$ 2.5 GHz. The coarse model is implemented in a circuit simulator Agilent ADS (2011), Fig. 8.3b. The initial design is the coarse model optimal solution $\boldsymbol{x}^{(0)} = [25.00\ 5.00\ 1.221\ 0.652\ 0.187\ 0.100]^T$ mm (minimax specification error +15.7 dB).

The first iteration of our optimization procedure already yielded a design satisfying the specifications, $\boldsymbol{x}^{(1)} = [23.79\ 5.00\ 1.00\ 0.694\ 0.192\ 0.10]^T$ mm (specification error −0.6 dB). After the second iteration, the design was further improved to $\boldsymbol{x}^{(2)} = [23.68\ 5.00\ 1.00\ 0.717\ 0.193\ 0.10]^T$ mm (specification error −1.7 dB). Figure 8.4 shows the fine and coarse model responses at $\boldsymbol{x}^{(0)}$ and the fine model response at the final design. For the sake of comparison, the filter was also optimized using the frequency SM algorithm (Koziel 2010b). The design obtained in three iterations, $[23.66\ 5.00\ 1.00\ 0.654\ 0.188\ 0.100]^T$ mm, satisfies the design specifications; however, it is not as good as the one obtained using AADS (specification error −0.8 dB).

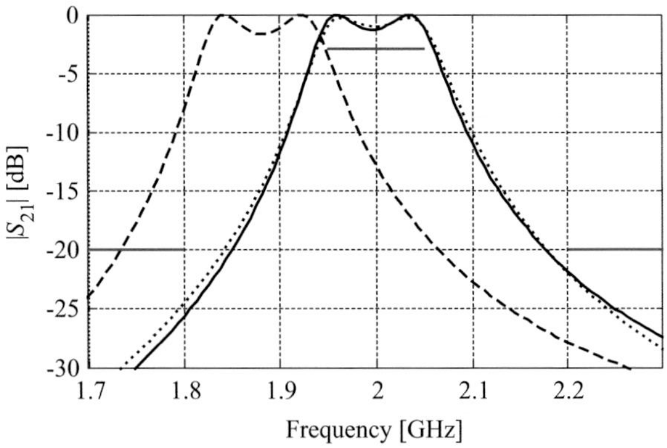

Fig. 8.4 Bandpass filter with open stub inverter: high-fidelity model response (*solid line*) at the final design obtained after two iterations of our optimization procedure; high- (*dashed line*) and low-fidelity (*dotted line*) model responses at the initial design

8.2.2 *Third-Order Chebyshev Bandpass Filter*

In the case of considerable misalignment between the low- and high-fidelity model responses, AADS may not work well. However, in some situations, using a different reference design for the high- and low-fidelity models may be helpful. In particular, the low-fidelity model can be optimized with respect to the modified specifications starting not from $\boldsymbol{x}^{(0)}$ (the optimal solution of $\boldsymbol{c}$ with respect to the original specifications), but from another design, say $\boldsymbol{x}_c^{(0)}$, at which the response of $\boldsymbol{c}$ is as similar to the response of $\boldsymbol{f}$ at $\boldsymbol{x}^{(0)}$ as possible. Such designs can be obtained as follows (Koziel 2010b, f):

$$\boldsymbol{x}_c^{(0)} = \arg\min_{z} ||\boldsymbol{f}(\boldsymbol{x}^{(0)}) - \boldsymbol{c}(z)||. \tag{8.1}$$

At the iteration i of the AADS procedure, the optimal design of the low-fidelity model $\boldsymbol{c}$ with respect to the modified specifications, $\boldsymbol{x}_c^{(i)}$, has to be translated to the corresponding fine model design, $\boldsymbol{x}^{(i)}$, as follows $\boldsymbol{x}^{(i)} = \boldsymbol{x}_c^{(i)} + (\boldsymbol{x}^{(0)} - \boldsymbol{x}_c^{(0)})$. Note that the preconditioning procedure (8.1) is performed only once for the entire optimization process. The idea of coarse model preconditioning is borrowed from space mapping (more specifically, from the original space mapping concept, Bandler et al. 2004a). In practice, the coarse model can be "corrected" to reduce its misalignment with the high-fidelity model using any available degrees of freedom, for example, preassigned parameters as in implicit space mapping (Cheng et al. 2008b).

For the sake of illustration, consider the third-order Chebyshev bandpass filter (Kuo et al. 2003) shown in Fig. 8.5a. The design variables are $\boldsymbol{x} = [L_1\, L_2\, S_1\, S_2]^T$ mm; $W_1 = W_2 = 0.4$ mm. The fine model is simulated in Sonnet ***em*** (Sonnet 2010).

The design specifications are $|S_{21}| \geq -3$ dB for 1.8 GHz $\leq \omega \leq$ 2.2 GHz, and $|S_{21}| \leq -20$ dB for 1.0 GHz $\leq \omega \leq$ 1.6 GHz and 2.4 GHz $\leq \omega \leq$ 3.0 GHz. The low-fidelity model is implemented in Agilent ADS (2011) as shown in Fig. 8.5b.

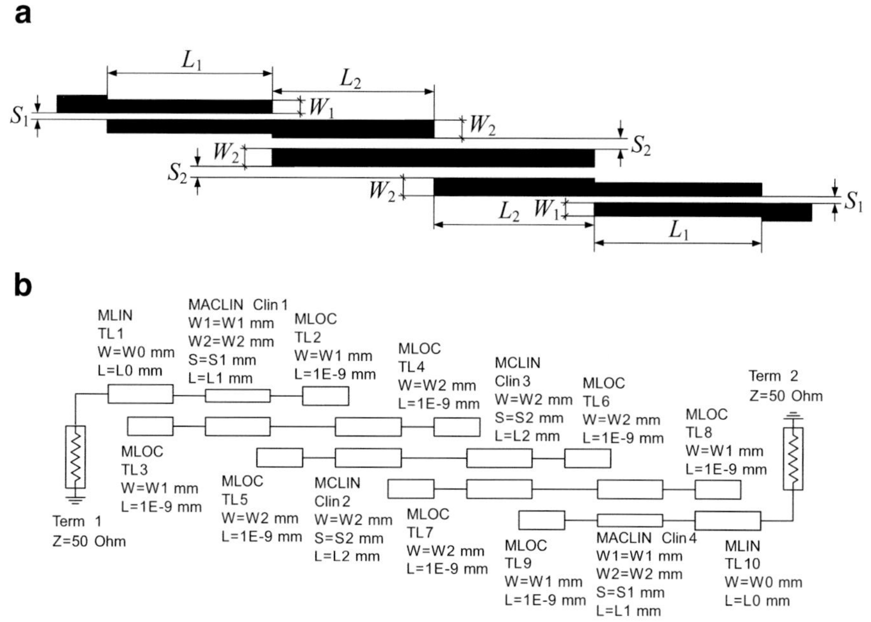

Fig. 8.5 Third-order Chebyshev bandpass filter: (**a**) geometry (Kuo et al. 2003); (**b**) low-fidelity model (Agilent ADS)

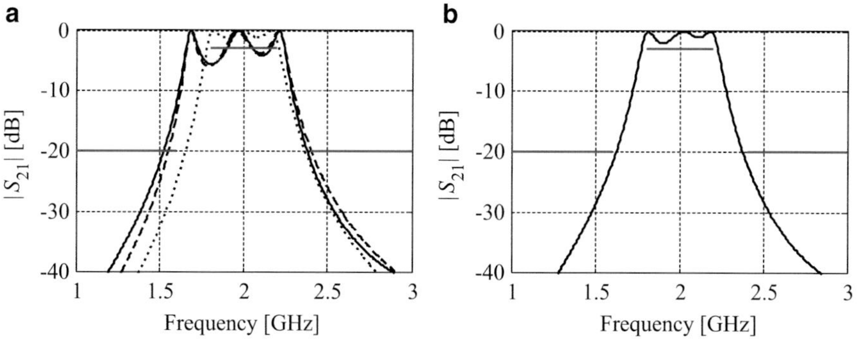

Fig. 8.6 Third-order Chebyshev filter (Kuo et al. 2003): (**a**) high- (*solid line*) and low-fidelity model (*dotted line*) responses at the initial design $\boldsymbol{x}^{(0)}$ and the low-fidelity model response (*dashed line*) design at $\boldsymbol{x}_c^{(0)}$; (**b**) high-fidelity model response at the design obtained after two iterations of generalized AADS

Figure 8.6a shows the high- and low-fidelity model responses at the initial design $\boldsymbol{x}^{(0)} = [14.6\ 15.3\ 0.56\ 0.53]^T$ mm. Large ripples in the passband of the fine model response and small ripples of the coarse model response prevent us from applying AADS directly. Using (8.1), a new design $\boldsymbol{x}_c^{(0)} = [14.56\ 15.8\ 0.687\ 0.316]^T$ mm was found so that $\boldsymbol{c}(\boldsymbol{x}_c^{(0)})$ is very similar to $\boldsymbol{f}(\boldsymbol{x}^{(0)})$ as shown in Fig. 8.6b. Using $\boldsymbol{x}_c^{(0)}$, the generalized AADS procedure was executed yielding $\boldsymbol{x}^{(1)} = [14.6\ 14.8\ 0.43\ 0.76]^T$ mm (specification error is −0.5 dB) and $\boldsymbol{x}^{(2)} = [14.6\ 14.8\ 0.42\ 0.82]^T$ mm (specification error −1.1 dB).

8.3 AADS for Design Optimization of Antennas

In this section, we demonstrate the use of AADS for design optimization of antenna structures. One of the issues here is that the low-fidelity antenna models are normally implemented through coarse-discretization EM simulations (Koziel and Ogurtsov 2014a). Consequently, they are relatively expensive. From the point of view of computational efficiency, simplicity of AADS (in particular, no need to extract any surrogate model parameters) is an important advantage.

8.3.1 Ultra-Wideband Monopole Antenna

As a first example consider the ultra-wideband (UWB) monopole antenna shown in Fig. 8.7. The monopole is on a 0.508 mm thick Rogers RO3203 substrate. Design variables are $\boldsymbol{x} = [h_0\ w_0\ a_0\ s_0\ h_1\ w_1\ l_{gnd}\ w_s]^T$. Other parameters are fixed: $l_s = 25$, $w_m = 1.25$, $h_p = 0.75$ (all in mm).

The microstrip input of the monopole is fed through an edge-mount SMA connector (SMA 2013) having a hex nut. The ground of the monopole has a profiled edge. Simulation time of $\boldsymbol{c}$ (152,640 mesh cells) is 2 min, and that of $\boldsymbol{f}$ (1,151,334 mesh cells) is 45 min (both at the initial design). Both models are evaluated using the transient solver of CST Microwave Studio (CST 2013). The design specifications for the reflection response are $|S_{11}| \leq -10$ dB for

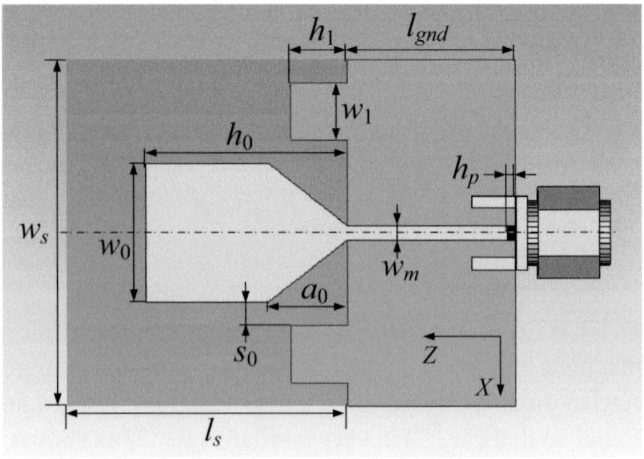

Fig. 8.7 UWB monopole: top view, substrate shown transparent. Magnetic-symmetry wall is shown with the *dash-dot line*

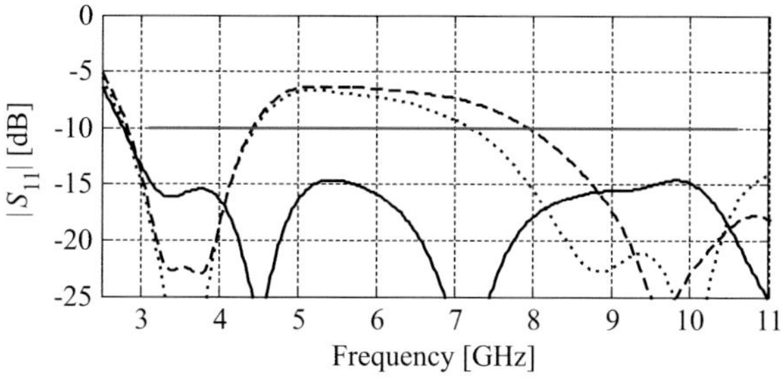

Fig. 8.8 UWB monopole: high-(*dashed line*) and low-fidelity (*dotted line*) model responses at the initial design $\boldsymbol{x}^{(0)}$, as well as fine (*solid line*) at the final design $\boldsymbol{x}^{(3)}$

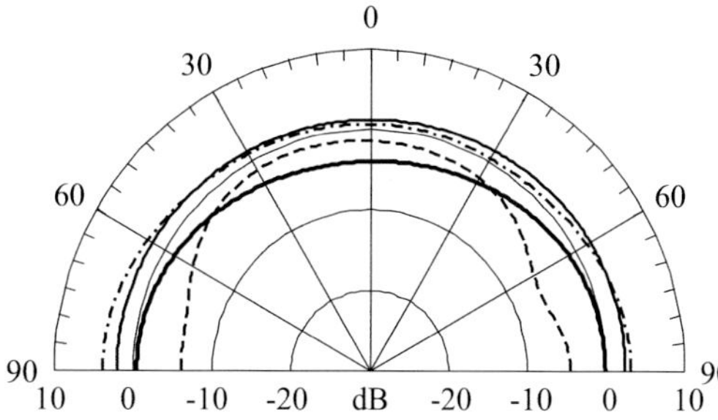

Fig. 8.9 Realized gain (x-pol.) of the UWB monopole: pattern cut in the *XOY* plane at 3 GHz (*solid line*), 5 GHz (*thick solid line*), 7 GHz (*dash-dotted line*), and 9 GHz (*dashed line*). 90° on the left, 0°, and 90° on the right are for *Y*, *X*, and −*Y* directions, respectively

Table 8.1 UWB monopole: optimization cost

		CPU time	
Algorithm component	Number of model evaluations	Absolute	Relative to $\boldsymbol{f}$
Optimizing $\boldsymbol{c}$	$252 \times \boldsymbol{c}$	8 h 24 min	11.2
Evaluation of $\boldsymbol{f}$	$3 \times \boldsymbol{f}$	2 h 15 min	3.0
Total cost[a]	N/A	10 h 39 min	14.2

[a]Excludes $\boldsymbol{f}$ evaluation at the initial design

3.1–10.6 GHz. In addition, the radiation pattern of the monopole is to be omnidirectional in the *XOY* plane.

The initial design is $\boldsymbol{x}^{(0)} = [18\ 12\ 2\ 0\ 5\ 1\ 15\ 40]^T$ mm. Optimization performed using AADS yields the final design $\boldsymbol{x}^{(3)} = [18.27\ 19.41\ 2.02\ 1.34\ 1.95\ 5.83\ 15.74\ 35.75]^T$ mm ($|S_{11}| < -14.5$ dB in the frequency band of interest) obtained after only three iterations of the procedure. Figure 8.8 shows the reflection responses of the high- and low-fidelity models at the initial design as well as the $\boldsymbol{f}$ response at the final design. The far-field response of the final design is shown in Fig. 8.9. The total number of evaluations of $\boldsymbol{c}$ in the optimization process is 252. Table 8.1 shows the computational cost of the optimization: the total optimization time corresponds to about 11 evaluations of the high-fidelity model.

8.3.2 Planar Yagi Antenna

Consider a planar Yagi antenna shown Fig. 8.10. The structure is supposed to operate in the 2.4–2.5 GHz frequency band (Koziel and Ogurtsov 2010). Optimization of planar Yagi antennas on finite substrate is a challenging task due to the finite substrate and proximity of the feeding circuitry to the radiators both introducing additional degrees of freedom to the design as well as complicate the use of methods developed for Yagi aerials (Chen and Cheng 1975; Cheng and Chen 1973). The antenna comprises three directors, one driving element of a modified shape consisting of partially overlapping strips, and the feeding microstrip ground plane serving also as the reflector. The presented antenna can be viewed as a planar realization of the five-element Yagi. The antenna components are defined on a single layer of 0.025″ thick Rogers RT6010 substrate which has the size of 100 mm × 160 mm. The ground dimensions are 100 mm × 40 mm. The input 50 Ω microstrip is to be interfaced to the terminals of the driving element through a section of the parallel strip transmission line in a way that provides the balanced input to the antenna. The antenna model is defined with CST (2013).

The primary design objective is maximization of the antenna directivity of the principal polarization (E-field is parallel to XOZ plane) in the 2.4–2.5 GHz. Furthermore, the following antenna figures are handled through design constraints (also in the 2.4–2.5 GHz band): the side lobe level relative to maximum (SLL < -10 dB), front-to-back ratio (FBR < -12 dB), direction of maximal radiation θ_m (elevation angle from the Z axis, $|\theta_m| < 1°$). The antenna should be interfaced to the 50 Ω environment so that $|S_{11}| < -10$ dB in the 2.4–2.5 GHz band.

As the input impedance of Yagi antennas is typically sensitive to variations of antenna dimensions (Balanis 2005), and since its value is not available prior to simulation while it is needed to define the feeding part of the antenna, the design

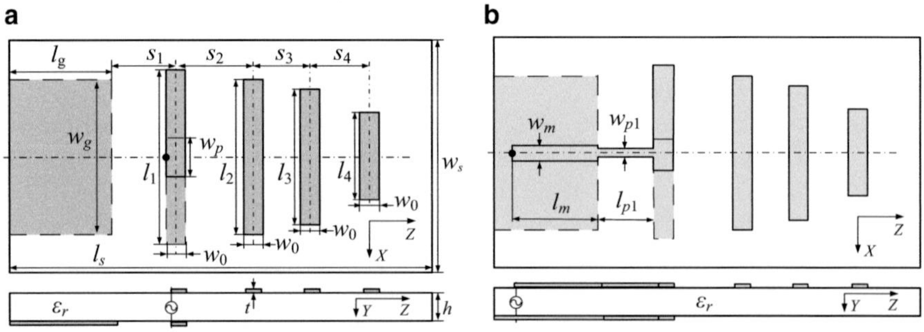

Fig. 8.10 Printed quasi-Yagi antenna: (**a**) the model used at the optimization stage of design (no feeding section), (**b**) the model updated with a feeding section starting from the 50 Ω microstrip. Source impedance is not shown at the diagrams. For simplicity, the feeding section at the panel (**b**) is shown as a simple two section structure: 50 Ω microstrip (dimensions l_m and w_m) and parallel strips (dimensions l_{p1} and w_{p1}). Detailed geometry of the optimized feeding section is given with Fig. 8.12a

optimization proceeds in two major steps: (1) the antenna is optimized for maximal directivity subject to the constraints on SLL, FBR, and θ_m; at this step the excitation is applied directly at the driving element's terminals (Fig. 8.10a); design optimization procedure is based on surrogate-based optimization and involves optimization of a coarse-discretization antenna model; (2) having the optimal antenna dimensions, the feed interfacing the 50 Ω input and the driving element terminals is designed.

For the sake of computational efficiency, optimization of the antenna is carried out at the coarse-discretization EM model level. However, the coarse-discretization model is of limited accuracy: the figures of interest (e.g., directivity, SLL, FBR) are shifted in frequency with respect to those of the high-fidelity model. The frequency relationship between the two models using characteristic points (e.g., local maxima, points of corresponding response levels) is captured using the AADS procedure as shown in Fig. 8.11. Using this relationship, the original frequency band of interest is mapped into the corresponding band that is used in the optimization of the coarse-discretization model.

The design variables when optimizing the antenna for maximal directivity are $\boldsymbol{x} = [l_1\, l_2\, l_3\, l_4\, s_1\, s_2\, s_3\, s_4\, w_0\, w_p]^T$ (Fig. 8.10a). Other parameters are fixed: $l_s = 160$, $w_s = 100$, $l_g = 40$, $w_g = 100$, and $h = 0.635$ (all in mm). The initial design is $\boldsymbol{x}^{(0)} = [40.42\; 35.7\; 31.5\; 27.3\; 17.85\; 22.05\; 22.05\; 22.05\; 2.35\; 1.5]^T$. Simulation time of the coarse-discretization model (51,580 cells at $\boldsymbol{x}^{(0)}$) is about 6 min, and it is about 2 h for the original, high-fidelity model (1,096,980 cells at $\boldsymbol{x}^{(0)}$). The optimum is found at $\boldsymbol{x}^* = [40.87\; 37.31\; 34.33\; 29.80\; 17.35\; 22.55\; 23.05\; 24.55\; 1.55\; 2.13]^T$.

Based on the optimum $\boldsymbol{x}^*$ and the antenna impedance at the driving element terminals Z_t (Fig. 8.10), a feed is designed (Fig. 8.12) using analytical formulas with a microstrip ($l_m = 35$ mm, $w_m = 0.586$ mm) and parallel strips ($l_{p1} = s_1 - w_0/2$, $w_{p1} = 0.36$ mm). The updated antenna model is then simulated. Its reflection does not meet the design specifications for frequency over 2.484 GHz. Therefore, the feed is redesigned with geometry of Fig. 8.12a through optimization of its full-wave model and a schematic of Fig. 8.12b. Dimensions of the simple feed are used as an initial guess. Optimal feed dimensions are found to be $[w_{p1}\, w_{p2}\, w_{p3}\, l_{p1}\, l_{p2}]^T = [0.428\; 0.275\; 0.245\; 0.575\; 8.08]^T$. The updated antenna model is then simulated,

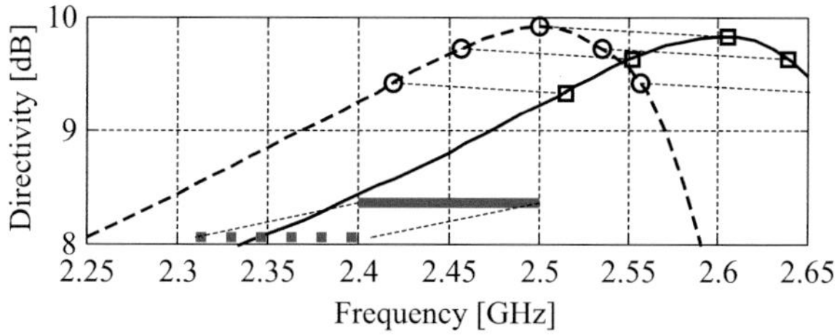

Fig. 8.11 Directivity versus frequency for the antenna structure (*solid line*) and its coarse-discretization model (*dashed line*). Characteristic points (*squares* and *circles*) are used to establish a frequency relationship between the two responses and to map the original frequency band of interest (2.4–2.5 GHz, *thick solid line*) into the corresponding band used in the coarse-discretization model optimization (*thick dashed line*) using the AADS principles

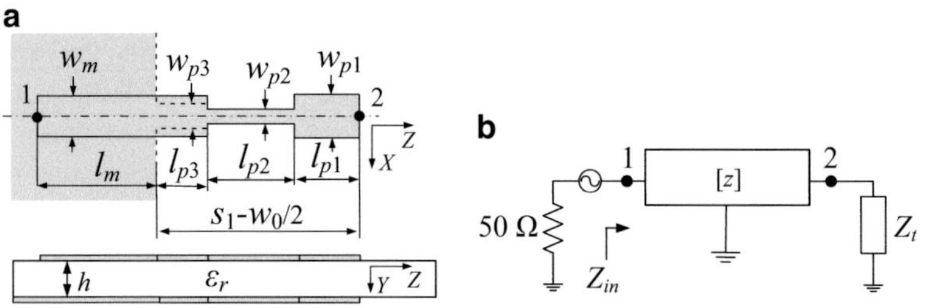

Fig. 8.12 A feed interfacing the 50 Ω input and the driving elements: (**a**) geometry of its full-wave model; (**b**) implemented schematic

Fig. 8.13 Reflection from the input of the Yagi antenna

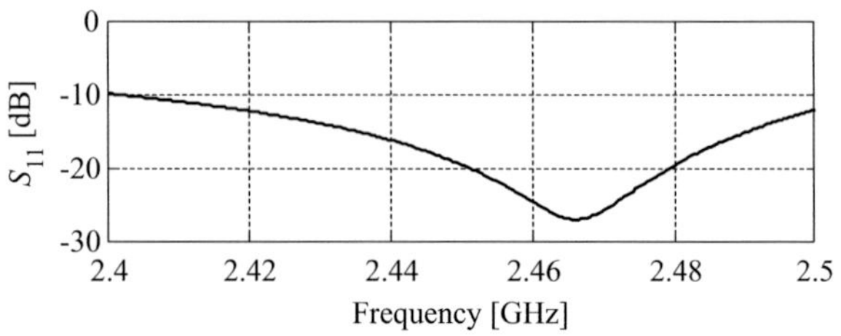

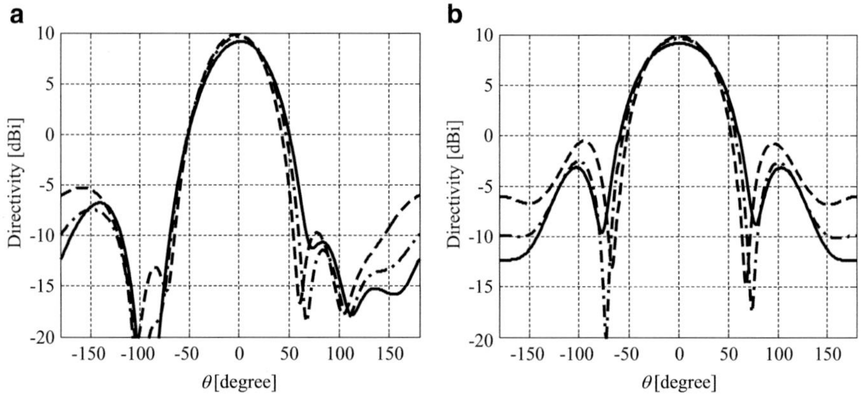

Fig. 8.14 Directivity pattern at 2.4 GHz (*solid*), 2.45 GHz (*dash-dot*), and 2.5 GHz (*dash*): (**a**) co-pol. in the E-plane (XOZ); (**b**) x-pol. in the H-plane (YOZ)

and responses are shown in Figs. 8.13, 8.14, and Table 8.2. Table 8.3 shows the computational cost of the optimization process, which corresponds to only 19 full-wave antenna simulations.

Table 8.2 Printed Yagi antenna: performance summary

Figure	Value
Directivity, maximum[a]	10 dBi
Directivity, end fire maximum[a]	9.85 dBi
IEEE gain, end fire maximum[a]	9.49 dBi
Radiation efficiency, minimum[a]	92 %
Front to back ratio (FBR), minimum[a]	15.5 dB
Side lobe level (SLL), maximum[a]	−10.2 dB
End-fire polarization purity, minimum[a]	40 dB
3 dB beamwidth at 2.45 GHz	E-plane: 59°
	H-plane: 74°
Relative bandwidth ($\|S_{11}\| < -10$ dB)	4.4 %
Input impedance at resonance (2.466 GHz)	54.6 Ω

[a]Maximum/minimum over 2.4–2.5 GHz

Table 8.3 Printed Yagi antenna: optimization cost summary

Algorithm component	# of model evaluations	Absolute time (h)	Relative time[a]
Coarse-discretization model optimization[b]	316	32	16
High-fidelity antenna simulation[c]	3	6	3
Total optimization time[d]	–	38	19

[a]Equivalent number of high-fidelity antenna simulations
[b]Total number of evaluations (coarse-discretization model is optimized once per iteration, two iterations were performed in total)
[c]Evaluation at the initial design and after each iteration
[d]Does not include the time necessary to design the antenna feed, which is negligible compared to the optimization time of the antenna itself

8.4 AADS for Design Optimization of High-Frequency Transition Structures

Substrate integrated circuits (SICs), including substrate integrated waveguides (SIWs) have become attractive solutions in contemporary microwave and millimeter wave engineering due to their ability to provide low cost realization of waveguide components, as well as integration of different components all in the frame of planar technology (Wu 2009; Deslandes and Wu 2001, 2005). Design of broadband, low loss, and well matched SIC transitions, such as microstrip-to-SIW and coplanar waveguide-to-SIW, is an important and still ongoing topic of engineering research (Deslandes 2010; Ogurtsov et al. 2010). One of the major tasks in SIC transition design is to meet given performance specifications for a given substrate, frequency band, and types of interfaced guide structures. Here, we illustrate the application of AADS for fast design of the microstrip-to-SIW transitions.

Consider the planar transition interfacing microstrip to SIW shown in Fig. 8.15. All components (microstrip, SIW, and transition) are on the same 0.787 mm

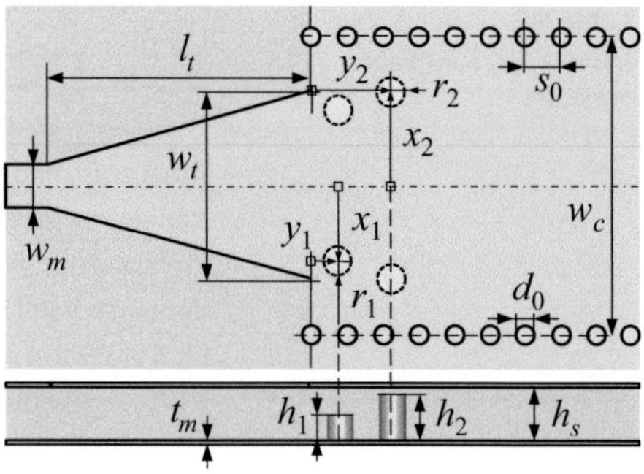

Fig. 8.15 Microstrip-to-SIW transition with two pairs of vias (four vias in total) added (Ogurtsov et al. 2010): top and side views. The fences of vias (SIW's side walls) are not shown on the side view

RT5880 substrate. The substrate is described by the first order Debye model in which the real part of permittivity, ε_r, is 3.5 at 10 GHz and the maximum value of loss tangent is 10^{-3} at 10 GHz. The substrate ground, which is the microstrip ground and the SIW bottom wall, and the SIW's top wall are of infinite lateral extent. All metal parts have conductivity of copper (5.8×10^7 S/m). The width of the 50 Ω input microstrip, w_m, is 2.25 mm. The SIW width, w_c (center to center distance between the rows of vias), is 15.95 mm. Via diameter, d, is 1 mm. Via spacing, s (center to center in a row), equals $2d$. The cutoff of the SIW's quasi TE_{10} dominant mode is at 6.55 GHz. The input microstrip is interfaced to the SIW through a linear taper. Design specifications are $|S_{11}|$, $|S_{22}| <= -20$ dB for the X-band (here 8.2–11.7 GHz).

The taper length in possible designs is constrained to 20 mm, i.e., 1.25 of the SIW's width, w_c. Then we introduce metalized vias partially protruding into substrate for the purpose of matching. Figure 8.15 gives a conceptual view of the transition with four extra vias. Via location, x_n, radii, r_n, and protruding depths, h_n, will be additional (to the taper length, l_t, and width, w_t) design variables. In the transition models, the port-to-port distance is 57 mm of which 32 mm is for the SIW section, and the SIW is excited through a 3.5 mm section of the equivalent rectangular waveguide (Ogurtsov et al. 2010).

Several design cases are considered that differ in the number of extra vias. More specifically, we present designs for two and four protruding vias in comparison to via-less case to demonstrate that enhanced designs (both in terms of bandwidth and matching level) can be obtained for the proposed modification of the microstrip-SIW transition by means of the surrogate based optimization.

Design variables for the first case are $\boldsymbol{x} = [w_t l_t]^T$. The design started from $\boldsymbol{x}^{in} = [10.00\ 14.25]^T$. The optimum was found at $\boldsymbol{x}^* = [5.91\ \ 15.01]^T$. Their reflection responses are shown in Fig. 8.16a. It can be observed that the optimum violates the design specifications at low frequencies of the bandwidth although the level of reflection is significantly decreased overall.

The next case involves a design with one pair of vias. These vias have the same protruding depth h_1, radius r_1, location on y-coordinate y_1. Via distance to the wall of symmetry (shown by the dash-dot line in Fig. 8.15) is x_1. Design variables here are $\boldsymbol{x} = [w_t\, l_t\, y_1\, x_1\, r_1\, h_1]^T$. The design started from $\boldsymbol{x}^{in}$

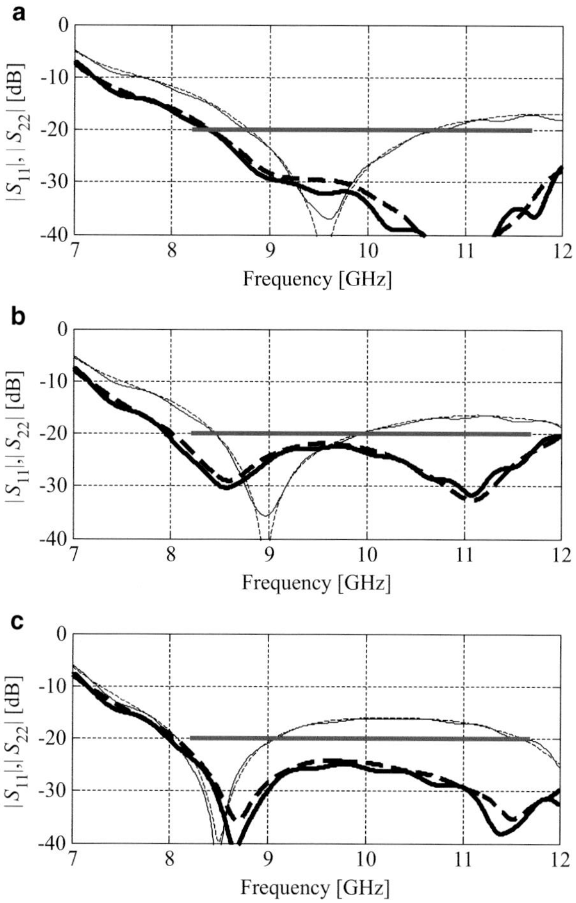

Fig. 8.16 $|S_{11}|$ (*solid*) and $|S_{22}|$ (*dash*) of the initial (*thin*) and optimized (*thick*) design (Ogurtsov et al. 2010): (**a**) original transition (without partially protruding vias); (**b**) transition with one pair of vias; and (**c**) transition with two pairs

$= [10.00\ 14.25\ 16.00\ 3.00\ 0.75\ 0.29]^T$. The optimum was found at $\boldsymbol{x}^* = [7.50\ 15.00\ 18.25\ 3.72\ 0.70\ 0.39]^T$. Their reflection responses are shown in Fig. 8.16b. One can notice that design specifications are met in this case.

The last case involves a structure with two pair of vias. The vias in the nth pair ($n = 1, 2$) have the same location on y-coordinate y_n. Via distance in the nth-pair to the wall of symmetry (shown by the dash-dot line in Fig. 8.15) is x_n. All vias have the same protruding depth ($h_1 = h_2$) and radius ($r_1 = r_2$). Design variables here are $\boldsymbol{x} = [w_t\ l_t\ y_1\ y_2\ x_1\ x_2\ r_1\ h_1]^T$. The design started from $\boldsymbol{x}^{in} = [10.00\ 14.25\ 16.00\ 18.00\ 2.00\ 3.00\ 0.75\ 0.29]^T$. The optimum was found at $\boldsymbol{x}^* = [7.30\ 15.00\ 16.34\ 19.63\ 0.94\ 2.94\ 0.49\ 0.26]^T$. Their reflection responses are shown in Fig. 8.16c. It is seen that the optimum found meets the design specifications, and the overall design quality is better than for other cases.

8.5 Summary and Discussion

In this chapter, a brief exposition of the adaptively adjusted design specifications technique has been provided. As opposed to most of surrogate-based optimization methods, including response correction techniques, AADS does not alter the low-fidelity model itself. Instead, it aims at modifying the design specifications in order to account for the discrepancy between the low- and high-fidelity models at hand. This results in a very simple implementation of the method while maintaining the overall efficiency at a level similar to other SBO methods, as demonstrated in Sects. 8.2 through 8.4 for a variety of microwave filters, antennas, and transition structures.

On the other hand, AADS is limited to minimax type of design specifications. Furthermore, the way of modifying the design specifications is problem dependent and it normally requires analysis of the model discrepancies in order to identify the characteristic point that should be utilized to implement such modifications. Consequently, it might be particularly useful if a problem-specific knowledge is available to the user that would help in establishing the aforementioned model relationships.

In general, AADS might be viewed as an interesting alternative to response-correction-based surrogate-assisted algorithms, which could be attractive for certain classes of design problems and certain groups of (rather experienced) users. An additional advantage is that, in many cases, modifications of the design specs can be done by hand (through visual inspection of the model responses), which allows for executing the AADS optimization procedure even when using built-in optimization capabilities of the simulation software utilized to evaluate the low- and high-fidelity models.

Chapter 9
Surrogate-Assisted Design Optimization Using Response Features

Highly nonlinear system responses, whether the nonlinearity is with respect to the adjustable parameters of the systems under design or with respect to certain free parameters (e.g., chord line (Leifsson and Koziel 2015a), time, or frequency (Koziel et al. 2006a, b)), are challenging to simulation-driven design optimization. The surrogate-based optimization techniques described in the previous chapters directly handled the relevant responses of the systems of interest (such as aerodynamic forces in case of aerodynamic shape optimization problems, or S-parameters versus frequency for microwave/photonic and antenna devices). One of the issues is the limited generalization capability of the surrogate models which may lead to a degradation of the convergence properties of the optimization algorithms. Fortunately, in many cases, it is possible to reformulate the optimization problem in terms of the so-called response features whose dependence on the optimization variables is much less nonlinear than for the original responses. Feature-based optimization executed at the level of these feature (or characteristic) points may be computationally much more efficient, even when realized using single-fidelity simulation models only. In this chapter, we formulate feature-based optimization, demonstrate its application for the design of microwave/photonic and antenna devices, as well as discuss its limitations and generalizations.

9.1 Optimization Using Response Features

Moving the optimization process into a domain in which the objective function landscape is less nonlinear may be critical for improving the computational efficiency of the optimization process. Here, we discuss the feature-based optimization (Glubokov and Koziel 2014a, b; Koziel and Bekasiewicz 2015), where instead of handling complete (often highly nonlinear) responses, only selected response features are processed. Dependence of those features on the designable parameters

S. Koziel, L. Leifsson, *Simulation-Driven Design by Knowledge-Based Response Correction Techniques*, DOI 10.1007/978-3-319-30115-0_9

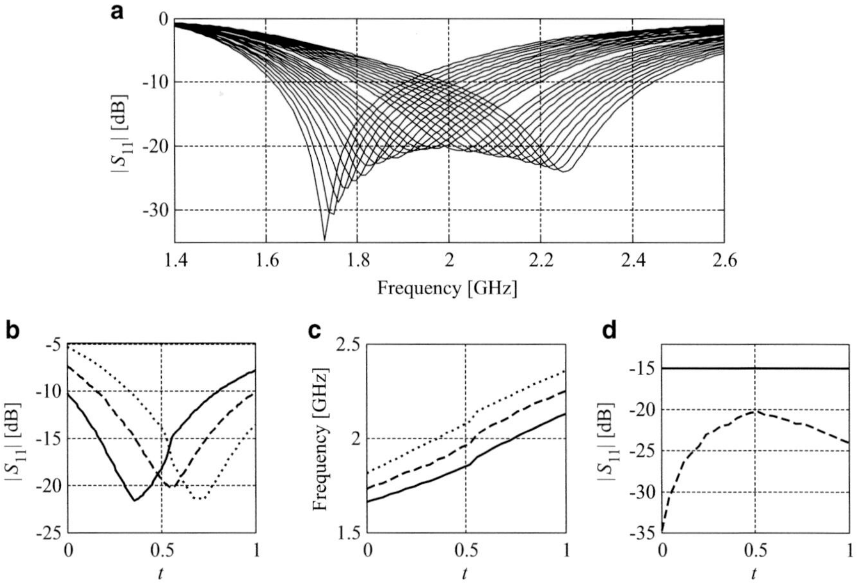

Fig. 9.1 Response features: (**a**) family of reflection responses of the example antenna structure, evaluated along certain line segment in the design space, example features correspond to -15 dB levels and the center frequency; (**b**) response variability at selected frequencies, 1.9 GHz (*solid line*), 2.0 GHz (*dashed line*), and 2.1 GHz (*dotted line*), indicating their high nonlinearity; (**c**) variability of the response features (frequency components), here, corresponding to -15 dB levels as well as the center frequency; (**d**) variability of the level components of the response features of (**c**)

is usually less nonlinear than for the entire responses, which speeds up the optimization process.

The concept is explained in Fig. 9.1 that shows a family of reflection responses of the antenna structure considered in Koziel (2014). The reflection response versus frequency is evaluated along a certain line segment in the design space, $\boldsymbol{x} = t\boldsymbol{x}_a + (1 - t)\boldsymbol{x}_b$, $0 \le t \le 1$, where $\boldsymbol{x}_a$ and $\boldsymbol{x}_b$ are arbitrarily selected vectors. The figure also shows the plots of the corresponding features as a function of the parameter t. Figure 9.1b shows the reflection characteristic $|S_{11}|$ versus the line segment parameter t for various frequencies, here, 1.9, 2.0, and 2.1 GHz. When creating a conventional response surface approximation model, the response at each frequency is modeled separately. As indicated in the plots, the responses are highly nonlinear. At the same time, the behavior of the response features (here, the three points corresponding to the -15 dB levels of $|S_{11}|$ and $|S_{11}|$ at the center frequency) is much more linear, and, consequently, easier to model. In the optimization context, estimation of the next candidate design is usually obtained from a local model established in the vicinity of the current design (e.g., first-order Taylor expansion) so that more linear behavior of the optimized function results in faster convergence of the optimization algorithm.

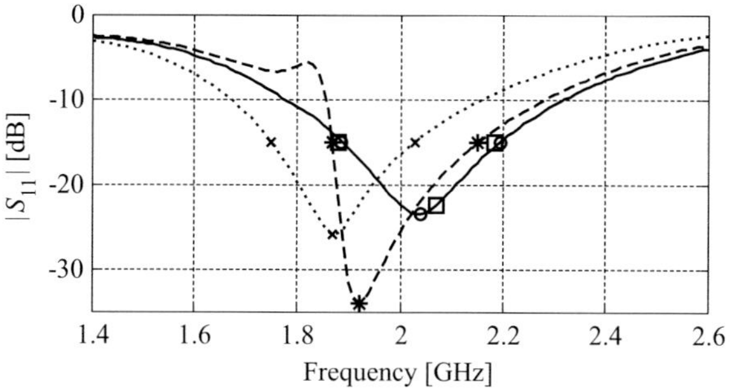

Fig. 9.2 Feature prediction using estimated feature gradients: high-fidelity model at the reference design $\boldsymbol{x}$ (*dotted line*) and corresponding feature points (*crosses*), high-fidelity model at the perturbed design $\boldsymbol{x}+d\boldsymbol{x}$ (*solid line*) and the corresponding feature points (*open circles*), linear expansion model constructed at the frequency-based response level and evaluated at $\boldsymbol{x}+d\boldsymbol{x}$ (*dashed line*) together with the corresponding feature points (*asterisks*), as well as the high-fidelity feature points predicted using feature gradients (*open squares*). It can be observed that the feature point prediction obtained using conventional linear expansion (at the frequency-based response level) is very inaccurate, whereas the feature prediction using feature gradients is much more reliable

We will use the symbols $f_k(\boldsymbol{x})$ and $l_k(\boldsymbol{x})$ to denote the frequency and the level of the kth feature, $k=1,\ldots,p$, where p is the number of features for a given problem. The response features should be selected in accordance to given performance specifications so that the original formulation of the design problem (3.1) can be replaced by an equivalent problem formulated for the features

$$\boldsymbol{x}_f^* = \arg\min_{\boldsymbol{x}} U_F\Big(\boldsymbol{f}_F(\boldsymbol{x})\Big), \tag{9.1}$$

where $\boldsymbol{f}_F \in R^{2p}$ denotes the feature-based model (as a function of the design variables $\boldsymbol{x}$), whereas U_F is the objective function at the feature level. Using the previously introduced notation, we have $\boldsymbol{f}_F(\boldsymbol{x}) = [f_1(\boldsymbol{x})\ f_2(\boldsymbol{x}) \ldots f_p(\boldsymbol{x})\ l_1(\boldsymbol{x})\ l_2(\boldsymbol{x}) \ldots l_p(\boldsymbol{x})]^T$.

Figure 9.2 shows a typical situation for a narrow-band antenna. In this specific example, the goal is maximization of the fractional bandwidth at -15 dB level. For illustration, the three response features are utilized corresponding to the above response level and to the response minimum (although only the first two features are necessary in the optimization process). It can be observed that the prediction of the feature location obtained using the feature model gradients $\nabla f_k(\boldsymbol{x})$ and $\nabla l_k(\boldsymbol{x})$ (i.e., $f_k(\boldsymbol{x}+d\boldsymbol{x}) \approx f_k(\boldsymbol{x}) + \nabla f_k(\boldsymbol{x})d\boldsymbol{x}$, and $l_k(\boldsymbol{x}+d\boldsymbol{x}) \approx l_k(\boldsymbol{x}) + \nabla l_k(\boldsymbol{x})d\boldsymbol{x}$) for a specific search step size $d\boldsymbol{x}$, versus the actual feature location verified by EM simulation, is very good. This is not the case when the same search step is utilized for the first-order Taylor model constructed for the frequency-based response, i.e., $\boldsymbol{f}(\boldsymbol{x}+d\boldsymbol{x}) \approx \boldsymbol{f}(\boldsymbol{x}) + \boldsymbol{J}_f(\boldsymbol{x})d\boldsymbol{x}$, where $\boldsymbol{J}_f$ denotes the model Jacobian.

The reason, again, is less nonlinear behavior of the response features with respect to the location in the design space. Perhaps even more important reason is that modeling of frequency coordinates o the features, i.e., $f_k(\boldsymbol{x})$, directly describe relocation of important parameters (such as bandwidth or a center frequency), whereas the conventional description does this indirectly through the modeling of the entire responses.

The feature-based optimization algorithm is formulated as an iterative process generating a series $\boldsymbol{x}^{(i)}$ of approximations to $\boldsymbol{x}^*$ (solution to the original problem (3.1)) as follows:

$$\boldsymbol{x}^{(i+1)} = \arg\min_{\boldsymbol{x}} U_F\left(\boldsymbol{l}^{(i)}(\boldsymbol{x})\right), \tag{9.2}$$

where $\boldsymbol{l}^{(i)}$ is a linear model of the response features set up at the current design $\boldsymbol{x}^{(i)}$ using finite differentiation; recall (cf. (9.1) and the text below it) that U_F is the objective function encoding the design specifications at the feature level. The model $\boldsymbol{l}^{(i)}$ is defined as

$$\boldsymbol{l}^{(i)}(\boldsymbol{x}) = \begin{bmatrix} \begin{bmatrix} f_1(\boldsymbol{x}^{(i)}) \\ \cdots \\ f_p(\boldsymbol{x}^{(i)}) \end{bmatrix} + \begin{bmatrix} \dfrac{f_1(\boldsymbol{x}_{d.1}^{(i)}) - f_1(\boldsymbol{x}^{(i)})}{d_1^{(i)}} \\ \cdots \\ \dfrac{f_p(\boldsymbol{x}_{d.p}^{(i)}) - f_p(\boldsymbol{x}^{(i)})}{d_p^{(i)}} \end{bmatrix} \circ (\boldsymbol{x} - \boldsymbol{x}^{(i)}) \\ \\ \begin{bmatrix} l_1(\boldsymbol{x}^{(i)}) \\ \cdots \\ l_p(\boldsymbol{x}^{(i)}) \end{bmatrix} + \begin{bmatrix} \dfrac{l_1(\boldsymbol{x}_{d.1}^{(i)}) - l_1(\boldsymbol{x}^{(i)})}{d_1^{(i)}} \\ \cdots \\ \dfrac{l_p(\boldsymbol{x}_{d.p}^{(i)}) - l_p(\boldsymbol{x}^{(i)})}{d_p^{(i)}} \end{bmatrix} \circ (\boldsymbol{x} - \boldsymbol{x}^{(i)}) \end{bmatrix} \tag{9.3}$$

where $\boldsymbol{x}^{(i)} = [x_1{}^{(i)} x_2{}^{(i)} \ldots x_n{}^{(i)}]^T$ is a current design, whereas $\boldsymbol{x}_{d.k}{}^{(i)} = [x_1{}^{(i)} \ldots x_k{}^{(i)} + d_k{}^{(i)} \ldots x_n{}^{(i)}]^T$, are the perturbed designs. Here, $\boldsymbol{d}^{(i)} = [d_1{}^{(i)} d_2{}^{(i)} \ldots d_n{}^{(i)}]^T$ is the perturbation size which may, in general, be iteration dependent. The symbol $\circ$ denotes component-wise multiplication. Note that frequency and level coordinates of the feature points are modeled independently. To ensure convergence, the algorithm is embedded in the trust-region framework (Conn et al. 2000), i.e., (9.2) is replaced by

$$\boldsymbol{x}^{(i+1)} = \arg \min_{||\boldsymbol{x} - \boldsymbol{x}^{(i)}|| \le \delta^{(i)}} U_F\left(\boldsymbol{l}^{(i)}(\boldsymbol{x})\right), \tag{9.4}$$

where $\boldsymbol{l}^{(i)}$ is optimized within the trust region defined as $||\boldsymbol{x} - \boldsymbol{x}^{(i)}|| \le \delta^{(i)}$; $\delta^{(i)}$ is the trust region size updated using conventional rules (Conn et al. 2000). The new design $\boldsymbol{x}^{(i+1)}$ produced by (9.4) is only accepted if it leads to the improvement of $f_F(\boldsymbol{x})$, i.e., $f_F(\boldsymbol{x}^{(i+1)}) < f_F(\boldsymbol{x}^{(i)})$.

Fig. 9.3 High- (*solid line*) and low-fidelity (*dashed line*) model responses at a selected design, as well as the corresponding feature points (−15 dB levels and center frequency), (*open circles*) and (*open squares*), respectively

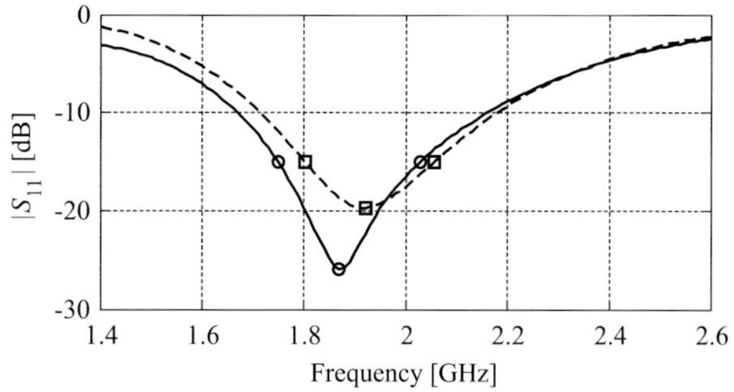

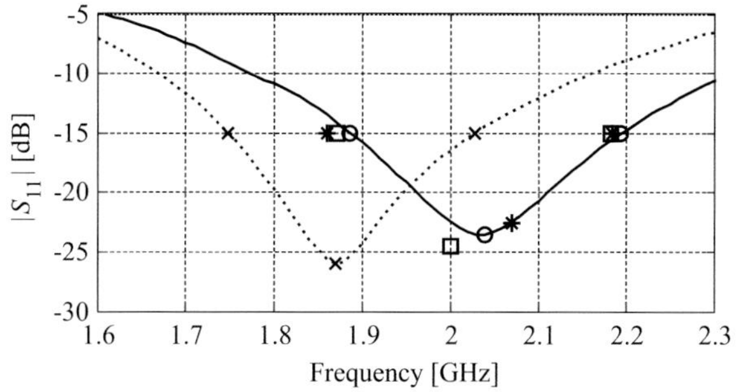

Fig. 9.4 Feature prediction using estimated low-fidelity feature gradients: high-fidelity model at the reference design $\boldsymbol{x}$ (*dotted line*) and corresponding feature points (*crosses*), high-fidelity model at the perturbed design $\boldsymbol{x} + d\boldsymbol{x}$ (*solid line*) and the corresponding feature points (*open circles*), feature points predicted using feature gradients estimated using high-fidelity model responses (*open squares*), feature points predicted using feature gradients estimated using low-fidelity model responses (*asterisks*). Note that the feature point prediction obtained using low-fidelity model is quite reliable and comparable to that obtained using high-fidelity model, particularly for −15 dB levels

In order to speed up the optimization process, the gradients of the features can be estimated (cf. (9.3)) using the low-fidelity model evaluations rather than the high-fidelity ones. In particular, the left hand side terms $f_k(\boldsymbol{x}^{(i)})$ and $l_k(\boldsymbol{x}^{(i)})$ come from the high-fidelity model responses, whereas the finite differentiation terms $[f_k(\boldsymbol{x}_{d.k}{}^{(i)}) - f_k(\boldsymbol{x}^{(i)})]$ and $[l_k(\boldsymbol{x}_{d.k}{}^{(i)}) - l_k(\boldsymbol{x}^{(i)})]$ come from the low-fidelity model. This is crucial to obtain considerable computational savings: gradient estimation using high-fidelity model requires n evaluations of $\boldsymbol{f}$, whereas the same estimation realized with the low-fidelity model requires n evaluations of $\boldsymbol{c}$, which normally corresponds (CPU time-wise) to much less than a single evaluation of $\boldsymbol{f}$. Although the low- and high-fidelity models are normally misaligned (see Fig. 9.3), the gradient estimations of the features are quite similar as illustrated in Fig. 9.4. In particular, the feature point prediction obtained using the low-fidelity model is comparable to that obtained using the high-fidelity one.

9.2 Feature-Based Tuning of Microwave Filters

In order to illustrate the application of feature-based optimization for microwave filter tuning, consider the third-order cross-coupled substrate-integrated waveguide (SIW) bandpass filter (Glubokov and Koziel 2014a) shown in Fig. 9.5. The design parameters are $\boldsymbol{x} = [L_{res1}\, L_{res2}\, W_{12}\, W_{13}\, W_{in}]^T$. The width of all the resonators is fixed to $W_{res} = 8$ mm. The substrate parameters are: height $H_{sub} = 1$ mm and $\varepsilon_r = 2.2$; these values remain constant throughout the entire optimization. The specifications of the filter are given as follows: $|S_{11}| \leq -20$ dB for 9.9 GHz $\leq f \leq$ 10.1 GHz, and $|S_{21}| \leq -30$ dB for $f \geq 10.25$ GHz. The feature concept is also applied to the transmission losses response by allocating an additional characteristic point representing the local maximum of the $|S_{21}|$ in the upper stopband.

Figure 9.6 shows the behavior of the return loss characteristic when the design is swept along a selected line segment within the design space. One can observe that the responses are highly nonlinear functions of frequency and they change considerably from one design to another (Fig. 9.6a). The behavior of the feature points selected for the sake of the optimization process is much less nonlinear (Fig. 9.6b).

Figure 9.7 shows the relationship between the low- and high-fidelity filter model responses and their corresponding feature points. Despite considerable response discrepancies, the feature point changes are consistent between the models which confirms that the low-fidelity model can be reliably used to estimate the feature point gradients with respect to the geometry parameters of the filter.

The initial design parameters $\boldsymbol{x}_{init} = [14\ 15.5\ 5\ 3\ 4.5]^T$ mm have been found by manual tuning of the surrogate model. It is extremely important that both components of $\boldsymbol{f}(\boldsymbol{x})$ reach the levels of interest, so the features can be automatically extracted at each iteration; this is also correct for all the responses $\boldsymbol{c}(\boldsymbol{x} + \Delta\boldsymbol{x})$ used for the calculation of the gradients. The initial experiments have indicated that the surrogate model with $LPW = 7$ (computation time: 30 s) is sufficient to meet all the surrogate model requirements. $LPW = 25$ has been chosen for the fine model.

Figure 9.8 shows the optimal response $\boldsymbol{f}(\boldsymbol{x}_{opt})$ compared to the $\boldsymbol{f}(\boldsymbol{x}_{init})$, where $\boldsymbol{x}_{opt} = [13.51\ 15.52\ 5.24\ 2.96\ 4.82]^T$ mm, together with the filter specifications. The optimum has been obtained for 6 iterations including 7 evaluations of the fine model and 30 evaluations of the surrogate model (less than $9 \times \boldsymbol{f}$ in total).

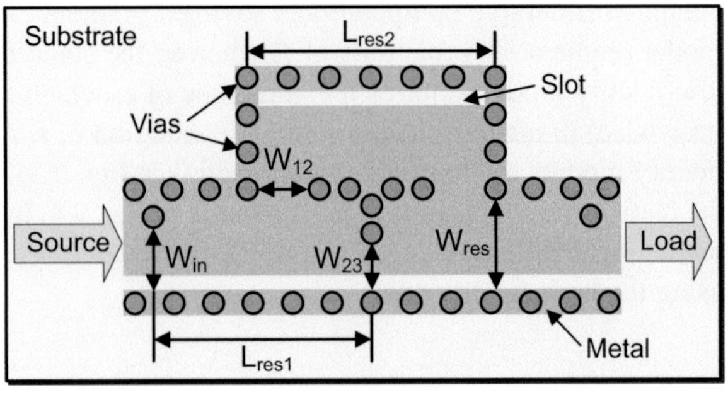

Fig. 9.5 Configuration of the third-order folded SIW filter (Glubokov and Koziel 2014a)

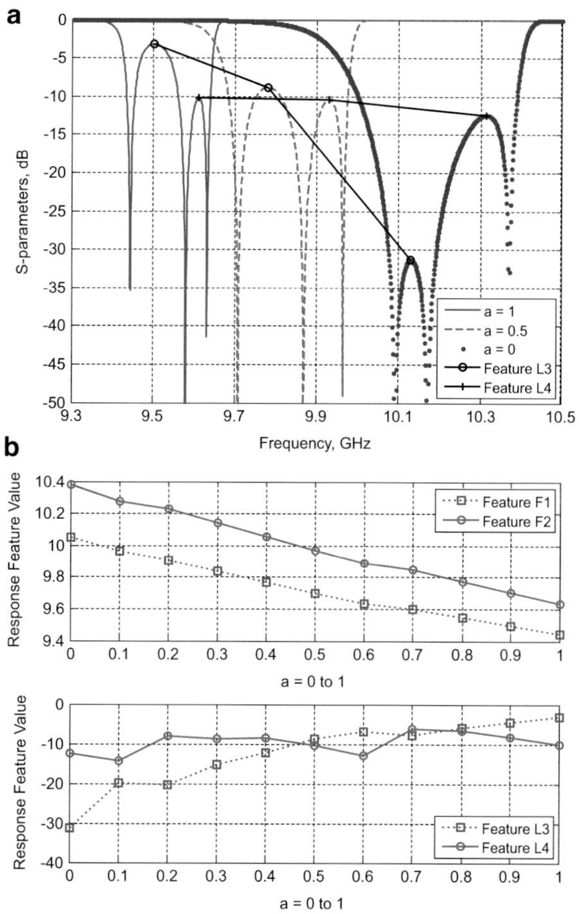

Fig. 9.6 Typical behavior of fine model responses $\boldsymbol{f}(\boldsymbol{x})$ (**a**) and corresponding response features (**b**) with respect to linear variation of the design parameters: $\boldsymbol{x} = a \cdot \boldsymbol{x}_A + (1 - a) \cdot \boldsymbol{x}_B$, $0 \leq a \leq 1$; $\boldsymbol{x}_A$ and $\boldsymbol{x}_B$ – arbitrary (preselected) points in the design space. The features F1 and F2 correspond to the −20 dB return loss levels determining the filter bandwidth, whereas the features L3 and L4 are the response levels corresponding to the local maxima within the pass-band

9.3 Feature-Based Optimization of Antennas

Consider the planar inverted-F antenna (PIFA) (Koziel 2014) shown in Fig. 9.9. The design variables are $\boldsymbol{x} = [v_0\, v_1\, v_2\, v_3\, v_4\, v_5\, v_6]^T$. The initial design is $\boldsymbol{x}^{(0)} = [2\,8\,5\,2\,10\,25\,45]^T$ mm. Fixed parameters are $[u_0\, u_1\, u_2\, u_3\, u_4\, u_5\, u_6\, u_7\, w_0\, r_0]^T = [6.15\; 50.0\; 15.0\; 10.5\; 29.35\; 11.65\; 5.0\; 1.0\; 0.5]^T$ mm; 0.508 mm substrate (Fig. 9.9a on the right) and the box (on the left) are of Rogers TMM4 and TMM6.

The high-fidelity model $\boldsymbol{f}$ is evaluated in CST Microwave Studio (~650,000 mesh cells, simulation time 13 min). The low-fidelity model $\boldsymbol{c}$ is also evaluated in CST (~22,000 mesh cells, simulation time 50 s). The antenna was optimized using the feature-based algorithm, assuming two scenarios: (1) obtaining as wide fractional bandwidth as possible at $|S_{11}| = -15$ dB (symmetrically) around 2.0 GHz (Case 1), and (2) minimizing $|S_{11}|$ at the center frequency of 2.0GHz (Case 2).

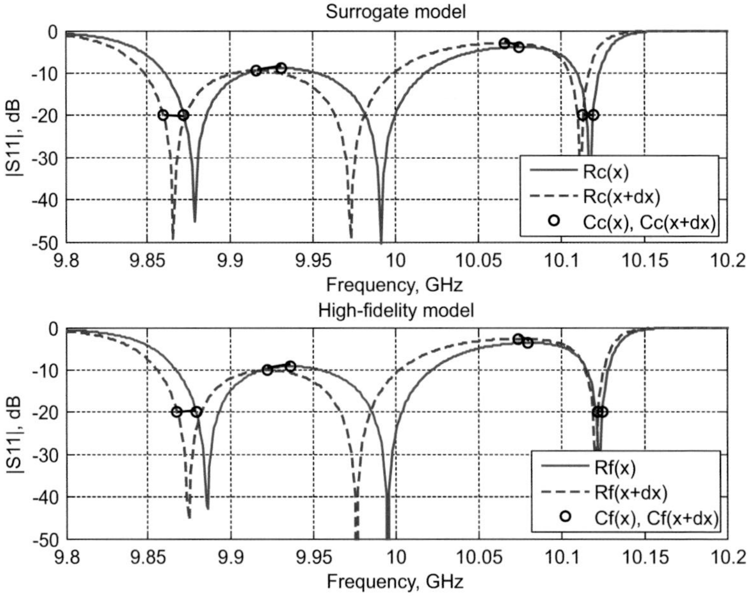

Fig. 9.7 Responses ($\boldsymbol{c}(\boldsymbol{x})$, $\boldsymbol{f}(\boldsymbol{x})$), corresponding characteristic points of a SIW filter, and their behavior with respect to small perturbation of the design variable $d\boldsymbol{x}$

For the first scenario, the final design $\boldsymbol{x}^* = [2.028\ 8.226\ 6.587\ 2.618\ 9.134\ 22.87\ 45.24]^T$ mm was obtained in 6 iterations of the algorithm (9.2), with the total cost corresponding to about 11 evaluations of the high-fidelity model (8 $\times \boldsymbol{f}$ and 50 $\times$ $\boldsymbol{c}$). Figure 9.10 shows the high-fidelity model responses at the initial and the final designs, as well as the evolution of the fractional bandwidth and the convergence plot. It should be noted that the presented approach is also efficient in terms of low-fidelity model evaluations, which is important when the time evaluation ratio between the $\boldsymbol{f}$ and $\boldsymbol{c}$ is low as in the present example.

For the second design scenario, the final design $\boldsymbol{x}^* = [2.116\ 8.186\ 6.454\ 2.341\ 8.8123\ 22.72\ 45.27]^T$ mm (Fig. 9.11) was obtained in 7 iterations at the total cost of about 13 evaluations of $\boldsymbol{f}$ (9 $\times \boldsymbol{f}$ and 58 $\times$ $\boldsymbol{c}$).

For the sake of comparison the antenna was also optimized using the two benchmark methods: (1) space mapping (SM) (Koziel et al. 2008a), and (2) the pattern search algorithm (Kolda et al. 2003). The SM algorithm utilized here exploits the low-fidelity model $\boldsymbol{c}$ as the underlying coarse model, and two types of model correction, specifically, frequency scaling (Koziel and Ogurtsov 2014a) and additive response correction (Koziel and Leifsson 2012e). The SM surrogate is reset at every iteration using the high-fidelity model data from the most current design; it is subsequently reoptimized using pattern search. The pattern search algorithm used in both direct search and space mapping procedure is based on the implementation described in Koziel (2010g).

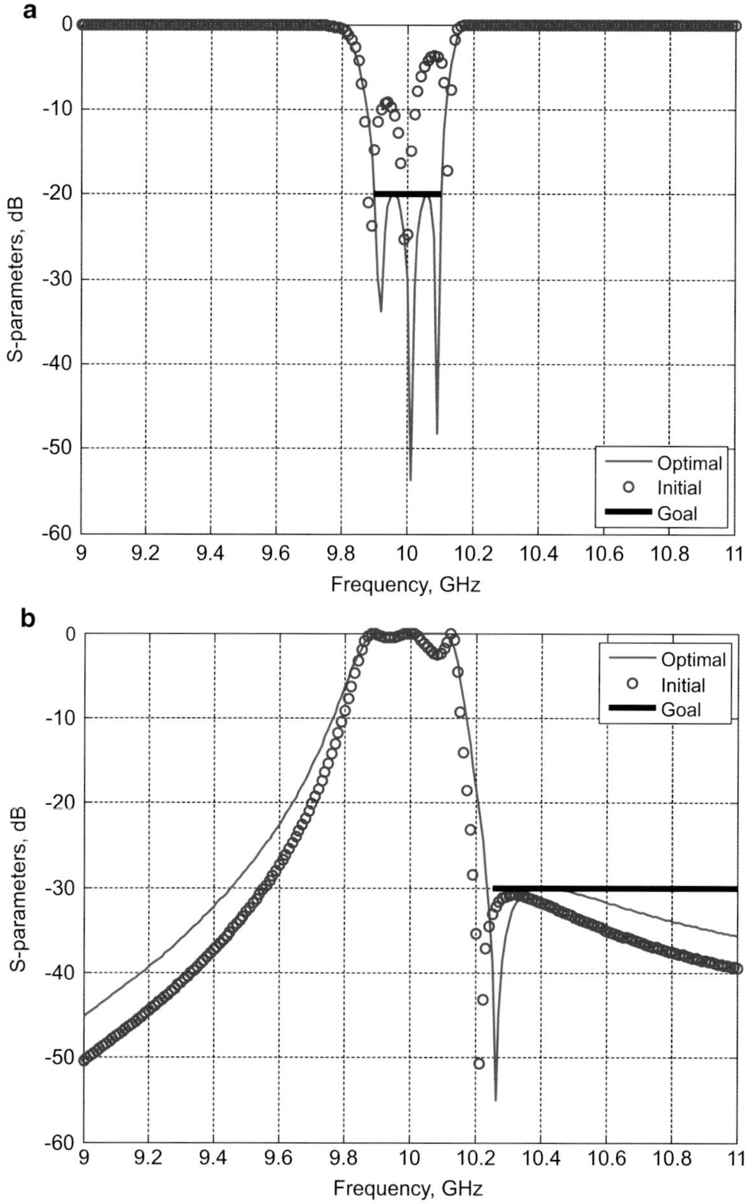

Fig. 9.8 Comparison of initial $\boldsymbol{f}(\boldsymbol{x}_{init})$ and optimal $\boldsymbol{f}(\boldsymbol{x}_{opt})$ filter responses: (**a**) $|S_{11}|$; (**b**) $|S_{21}|$

The final designs obtained using all three methods are comparable; however, the design costs differ considerably. The design cost breakdown for both cases is shown in Tables 9.1 and 9.2, respectively. It can be observed that direct optimization of the high-fidelity model is clearly the most expensive approach ($205 \times \boldsymbol{f}$ and $141 \times \boldsymbol{f}$ for

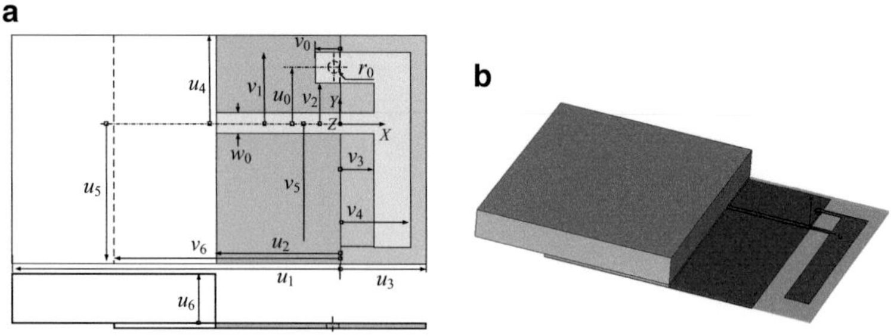

Fig. 9.9 PIFA geometry (Koziel 2014): (**a**) top and side view, substrate shown transparent; (**b**) perspective view

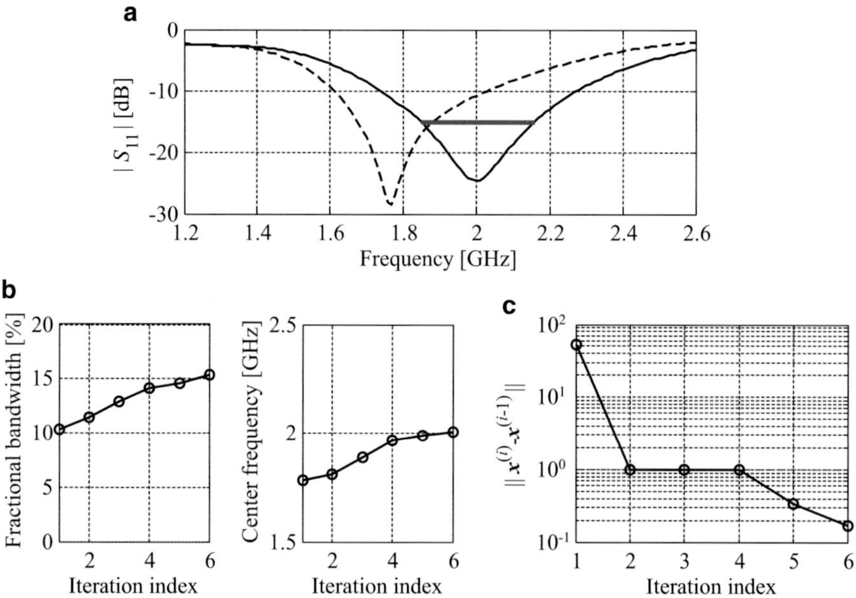

Fig. 9.10 PIFA optimization (Case 1): (**a**) initial (*dashed line*) and optimized (*solid line*) response (fractional bandwidth around 2 GHz of 15 %); (**b**) evolution of the fractional bandwidth and the center frequency vs. iteration index; (**c**) convergence plot

Fig. 9.11 PIFA optimization (Case 2): initial (*dashed line*) and optimized (*solid line*) response ($|S_{11}| = -32$ dB at the center frequency)

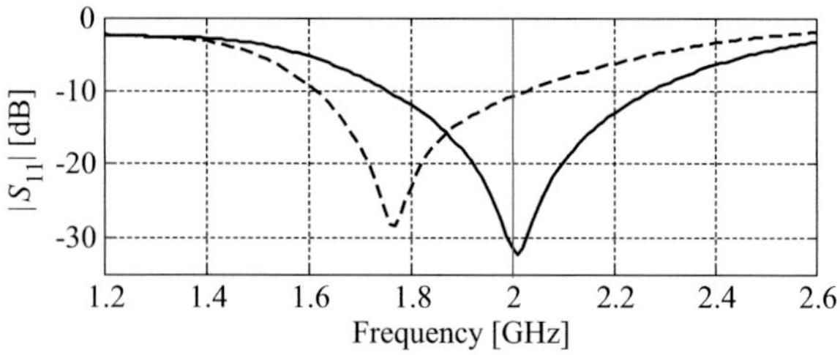

Table 9.1 Planar interted-F antenna: design optimization cost for Case 1

Algorithm	Algorithm component	Number of model evaluations	CPU Cost Absolute (min)	CPU Cost Relative to f
Direct f optimization (pattern search)	Evaluating f	$205 \times f$	2665	205.0
Space mapping	Evaluating c	$345 \times c$	288	22.2
	Evaluating f	$5 \times f$	65	5.0
	Total cost	N/A	353	27.2
Feature-based optimization	Evaluating c	$50 \times c$	42	4.0
	Evaluating f	$8 \times f$	104	9.0
	Total cost	N/A	146	13.0

Table 9.2 Planar interted-F antenna: design optimization cost for Case 2

Algorithm	Algorithm component	Number of model evaluations	CPU Cost Absolute (min)	CPU Cost Relative to f
Direct f optimization (pattern search)	Evaluating f	$141 \times f$	1833	141.0
Space mapping	Evaluating c	$350 \times c$	292	22.4
	Evaluating f	$6 \times f$	78	6.0
	Total cost	N/A	370	28.4
Feature-based optimization	Evaluating c	$58 \times c$	48	3.7
	Evaluating f	$9 \times f$	117	9.0
	Total cost	N/A	165	12.7

Case 1 and 2, respectively). Space mapping offers significant reduction of the design cost (to less than 30 equivalent evaluations of the high-fidelity model). However, due to relatively low time evaluation ration between the high- and low-fidelity model (only about 16:1), the computational overhead related to surrogate model optimization (around $350 \times c$) becomes the major contributor (~80 %) to the total cost. The feature-based optimization algorithm is much more efficient with this respect (the low-fidelity model evaluation corresponds to only about 30 % of the total cost), which results in substantial savings compared to SM even though the number of algorithm iterations is larger.

9.4 Feature-Based Optimization of Photonic Devices

Simulation-driven optimization of integrated photonic components is a challenging task because of very long simulation time of their EM models. In this section we demonstrate the use of feature-based optimization algorithm for the design of a silicon-on-insulator (SOI) add-drop microring resonator. The structure is shown in

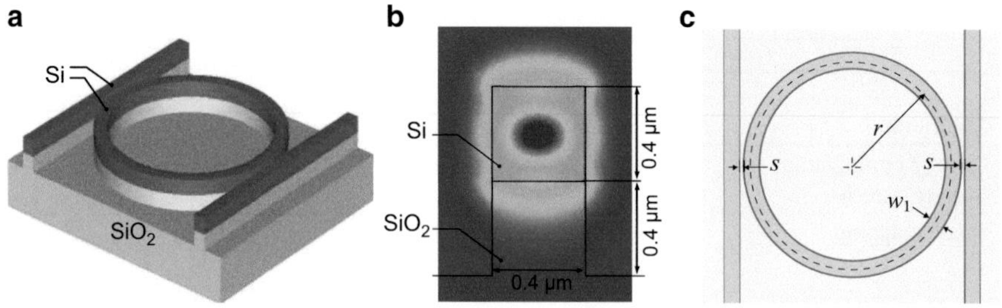

Fig. 9.12 Microring resonator: (**a**) 3D view; (**b**) E-field amplitude distribution of the used mode (hybrid TM_{00}) at the input/output waveguide cross sections; (**c**) layout

Fig. 9.12. In the EM model, the refractive index of the silicon core is 3.45 and that of the insulator is 1.45 over the simulation frequencies so that all dispersion effects come from the EM phenomena. The upper cladding is air. Cross sections of the input–output rib waveguides are shown in Fig. 9.12b with the E-field distribution of the used vertical polarization/mode. The use of horizontal polarization (TE modes) have been the most common approach for photonic components comprising microrings and bus waveguides (Bogaerts et al. 2012; Little et al. 1997; Prabhu et al. 2009). However, it has been demonstrated that vertical polarization (TM modes) can substantially alleviate the issue of excessive losses associated with manufacturing non-idealities of the optical waveguides (De Heyn et al. 2010). With the material parameters and cross-sectional dimensions of the input–output rib waveguides the hybrid TM_{00} mode (vertical polarization) is dominant according to the numerical analysis although single mode propagation is not provided over the simulated bandwidth (1.5 μm-to-1.63 μm) so the hybrid TE_{00} mode can propagate as well.

The layout of the resonator is shown in Fig 9.12c where $\boldsymbol{x} = [s\ r\ w_1]^T$ mm are the design variables, namely, the ring-to-bus spacing, the ring radius, and width of the ring (allowed to be different from the fixed width of the input/output waveguides). The EM model of the microring resonator is defined with CST MWS software (CST 2013) and simulated using the CST MWS transient solver. In this case, only one EM model is utilized (at medium discretization level): the high-fidelity model was too expensive to handle, whereas the low-fidelity model was too unreliable to be utilized in the optimization process. The typical microring responses are shown in Fig. 9.13.

There are several design objectives in our problem:

1. $FSR \geq 3.4$ THz, where FSR (free spectral range) is the peak-to-peak distance for $|S_{41}|$ curve
2. $\delta f/FSR \leq 0.06$, where δf is the -3 dB (half-power) bandwidth (relative to the maximum) in the $|S_{41}|$ curve
3. $|S_{41}| \geq -0.5$ dB at its maxima
4. $|S_{41}| \leq -20$ dB at its minima
5. $|S_{21}| \leq -20$ dB at its minima

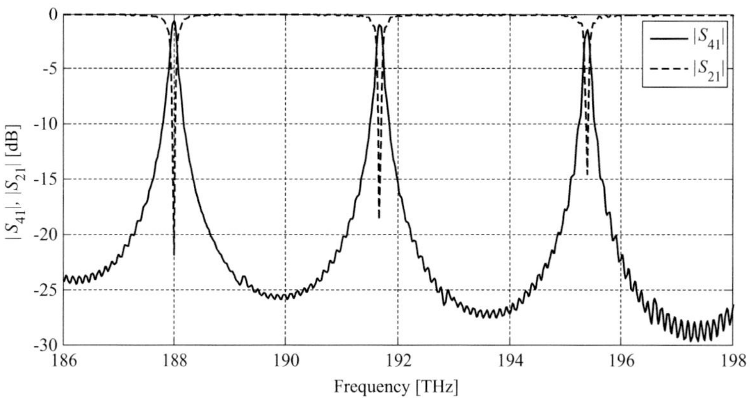

Fig. 9.13 Typical $|S_{21}|$ (through) and $|S_{41}|$ (add/drop) response of a microring under design (Koziel and Ogurtsov 2014b)

where $|S_{41}|$ stands for drop/add transmission response and $|S_{21}|$ stands for through (pass) transmission according to contemporary lightwave terminology (Bogaerts et al. 2012).

The feature vector is defined as $\boldsymbol{f}_F(\boldsymbol{x}) = [f_{S41.\max.1}(\boldsymbol{x})\, f_{S41.\max.2}(\boldsymbol{x})\; l_{S41.\max.1}(\boldsymbol{x})\; l_{S41.\max.2}(\boldsymbol{x})\; b_{S41.1}(\boldsymbol{x})\; b_{S41.2}(\boldsymbol{x})\; l_{S41.\min}(\boldsymbol{x})\; l_{S21.\min.1}(\boldsymbol{x})\; l_{S21.\min.2}(\boldsymbol{x})]^T$, where the first four components describe the two $|S_{41}|$ maxima closest to 190 THz (frequency and levels), the next two components are -3 dB bandwidths of $|S_{41}|$, whereas the remaining three are the minimum of $|S_{41}|$ and the two minima of $|S_{21}|$ (levels only). Figure 9.14 shows the response features of $\boldsymbol{f}_F$ for the case considered. All the objectives as listed above can be calculated from the components of $\boldsymbol{f}_F(\boldsymbol{x})$. The objective function for the feature-based optimization process is defined as

$$\begin{aligned} U_F\left(\boldsymbol{f}_F(\boldsymbol{x})\right) &= \max\{l_{S21.\max.1}(\boldsymbol{x}), l_{S21.\max.2}(\boldsymbol{x})\} + \beta_1 \cdot \left\{\min\left(FSR(\boldsymbol{x}) - 3.4, 0\right)/3.4\right\}^2 + \\ &+ \beta_2 \cdot [\max(\delta f/FSR(\boldsymbol{x}) - 0.06, 0\big)/0.06]^2 + \\ &+ \beta_3 \cdot [\min\left(\min\{l_{S41.\max.1}(\boldsymbol{x}), l_{S41.\max.2}(\boldsymbol{x})\} + 0.5, 0\right)/0.5]^2 + \\ &+ \beta_4 \cdot [\max\left(l_{S41.\min}(\boldsymbol{x}) + 20, 0\right)/20]^2 \end{aligned} \tag{9.5}$$

Here, $FSR(\boldsymbol{x}) = f_{S41.\max.2}(\boldsymbol{x}) - f_{S41.\max.1}(\boldsymbol{x})$, and $\delta f = \max\{b_{S41.1}(\boldsymbol{x}), b_{S41.2}(\boldsymbol{x})\}$. Here, we use large values of the penalty factors $\beta_1 = \beta_2 = \beta_3 = \beta_4 = 1{,}000$, which essentially sets a hard limit for all the objectives.

The initial design is $\boldsymbol{x}^{(0)} = [0.15\; 2.80\; 0.40]^T$ mm. The values of all five objectives are gathered in Table 9.3. The final design $\boldsymbol{x}^* = [0.1051\; 3.0808\; 0.4085]^T$ mm has been obtained at the total cost of 19 only EM simulations of the microring (total CPU time 28 h). Figure 9.15 shows the microring responses at the initial and at the final designs, as well as the convergence plot of the algorithm and the evolution of the objective function (9.5). It can be observed that two of the objectives,

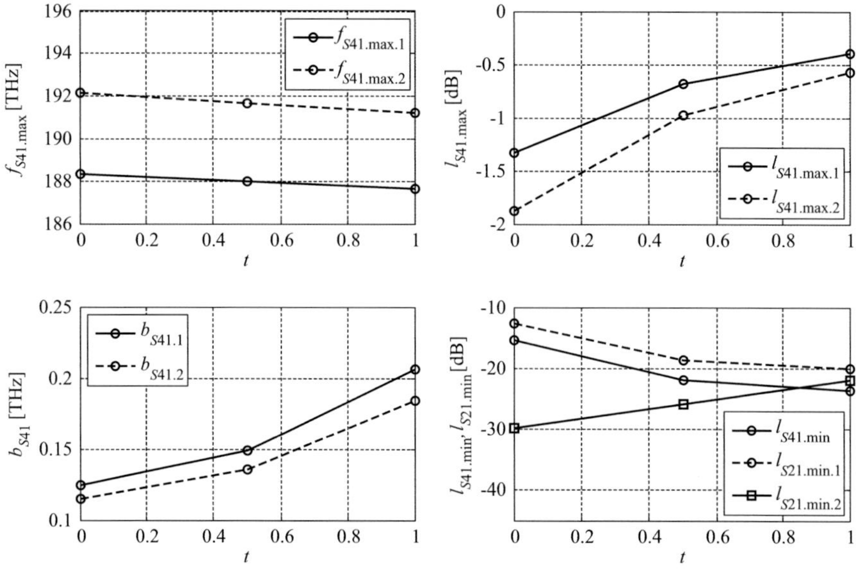

Fig. 9.14 Plots of the microring response features F evaluated for three different designs allocated along a straight line in the design space and parameterized by a parameter $0 \leq t \leq 1$. The plots indicate simple relationships between the features and the designable parameters of the microring (Koziel and Ogurtsov 2014b)

Table 9.3 Optimization results for the microring of Fig. 9.12

Objective description	Design requirements	Objective values at the initial design	Objective values at the final design
FSR	≥3.4 THz	3.82 THz	3.46 THz
$\delta f/FSR$	≤0.06	0.033	0.061
$\|S_{41}\|$ maximum	≥−0.5 dB	−1.9 dB	−0.45 dB
$\|S_{41}\|$ minimum	≤−20 dB	−29.7 dB	−21.3 dB
$\|S_{21}\|$ minimum	≤−20 dB	−12.6 dB	−21.8 dB

specifically those regarding $|S_{41}|$ maximum and $|S_{21}|$ minimum, are not satisfied at the initial design. At the final design, both of these objectives are greatly improved, whereas all the others are satisfied (except some small violation of the second one).

9.5 Limitations and Generalizations

An important prerequisite of the feature-based optimization process is that the feature point set is consistent along the optimization path. In particular, it is necessary to establish the linear model (9.3) as well as to compare the quality of the designs at subsequent iterations.

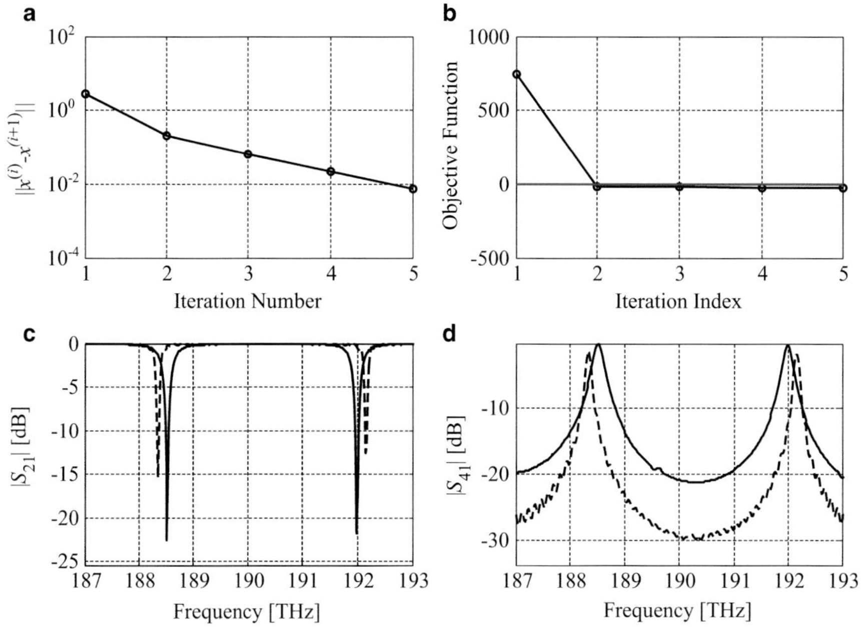

Fig. 9.15 Microring optimization results: (**a**) convergence plot, (**b**) evolution of the objective function (2), (**c**) initial (*dashed line*) and optimized (*solid line*) $|S_{21}|$ response, (d) initial (*dashed line*) and optimized (*solid line*) $|S_{41}|$ response

For certain problems, the above consistency may be difficult to maintain due to considerable changes of the system response during the optimization process. In some situations, it is possible to work around this issue, e.g., by neglecting the feature points that are not present in one (or more) of the designs that are being compared (e.g., the new candidate solution versus the current design). For example, the typical cost function may be replaced by the total discrepancy between the extracted features vector and the objective averaged with respect to the number of available features $p^{(i)}$ at iteration i. In this case, the iterative optimization process is reformulated as follows (here, assuming norm minimization with respect to the target feature set $\boldsymbol{f}_{F.\mathrm{target}}$):

$$\boldsymbol{x}^{(i+1)} = \arg \min_{||\boldsymbol{x}-\boldsymbol{x}^{(i)}|| \le \delta^{(i)}} \| \boldsymbol{f}_F(\boldsymbol{x}) - \boldsymbol{f}_{F.\mathrm{target}} \| \tag{9.6}$$

and the new design $\boldsymbol{x}^{(i+1)}$ is accepted if

$$\frac{1}{p^{(i+1)}} \left\| \boldsymbol{f}_F(\boldsymbol{x}^{(i+1)}) - \boldsymbol{f}_{F.\mathrm{target}} \right\|^2 < \frac{1}{p^{(i)}} \left\| \boldsymbol{f}_F(\boldsymbol{x}^{(i)}) - \boldsymbol{f}_{F.\mathrm{target}} \right\|^2. \tag{9.7}$$

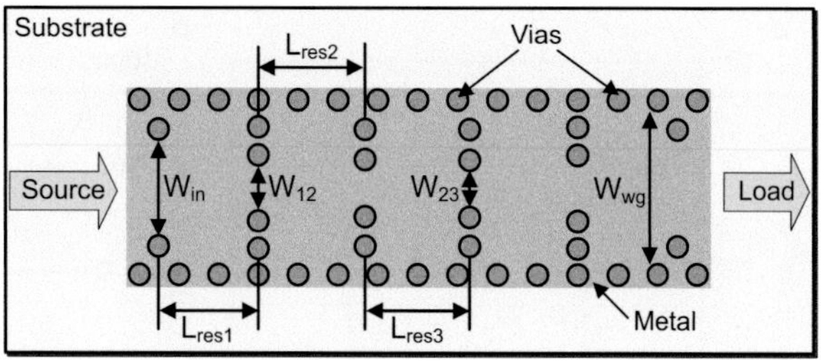

Fig. 9.16 Configuration of the fifth-order SIW filter

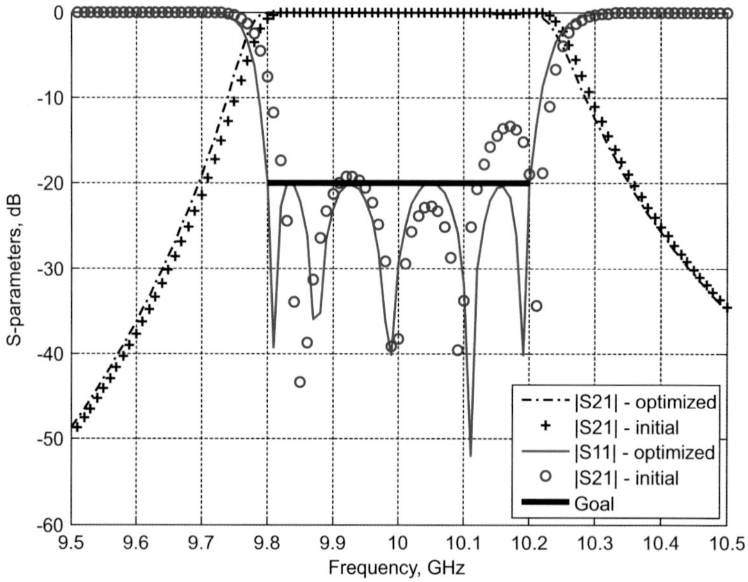

Fig. 9.17 Initial $f(\boldsymbol{x}_{init})$ and optimal $f(\boldsymbol{x}_{opt})$ responses of the fifth-order substrate-integrated waveguide filter of Fig. 9.16

Figure 9.16 shows the configuration of the fifth-order direct-coupled SIW filter (Glubokov and Koziel 2014b) utilized to illustrate this concept. Its design parameters are $\boldsymbol{x} = [L_{res1}\, L_{res2}\, L_{res3}\, W_{12}\, W_{23}\, W_{in}]^T$. The width of the waveguide is $W_{res} = 14$ mm, the substrate height is $H_{sub} = 1$ mm and permittivity is $\varepsilon_r = 2.2$. The specifications of the filter are given as follows: $|S_{11}| \leq -20$ dB for 9.8 GHz $\leq f \leq$ 10.2 GHz.

The initial design, found by optimization using coupling matrices (Glubokov and Koziel 2014b), is $\boldsymbol{x}_{init} = [13.182\; 14.536\; 14.664\; 5.96\; 5.477\; 8.099]^T$ mm. It should be emphasized that the initial response, shown in Fig. 9.17, has less features than at the optimum design, which is handled by the feature-based algorithm using (9.16) and (9.17). The optimized response $\boldsymbol{f}(\boldsymbol{x}_{opt})$ of the filter, where $\boldsymbol{x}_{opt} = [13.168\; 14.553\; 14.725\; 5.983\; 5.495\; 8.097]^T$ mm, is shown in Fig. 9.17. The total design cost (four iterations of (9.6)) corresponds to less than $6 \times \boldsymbol{f}$.

9.6 Summary

Feature-based optimization presented in this chapter attempts to move the optimization process from the original domain, in which the system responses are presumably highly nonlinear, to the feature space, where the figures of interest are much less nonlinear as functions of the designable parameters of the system at hand. Owing to that, the computational cost of the optimization process can be greatly reduced. If available, the low-fidelity model can also be used to further reduce the numerical expenses, specifically by estimating the response feature gradients at the level of the low- rather than the high-fidelity model. As demonstrated using a number of examples from the areas of microwave and antenna engineering as well as photonics, feature-based optimization can be very efficient.

At the same time, one needs to remember about an important limitation of feature-based optimization, which is the necessity of maintaining consistency of the feature set along the entire optimization path so that various designs can be compared to each other in the feature space. Furthermore, selection of the feature points is problem dependent and although straightforward in most cases, it may require some engineering insight. Consequently, feature-based optimization is a good choice for problems with well-defined system response shapes. However, it is not a general-purpose method.

Chapter 10
Enhancing Response Correction Techniques by Adjoint Sensitivity

Utilization of adjoint sensitivity techniques allows us to obtain both the response and its derivatives with respect to the geometry and/or material parameters of the structure of interest with a relatively small extra computational cost (El Sabbagh et al. 2006). In electromagnetic simulations, the process of obtaining the derivatives may not require additional simulations to the simulation for obtaining the figures of merit (El Sabbagh et al. 2006). In computational fluid dynamics (CFD), the cost of obtaining the gradients is almost equivalent to one additional flow solution (Jameson 1988). In both cases, the cost of obtaining the derivatives is independent of the number of design variables. Needless to say, the addition of adjoint sensitivities to simulation-based design and optimization has been transformative. In this chapter, we illustrate how surrogate-based modeling and optimization using response correction techniques can be enhanced with adjoint sensitivities. In particular, we start by discussing how to incorporate derivative data into the surrogate modeling and optimization process. Then, we provide the formulations for adjoint-enhanced versions of space mapping (Chap. 6), manifold mapping (Chap. 6), and shape-preserving response prediction (Chap. 7). For each technique, we provide example applications involving simulation-based design of several complex engineering systems including filters, and transonic airfoils.

10.1 Surrogate-Based Modeling and Optimization with Adjoint Sensitivity

A generic surrogate-based optimization (SBO) algorithm generates a sequence of approximate solutions to the original problem (3.1), $\boldsymbol{x}^{(i)}$, as follows (we repeat the problem formulation (4.1) here for the convenience of the reader):

S. Koziel, L. Leifsson, *Simulation-Driven Design by Knowledge-Based Response Correction Techniques*, DOI 10.1007/978-3-319-30115-0_10

$$\boldsymbol{x}^{(i+1)} = \arg\min_{\boldsymbol{x}} U\left(\boldsymbol{s}^{(i)}(\boldsymbol{x})\right), \tag{10.1}$$

where $\boldsymbol{s}^{(i)}$ is the surrogate model at iteration i and is assumed to be a computationally less expensive yet a reliable representation of $\boldsymbol{f}$, particularly in the neighborhood of $\boldsymbol{x}^{(i)}$. Robustness of the SBO process (10.1) depends on the quality of the surrogate model $\boldsymbol{s}^{(i)}$. To ensure convergence of the algorithm (10.1) to at least a local optimum of the high-fidelity model, first-order consistency conditions should be met (Alexandrov and Lewis 2001), i.e., $\boldsymbol{s}^{(i)}(\boldsymbol{x}^{(i)}) = \boldsymbol{f}(\boldsymbol{x}^{(i)})$ and $\boldsymbol{J}_{s(i)}(\boldsymbol{x}^{(i)}) = \boldsymbol{J}_f(\boldsymbol{x}^{(i)})$, where $\boldsymbol{J}$ stands for the Jacobian of the respective model. Also, the process (10.1) has to be embedded in the trust-region (TR) framework (Conn et al. 2000) as in (4.2).

Availability of adjoint sensitivities (Toivanen et al. 2010; Soliman et al. 2004) makes it possible to satisfy the consistency conditions in an easy way without excessive computations. From the selected previous works (Khalatpour et al. 2011; Cheng et al. 2011, 2012; Koziel et al. 2006a, b; Bakr et al. 1999a, b; Bandler et al. 2002), we summarize the ways in which model sensitivities can be used:

1. *Surrogate mapping*: use sensitivity to improve a mapping between the low- and high-fidelity models.
2. *Parameter extraction*: match the low- and the high-fidelity models using both the responses and the corresponding derivatives.
3. *Surrogate enhancement*: use sensitivity to enhance the surrogate model by enforcing first-order consistency.

In this chapter, we describe how (1), (2), and (3) can be applied to space mapping (Bandler et al. 1994), manifold mapping (Echeverria and Hemker 2005), and shape-preserving response prediction (Koziel 2010a). But first, we will illustrate how a linear approximation of the high-fidelity model with first-order derivatives can accelerate the optimization process when using only that model. This is the conventional way of using derivative information to expedite the optimization process.

Consider the first-order Taylor model:

$$\boldsymbol{s}^{(i)}(\boldsymbol{x}) = \boldsymbol{f}\left(\boldsymbol{x}^{(i)}\right) + \boldsymbol{J}_f\left(\boldsymbol{x}^{(i)}\right) \cdot \left(\boldsymbol{x} - \boldsymbol{x}^{(i)}\right), \tag{10.2}$$

where $\boldsymbol{J}_f$ is the Jacobian of $\boldsymbol{f}$ obtained by means of the adjoint sensitivity technique. Using the surrogate model (10.2) within the generic SBO scheme (10.1) with trust regions (as given in (4.2)) forms a first-order consistent approximation of the original problem (3.1). The key mechanisms of the adjoint-enhanced algorithm is how the search for the new design $\boldsymbol{x}^{(i)}$ is performed as well as how the search radius $\delta^{(i)}$ is updated. Instead of the standard updating rules, we can use the following strategy ($\boldsymbol{x}^{(i-1)}$ and $\delta^{(i-1)}$ are the previous design and the search radius, respectively):

1. For $\delta_k = k \cdot \delta^{(i-1)}$, $k = 0, 1, 2$, solve $\boldsymbol{x}^k = \arg \min_{\boldsymbol{x}: \left\| \boldsymbol{x} - \boldsymbol{x}^{(i)} \right\| \leq \delta_k} U\left(\boldsymbol{s}^{(i)}(\boldsymbol{x})\right)$. Note that $\boldsymbol{x}^0 = \boldsymbol{x}^{(i-1)}$. The values of δ_k and $U_k = U(\boldsymbol{s}^{(i)}(\boldsymbol{x}^k))$ are interpolated with a second-order polynomial to find δ^* that gives the smallest (estimated) value of the specification error (δ^* is limited to $3 \cdot \delta^{(i-1)}$). Set $\delta^{(i)} = \delta^*$.
2. Find a new design $\boldsymbol{x}^{(i)}$ by solving (10.1) constrained to $\|\boldsymbol{x} - \boldsymbol{x}^{(i)}\| \leq \delta^{(i)}$.
3. Calculate the gain ratio $\rho = [U(\boldsymbol{f}(\boldsymbol{x}^{(i)})) - U_0]/[U(\boldsymbol{s}^{(i)}(\boldsymbol{x}^{(i)})) - U_0]$. If $\rho < 0.25$ then $\delta^{(i)} = \delta^{(i)}/3$; else if $\rho > 0.75$ then $\delta^{(i)} = 2 \cdot \delta^{(i)}$.
4. If $\rho < 0$ go to 2.
5. Return $\boldsymbol{x}^{(i)}$ and $\delta^{(i)}$.

The trial points $\boldsymbol{x}^k$ are used to find the best value of the search radius, which is further updated based on the gain ratio ρ which is the actual versus the expected objective function improvement. If the new design is worse than the previous one, the search radius is reduced to find $\boldsymbol{x}^{(i)}$ again, which eventually will bring the improvement of U because the models $\boldsymbol{s}^{(i)}$ and $\boldsymbol{f}$ are first-order consistent (Alexandrov and Lewis 2001). This precaution is necessary because the procedure in Step 1 only gives an estimation of the search radius.

As an example, consider a wideband hybrid antenna (Petosa 2007) shown in Fig. 10.1, which is a quarter-wavelength monopole loaded by a dielectric ring resonator. The design goal is to have $|S_{11}| \leq -20$ dB for 8–13 GHz. The design variables are $\boldsymbol{x} = [h_1 h_2 r_1 r_2 g]^T$. The initial design is $\boldsymbol{x}^{(0)} = [2.5\ 9.4\ 2.3\ 3.0\ 0.5]^T$ mm. Other parameters are fixed. The final design with the proposed algorithm is $\boldsymbol{x}^{(0)} = [3.94\ 10.01\ 2.23\ 3.68\ 0.0]^T$ mm. Table 10.1 and Fig. 10.2 compare the design cost and quality of the final design found by the algorithm described above and Matlab's *fminimax*. It can be observed that the adjoint-based algorithm described above yields better design at significantly smaller computational cost (75 % design time reduction).

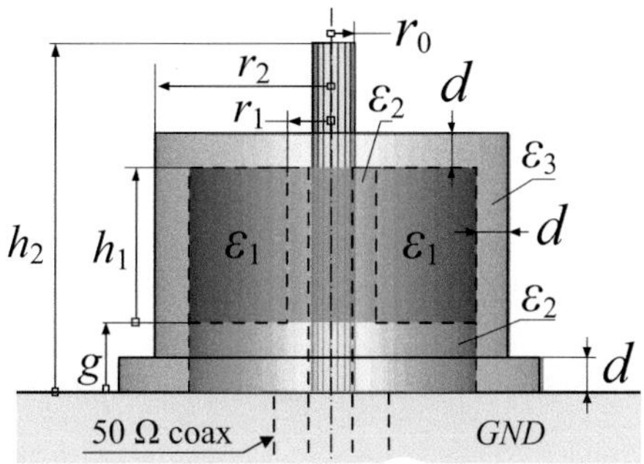

Fig. 10.1 Wideband hybrid antenna: geometry

Table 10.1 Wideband hybrid antenna: design results

Algorithm	max$\|S_{11}\|$ for 8–13 GHz at final design	Design cost (number of EM analyses)
Matlab's *fminimax*	–22.6 dB	98
SBO (10.1) with trust region and surrogate model (10.2)	–24.6 dB	24

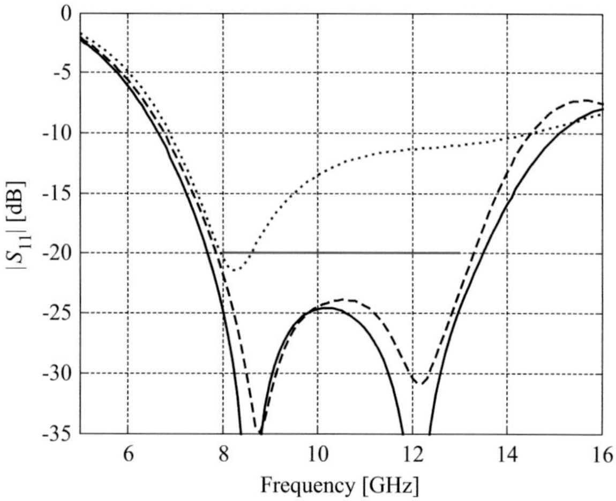

Fig. 10.2 Wideband hybrid antenna: reflection response at the initial design (· · ·), at the final design by Matlab's *fminimax* (- - -), and by the algorithm (10.1) with the surrogate (10.2) embedded in the trust region framework (—)

10.2 Space Mapping with Adjoints

This section provides the details of using adjoint sensitivity information to enhance space mapping (SM) in terms of surrogate modeling, parameter extraction, and optimization. The SM procedure is then demonstrated on design problems involving filters, antennas, and airfoils.

10.2.1 SM Surrogate Modeling and Optimization

An SM surrogate enhanced with sensitivities is constructed using input and output SM (El Sabbagh et al. 2006; Koziel et al. 2013c) of the form

$$\boldsymbol{s}^{(i)}(\boldsymbol{x}) = \boldsymbol{c}\left(\boldsymbol{x} + \boldsymbol{p}^{(i)}\right) + \boldsymbol{d}^{(i)} + \boldsymbol{E}^{(i)}\left(\boldsymbol{x} - \boldsymbol{x}^{(i)}\right), \tag{10.3}$$

where $\boldsymbol{c}$ is the response of a low-fidelity model, $\boldsymbol{p}^{(i)}$ is the input SM vector, $\boldsymbol{d}^{(i)}$ is the additive term of output SM, $\boldsymbol{E}^{(i)}$ is the linear term of output SM, $\boldsymbol{x}$ is a vector with the current design, and $\boldsymbol{x}^{(i)}$ is a vector with the reference design. In (10.3), only the input SM vector $\boldsymbol{p}^{(i)}$ is obtained through the nonlinear parameter extraction process:

$$\boldsymbol{p}^{(i)} = \arg\min_{\boldsymbol{c}} \left\|\boldsymbol{f}\left(\boldsymbol{x}^{(i)}\right) - \boldsymbol{c}\left(\boldsymbol{x}^{(i)} + \boldsymbol{p}\right)\right\|. \tag{10.4}$$

The output SM parameters are calculated as

$$\boldsymbol{d}^{(i)} = \boldsymbol{f}\left(\boldsymbol{x}^{(i)}\right) - \boldsymbol{c}\left(\boldsymbol{x}^{(i)} + \boldsymbol{p}^{(i)}\right), \tag{10.5}$$

and

$$\boldsymbol{E}^{(i)} = \boldsymbol{J}_f\left(\boldsymbol{x}^{(i)}\right) - \boldsymbol{J}_c\left(\boldsymbol{x}^{(i)} + \boldsymbol{p}^{(i)}\right), \tag{10.6}$$

where $\boldsymbol{J}$ denotes the Jacobian of the respective model obtained using adjoint sensitivities. The formulation (10.3)–(10.6) ensures zero- and first-order consistency (Alexandrov and Lewis 2001) between the surrogate and the high-fidelity model, i.e., $\boldsymbol{s}^{(i)}(\boldsymbol{x}^{(i)}) = \boldsymbol{f}(\boldsymbol{x}^{(i)})$ and $\boldsymbol{J}_{\boldsymbol{s}^{(i)}}(\boldsymbol{x}^{(i)}) = \boldsymbol{J}_f(\boldsymbol{x}^{(i)})$, which substantially improves the ability of the SM algorithm to locate the high-fidelity model optimum (Koziel et al. 2006a, b).

Zero- and first-order consistency alone does not guarantee that the SM algorithm converges to the fine model optimum. Convergence can be ensured by embedding the SM optimization process (10.1) in the trust region (TR) framework (4.2) when using formulation (10.3)–(10.6).

The value of the TR radius is updated based on the gain ratio $\rho = (U(\boldsymbol{f}(\boldsymbol{x}^{(i+1)}) - U(\boldsymbol{f}(\boldsymbol{x}^{(i)}))/(U(\boldsymbol{s}(\boldsymbol{x}^{(i+1)}) - U(\boldsymbol{s}(\boldsymbol{x}^{(i)}))$. The designs obtained from unsuccessful iterations are rejected. First-order consistency ensures positive gain ratios for sufficiently small values of $\delta^{(i)}$. This, together with the smoothness of $\boldsymbol{s}^{(i)}(\boldsymbol{x})$ and $\boldsymbol{f}$, ensures convergence to a local high-fidelity model optimum. Because of numerical noise inherent to simulation-based models the convergence conditions are unlikely to be satisfied exactly. Therefore, the accuracy of locating the minimum by the SM algorithm may be limited. Embedding the SM algorithm in the trust region framework allows us to handle design variable constraints by adding them to the TR, i.e., $\|\boldsymbol{x} - \boldsymbol{x}^{(i)}\| \leq \delta^{(i)}$.

To speed up the parameter extraction process (10.4) involving coarse mesh low-fidelity models their adjoint sensitivities can be exploited. A simple trust region-based (Conn et al. 2000) algorithm can be used where the approximate solution $\boldsymbol{p}^{(i.k+1)}$ of $\boldsymbol{p}^{(i)}$ is found as (Koziel et al. 2013c)

$$\boldsymbol{p}^{(i.k+1)} = \arg \min_{\left\|\boldsymbol{p} - \boldsymbol{p}^{(i.k)}\right\| \leq \delta_{PE}^{(k)}} \left\|\boldsymbol{f}\left(\boldsymbol{x}^{(i)}\right) - \boldsymbol{L}_{c.p}{}^{(i.k)}(\boldsymbol{p})\right\|, \tag{10.7}$$

where k is the iteration index for this parameter extraction process. Here, $\boldsymbol{L}_{c.p}{}^{(i.k)}(\boldsymbol{p}) = \boldsymbol{c}(\boldsymbol{x}^{(i)} + \boldsymbol{p}^{(i.k)}) + \boldsymbol{J}_c(\boldsymbol{x}^{(i)} + \boldsymbol{p}^{(i.k)}) \cdot (\boldsymbol{p} - \boldsymbol{p}^{(i.k)})$ is a linear approximation of $\boldsymbol{c}(\boldsymbol{x}^{(i)}$

$+\boldsymbol{p}$) in the vicinity of $\boldsymbol{p}^{(i.k)}$. The TR radius $\delta_{PE}{}^{(k)}$ is updated according to standard rules (Conn et al. 2000). The parameter extraction is terminated upon convergence or exceeding the maximum number of coarse model evaluations.

Low-cost surrogate optimization with adjoint sensitivities can be realized using a TR-based algorithm to produce a sequence of approximations $\boldsymbol{x}^{(i+1.k)}$ of the solution $\boldsymbol{x}^{(i+1)}$ to (4.2) as follows (k is the iteration index for the surrogate model optimization process) (Koziel et al. 2013c):

$$\boldsymbol{x}^{(i+1.k+1)} = \arg \min_{\left\|\boldsymbol{x}-\boldsymbol{x}^{(i+1.k)}\right\| \le \delta_{SO}^{(k)}} U\left(\boldsymbol{L}_{c.x}{}^{(i.k)}(\boldsymbol{x})\right), \tag{10.8}$$

where $\boldsymbol{L}_{c.x}{}^{(i.k)}(\boldsymbol{x}) = \boldsymbol{s}^{(i)}(\boldsymbol{x}^{(i+1.k)}) + \boldsymbol{J}_{\boldsymbol{S}^{(i)}}(\boldsymbol{x}^{(i+1.k)})\cdot(\boldsymbol{x} - \boldsymbol{x}^{(i+1.k)})$ is a linear approximation of $\boldsymbol{s}^{(i)}(\boldsymbol{x})$ at $\boldsymbol{x}^{(i+1.k)}$. The TR radius $\delta_{SO}{}^{(k)}$ is updated according to the standard rules.

Note that the sensitivities of the surrogate model can be calculated using the sensitivities of both $\boldsymbol{f}$ and $\boldsymbol{c}$ as follows: $\boldsymbol{J}_{\boldsymbol{S}^{(i)}}(\boldsymbol{x}) = \boldsymbol{J}_c(\boldsymbol{x} + \boldsymbol{p}^{(i)}) + [\boldsymbol{J}_f(\boldsymbol{x}^{(i)}) - \boldsymbol{J}_c(\boldsymbol{x}^{(i)} + \boldsymbol{p}^{(i)})]$.

It should be observed that the adjoint sensitivities enhance the match between the surrogate and high-fidelity models through the linear term in (10.3), and also speed up the parameter extraction process (10.4), as well as the surrogate model optimization (10.1). These may not be critical if a well-behaved and fast equivalent circuit model is available but it is desirable for a coarsely discretized low-fidelity model. Adjoint-enhanced SBO with SM will now be illustrated through three design examples.

10.2.2 Design Example: UWB Monopole Antenna

Consider the monopole antenna (Koziel et al. 2012) shown in Fig. 10.3. The design variables are $\boldsymbol{x} = [l_1 l_2 l_3 w_1]^T$. The input microstrip of the monopole is fed through an edge mount SMA (SubMiniature version A) connector. Simulation time of the

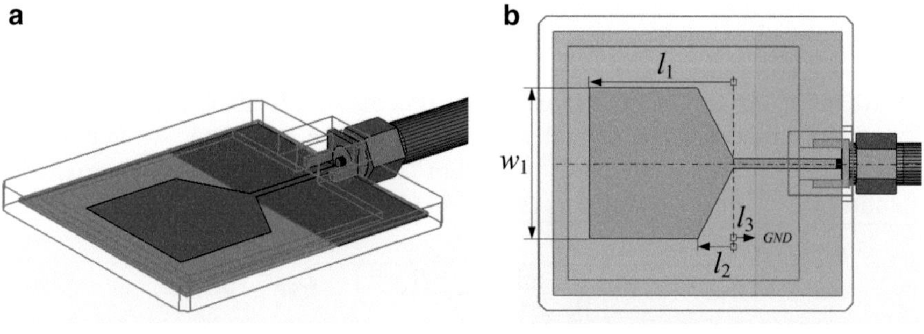

Fig. 10.3 UWB monopole (Koziel et al. 2012): (**a**) 3D view; (**b**) top view. The housing is shown transparent

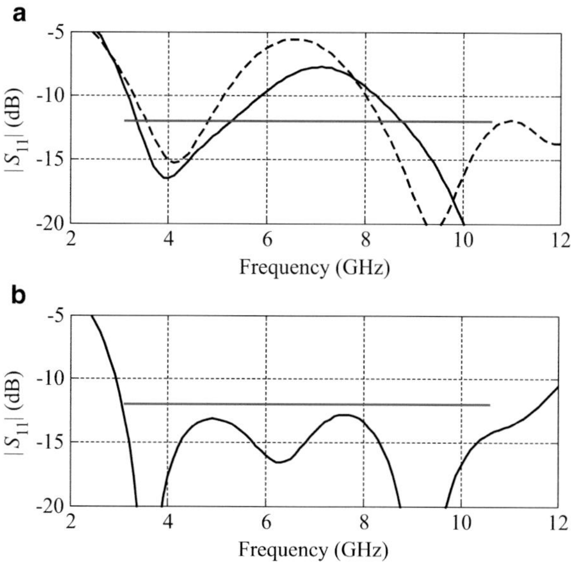

Fig. 10.4 UWB monopole (Koziel et al. 2012): (**a**) responses of $\boldsymbol{f}$ (—) and $\boldsymbol{c}$ (- - -) at the initial design $\boldsymbol{x}^{init}$; (**b**) response of $\boldsymbol{f}$ (—) at the final design

low-fidelity model $\boldsymbol{c}$ (156,000 mesh cells) is 1 min, and that of the high-fidelity model $\boldsymbol{f}$ (1,992,060 mesh cells) is 40 min (both at the initial design). Both models are evaluated using the transient solver of CST MWS. Simulation time per low/high-fidelity model evaluation is reduced by approximately 10 % without adjoint sensitivity evaluation. The design specifications for the reflection are $|S_{11}| \leq -12$ dB for 3.1–10.6 GHz. The initial design is $\boldsymbol{x}^{init} = [20\ 2\ 0\ 25]^T$ mm.

The antenna was optimized using the space mapping algorithm of Sect. 10.2.1. Figure 10.4a shows the responses of $\boldsymbol{f}$ and $\boldsymbol{c}$ at $\boldsymbol{x}^{init}$, and Fig. 10.4b shows the response of the high-fidelity model at the final design $\boldsymbol{x}^{(3)} = [20.29\ 2.27\ 0.058\ 19.63]^T$ obtained after three SM iterations.

Table 10.2 summarizes the total optimization cost. Note that using adjoint sensitivities allows us to greatly improve the efficiency of parameter extraction and surrogate model optimization in the design process. Otherwise, a less efficient non-gradient method has to be used. The average cost of parameter extraction and surrogate optimization processes is only about five evaluations of $\boldsymbol{c}$. The evolution of the specification error is shown in Fig. 10.5.

The results are compared with those obtained by direct optimization (Matlab's *fminimax* using adjoint sensitivity) and a space mapping method without using adjoint sensitivity (input and output space mapping of the form $\boldsymbol{s}(\boldsymbol{x}) = \boldsymbol{c}(\boldsymbol{x} + \boldsymbol{p}) + \boldsymbol{d}$, Matlab's *lsqnonlin* for parameter extraction, and a pattern search algorithm for coarse/surrogate model optimization). The space mapping method with adjoints provides results faster and with better accuracy (see Table 10.2).

Table 10.2 Monopole antenna: SM optimization results (Koziel et al. 2013c)[a]

Algorithm	SM with adjoint sensitivity	SM without adjoint sensitivity	Direct optimization
Coarse model evaluation	45	305	–
Fine model evaluation[b]	4	6	62
Coarse model actual CPU time	45 min	275 min[c]	–
Fine model actual CPU time[b]	160 min	220 min[c]	2,230 min[c]
Total CPU time[b]	205 min	495 min[c]	2,230 min[c]
Coarse model relative (w.r.t. *f*) CPU time	1.1	6.9	–
Fine model relative (w.r.t. *f*) CPU time[b]	4.0	5.5	56.0
Total relative CPU time[b]	5.1	12.4	56.0

[a]Obtained with Intel Xeon 2.53 GHz CPU, 6 GB RAM, and 500 GB hard drive
[b]Excludes *f* evaluation at the initial design
[c]Simulation time per fine/coarse model evaluation is reduced by approximately 10 % without adjoint sensitivity evaluation

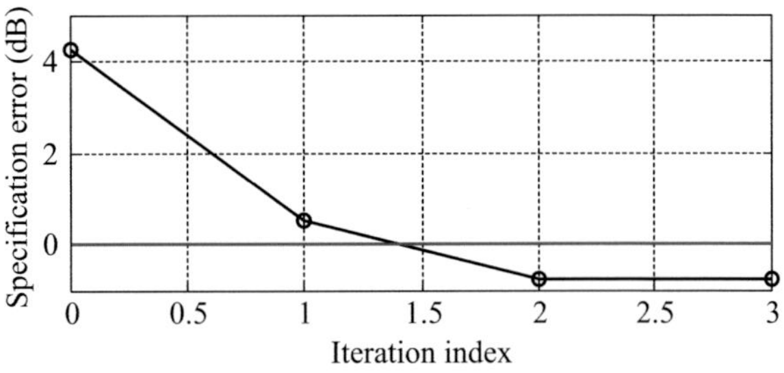

Fig. 10.5 UWB monopole (Koziel et al. 2012): minimax specification error versus SM iteration index

10.2.3 Design Example: Dielectric Resonator Filter

The dielectric resonator filter shown in Fig. 10.6 is considered. The design variables are $\boldsymbol{x} = [u_1 u_2 u_3 v_1 v_2]^T$ mm. The relative permittivity of the dielectric resonators (DR) is $\varepsilon_{r1} = 38$. The relative permittivity of the DR supports and the coax filling is $\varepsilon_{r2} = 2.1$. The materials are considered lossless. We have $h_3 = h_4 = h_5 = 1$ mm, $r_3 = r_2$ (the inner radius of the DR), and $r_4 = r_2 + 1$ mm, and $\rho_1 = 2.05$ mm. The other dimensions are fixed as in dielectric resonator filter (2011).

The filter models are simulated by the CST MWS transient solver. The low-fidelity model $\boldsymbol{c}$ comprises 38,220 hexahedral cells and runs for 1 min, and the high-fidelity model $\boldsymbol{f}$ comprises 333,558 cells and runs for 28 min, both at the initial design, which is $\boldsymbol{x}^{(ini)} = [8\ 12\ 25\ 25\ {-10}]^T$ mm. Simulation time per model evaluation is reduced by approximately 10 % without adjoint sensitivity evaluation. The design specifications are $|S_{21}| \geq -0.75$ dB for 4.52 GHz $\leq \omega \leq$ 4.54 GHz, and $|S_{21}| \leq -20$ dB for 4.4 GHz $\leq \omega \leq$ 4.47 GHz and 4.59 GHz $\leq \omega \leq$ 4.63 GHz.

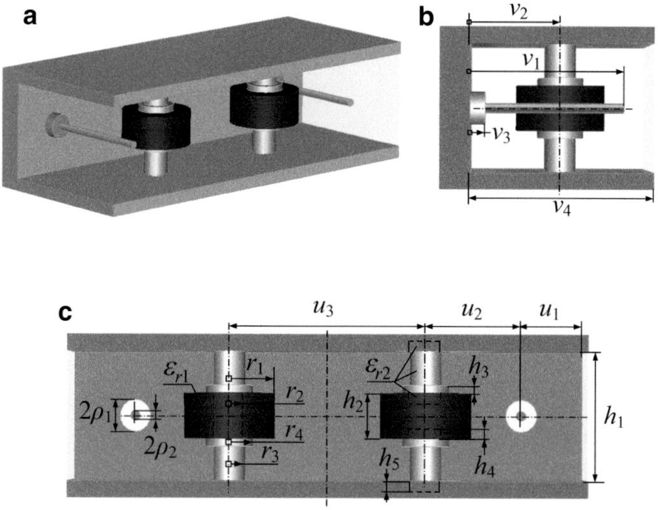

Fig. 10.6 Dielectric resonator filter (2011): (**a**) 3D view; (**b**) side view, and (**b**) front view. Front and side metal walls of the waveguide section are shown transparent

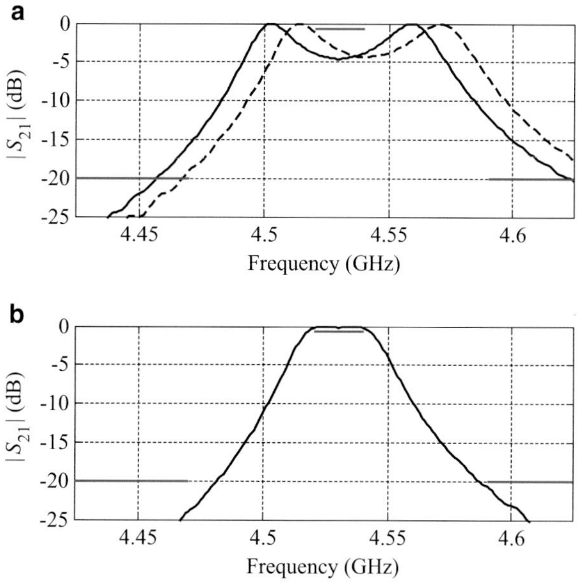

Fig. 10.7 Dielectric resonator filter (Koziel et al. 2013c): (**a**) responses of $\boldsymbol{f}$ (—) and $\boldsymbol{c}$ (- - -) at the initial design $\boldsymbol{x}^{init}$; (**b**) response of $\boldsymbol{f}$ (—) at the final design

The final design is found to be $\boldsymbol{x}^{(*)} = [7.28\ 11.06\ 24.81\ 26.00\ -11.07]^T$ mm after ten SM iterations. Figure 10.7 shows the responses at the initial and final designs. The algorithm performance is consistent with the previous examples with the design being improved at each iteration (Fig. 10.8), and overall low optimization cost (Table 10.3).

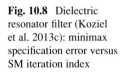

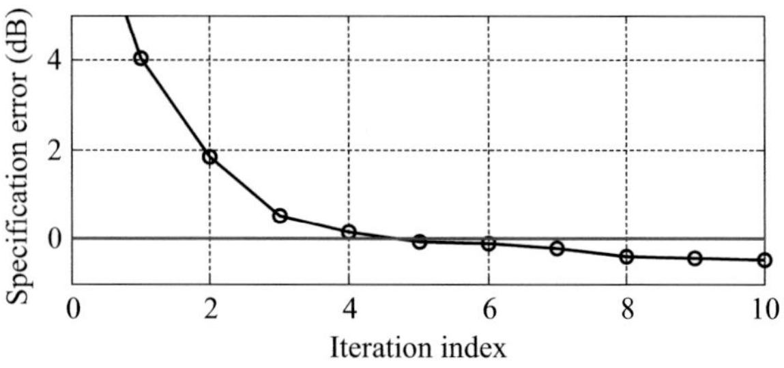

Fig. 10.8 Dielectric resonator filter (Koziel et al. 2013c): minimax specification error versus SM iteration index

Table 10.3 Dielectric resonator filter: SM optimization results (Koziel et al. 2013c)[a]

Algorithm	SM with adjoint sensitivity	SM without adjoint sensitivity	Direct optimization
Coarse model evaluation	80	346	–
Fine model evaluation[b]	11	7	67
Coarse model actual CPU time	80 min	315 min[c]	–
Fine model actual CPU time[b]	308 min	178 min[c]	1,690 min[c]
Total CPU time[b]	398 min	493 min[c]	1,690 min[c]
Coarse model relative (w.r.t. ***f***) CPU time	2.9	11.3	–
Fine model relative (w.r.t. ***f***) CPU time[b]	11	6.4	61.0
Total relative CPU time[b]	13.9	17.7	61.0

[a]Obtained using a computer with Intel Xeon 2.53GHz CPU, 6 GB RAM, and 500 GB hard drive
[b]Excludes ***f*** evaluation at the initial design
[c]Simulation time per fine/coarse model evaluation is reduced by approximately 10 % without adjoint sensitivity evaluation

We compare the results to those obtained by direct optimization (Matlab's *fminimax*) and space mapping without using adjoint sensitivity (input and output space mapping of the form $\boldsymbol{s}(\boldsymbol{x}) = \boldsymbol{c}(\boldsymbol{x} + \boldsymbol{p}) + \boldsymbol{d}$, Matlab's *lsqnonlin* for parameter extraction, and a pattern search algorithm for coarse/surrogate model optimization). The SM method with adjoint sensitivity converges faster with improved accuracy compared to the other methods (see Table 10.3).

10.2.4 Design Example: Transonic Airfoils

Here, we apply the adjoint-enhanced SM algorithm of Sect. 10.2.1 to the design optimization of transonic airfoil shapes. The objective is to minimize the drag coefficient (C_d) of a modified symmetric NACA 0012 airfoil section at a free-stream Mach number of $M_\infty = 0.85$ and an angle of attack $\alpha = 0°$ subject to a minimum thickness constraint. The optimization problem is stated as

$$\min_{l \le x \le u} C_d, \tag{10.9}$$

where $\boldsymbol{x}$ is the vector of design variables, and $\boldsymbol{l}$ and $\boldsymbol{u}$ are the lower and upper bounds, respectively, with the thickness constraint stated as

$$z(x) \ge z(x)_{baseline}, \tag{10.10}$$

where $z(x)$ is the airfoil thickness, $x \in [0,1]$ is the chord-wise location, and $z(x)_{baseline}$ is the thickness of the baseline airfoil, which is a modified version of the NACA 0012, defined as

$$z(x)_{baseline} = \pm 0.6 \begin{pmatrix} 0.2969\sqrt{x} - 0.1260x - 0.3516x^2 \\ +0.2843x^3 - 0.1036x^4 \end{pmatrix}. \tag{10.11}$$

In this case, only output space mapping is utilized (cf. (10.3)) and the space mapping surrogate model, $\boldsymbol{s}$, takes the following form (Tesfahunegn et al. 2015a, b):

$$\boldsymbol{s}^{(i)}(\boldsymbol{x}) = \boldsymbol{A}^{(i)} \circ \boldsymbol{c}(\boldsymbol{x}) + \boldsymbol{D}^{(i)} + \boldsymbol{q}^{(i)} + \boldsymbol{E}^{(i)} \cdot \left(\boldsymbol{x} - \boldsymbol{x}^{(i)}\right), \tag{10.12}$$

The multiplicative and additive response correction is defined, at the component level, as

$$\begin{aligned} &\boldsymbol{A}^{(i)} \circ \boldsymbol{c}(\boldsymbol{x}) + \boldsymbol{D}^{(i)} + \boldsymbol{q}^{(i)} \\ &= \left[a_l^{(i)} C_{l.c}(\boldsymbol{x}) + d_l^{(i)} + q_l^{(i)} \quad a_d^{(i)} C_{d.c}(\boldsymbol{x}) + d_d^{(i)} + q_d^{(i)} \quad A_c^{(i)}(\boldsymbol{x})\right]^T, \end{aligned} \tag{10.13}$$

where $\circ$ denotes a component-wise multiplication, $C_{l.c}$ is the lift coefficient obtained by the low-fidelity model, $C_{d.c}$ is the drag coefficient obtained by the low-fidelity model, A_c is the cross-sectional area of the low-fidelity mode (which is the same as the high-fidelity one), and a_i, d_i, and q_i, where $i = l, d$, are components of $\boldsymbol{A}$, $\boldsymbol{D}$, and $\boldsymbol{q}$, respectively. Note that there is no need to map A_c because $A_c(\boldsymbol{x}) = A_f(\boldsymbol{x})$ for all $\boldsymbol{x}$.

The matrix $\boldsymbol{E}^{(i)}$ is the linear part of the correction (10.12) and is defined as (Tesfahunegn et al. 2015a, b)

$$\boldsymbol{E}^{(i)} = \begin{bmatrix} \left[\nabla C_{l.f}\left(\boldsymbol{x}^{(i)}\right) - a_l^{(i)} \nabla C_{l.c}\left(\boldsymbol{x}^{(i)}\right)\right]^T \\ \left[\nabla C_{d.f}\left(\boldsymbol{x}^{(i)}\right) - a_d^{(i)} \nabla C_{d.c}\left(\boldsymbol{x}^{(i)}\right)\right]^T \\ 0 \end{bmatrix}. \tag{10.14}$$

The response correction parameters $\boldsymbol{A}^{(i)}$ and $\boldsymbol{D}^{(i)}$ are obtained by solving

$$\left[\boldsymbol{A}^{(i)}, \boldsymbol{D}^{(i)}\right] = \arg\min_{[\boldsymbol{A},\boldsymbol{D}]} \sum\nolimits_{k=0}^{i} \left\| \boldsymbol{f}\left(\boldsymbol{x}^{(k)}\right) - \boldsymbol{A} \circ \boldsymbol{c}\left(\boldsymbol{x}^{(k)}\right) + \boldsymbol{D} \right\|^2, \tag{10.15}$$

i.e., the response scaling is supposed to (globally) improve the matching for all previous iteration points. The additive response correction term $\boldsymbol{q}^{(i)}$ is defined as

$$\boldsymbol{q}^{(i)} = \boldsymbol{f}\left(\boldsymbol{x}^{(i)}\right) - \left[\boldsymbol{A}^{(i)} \circ \boldsymbol{c}\left(\boldsymbol{x}^{(i)}\right) + \boldsymbol{D}^{(i)}\right], \tag{10.16}$$

i.e., it ensures perfect matching between the surrogate and the high-fidelity model at the current design $\boldsymbol{x}^{(i)}$, $\boldsymbol{s}^{(i)}(\boldsymbol{x}^{(i)}) = \boldsymbol{f}(\boldsymbol{x}^{(i)})$ (the so-called zero-order consistency).

$\boldsymbol{A}^{(i)}$, $\boldsymbol{D}^{(i)}$, and $\boldsymbol{q}^{(i)}$ can be calculated analytically so that there is no need to carry out nonlinear minimization. Terms $\boldsymbol{A}^{(i)}$ and $\boldsymbol{D}^{(i)}$ can be obtained from Koziel and Leifsson (2012b):

$$\begin{bmatrix} a_l^{(i)} \\ d_l^{(i)} \end{bmatrix} = \left(\boldsymbol{C}_l^T \boldsymbol{C}_l\right)^{-1} \boldsymbol{C}_l^T \boldsymbol{F}_l, \quad \begin{bmatrix} a_d^{(i)} \\ d_d^{(i)} \end{bmatrix} = \left(\boldsymbol{C}_d^T \boldsymbol{C}_d\right)^{-1} \boldsymbol{C}_d^T \boldsymbol{F}_d, \tag{10.17}$$

where

$$\boldsymbol{C}_l = \begin{bmatrix} C_{l.c}\left(\boldsymbol{x}^{(0)}\right) & C_{l.c}\left(\boldsymbol{x}^{(1)}\right) & \cdots & C_{l.c}\left(\boldsymbol{x}^{(i)}\right) \\ 1 & 1 & \cdots & 1 \end{bmatrix}^T, \tag{10.18}$$

$$\boldsymbol{F}_l = \left[C_{l.f}\left(\boldsymbol{x}^{(0)}\right) \quad C_{l.c}\left(\boldsymbol{x}^{(1)}\right) \quad \cdots \quad C_{l.f}\left(\boldsymbol{x}^{(i)}\right)\right]^T, \tag{10.19}$$

$$\boldsymbol{C}_d = \begin{bmatrix} C_{d.c}\left(\boldsymbol{x}^{(0)}\right) & C_{d.c}\left(\boldsymbol{x}^{(1)}\right) & \cdots & C_{d.c}\left(\boldsymbol{x}^{(i)}\right) \\ 1 & 1 & \cdots & 1 \end{bmatrix}^T, \tag{10.20}$$

$$\boldsymbol{F}_d = \left[C_{d.f}\left(\boldsymbol{x}^{(0)}\right) \quad C_{d.c}\left(\boldsymbol{x}^{(1)}\right) \quad \cdots \quad C_{d.f}\left(\boldsymbol{x}^{(i)}\right)\right]^T, \tag{10.21}$$

which is a least-square optimal solution to the linear regression problems $\boldsymbol{C}_l a_l^{(i)} + d_l^{(i)} = \boldsymbol{F}_l$ and $\boldsymbol{C}_d\ a_d^{(i)} + d_d^{(i)} = \boldsymbol{F}_d$, equivalent to (10.15). Note that the matrices $\boldsymbol{C}_l^T \boldsymbol{C}_l$ and $\boldsymbol{C}_d^T \boldsymbol{C}_d$ are non-singular for $i > 1$. For $i = 1$, only the multiplicative correction with $\boldsymbol{A}^{(i)}$ components is used (calculated in a similar way).

Note that the linear term $\boldsymbol{E}^{(i)} \cdot (\boldsymbol{x} - \boldsymbol{x}^{(i)})$ in (10.12), defined as in (10.14), will ensure first-order consistency between the surrogate and the high-fidelity model, that is, perfect agreement of the model gradients, both for the drag and the lift coefficient. While this requires derivative information, it is obtained at a small computational overhead using adjoint sensitivities.

Adjoint sensitivity is also utilized to speed up the surrogate model optimization process (10.1). More specifically, in order to obtain the design $\boldsymbol{x}^{(i+1)}$, the following trust region-based process is launched (Tesfahunegn et al. 2015a, b):

$$\boldsymbol{x}^{(i+1.k+1)} = \arg \min_{\boldsymbol{x},\ \left\| \boldsymbol{x} - \boldsymbol{x}^{(i+1.k)} \right\| \leq \delta^{(i.k)}} H\left(\boldsymbol{L}^{(i.k)}(\boldsymbol{x})\right), \tag{10.22}$$

where $\boldsymbol{x}^{(i+1.k)}$, $k = 0, 1, \ldots$, is the sequence approximating $\boldsymbol{x}^{(i+1)}$, with $\boldsymbol{x}^{(i+1.0)} = \boldsymbol{x}^{(i)}$; $\boldsymbol{L}^{(i.k)}$ is the linear approximation of the surrogate model $\boldsymbol{s}^{(i)}$, established at $\boldsymbol{x}^{(i+1.k)}$ of the form

$$\boldsymbol{L}^{(i.k)}(\boldsymbol{x}) = \boldsymbol{s}^{(i)}\left(\boldsymbol{x}^{(i+1.k)}\right) + \nabla \boldsymbol{s}^{(i)}\left(\boldsymbol{x}^{(i+1.k)}\right) \cdot \left(\boldsymbol{x} - \boldsymbol{x}^{(i+1.k)}\right), \tag{10.23}$$

Here, the gradient of $\boldsymbol{s}^{(i)}$ is calculated using adjoint sensitivity data of the low-fidelity model $\boldsymbol{c}$. $\delta^{(i.k)}$ is the trust region radius of the inner optimization loop, updated independently from the trust region radius $\delta^{(i)}$ in (4.2).

To parameterize the airfoil shape the Hicks-Henne bump functions (Hicks and Henne 1978) are used. In this approach, a baseline airfoil shape, $z(x)_{baseline}$, is deformed to yield a new airfoil shape, $z(x)$. The new airfoil shape can be written as

$$z(x) = z(x)_{baseline} + \Delta z(x), \tag{10.24}$$

where

$$\Delta z(x) = \sum_{n=1}^{N} \delta_n f_n(x) \tag{10.25}$$

is the total deformation with

$$f_n(x) = \sin^3(\pi x^{e_n}) \tag{10.26}$$

being the Hicks-Henne bump functions, δ_n the deformation amplitude, N the number of deformations, and

$$e_n = \frac{\log_{10}(0.5)}{\log_{10}(x_n)}, \tag{10.27}$$

where $x_n \in [0,1]$ is the location of the function maximum. Examples of a few Hicks-Henne bump function shapes are shown in Fig. 10.9.

During the shape optimization run, the number of bumps, N, and their locations of maxima, x_n, are, typically, held fixed. The designable parameters are then the amplitudes, δ_n, with the design variable vector written as $\boldsymbol{x} = [\delta_1 \delta_2 \cdots \delta_N]^T$. In this application example, we use $N = 15$ with the bumps equally spaced in $x_n \in [0.05, 0.95]$.

The flow past the airfoil is assumed to be steady, inviscid, and adiabatic, with no body forces. The compressible Euler equations are taken as the governing fluid flow equations. The high-fidelity model employs the Stanford University Unstructured

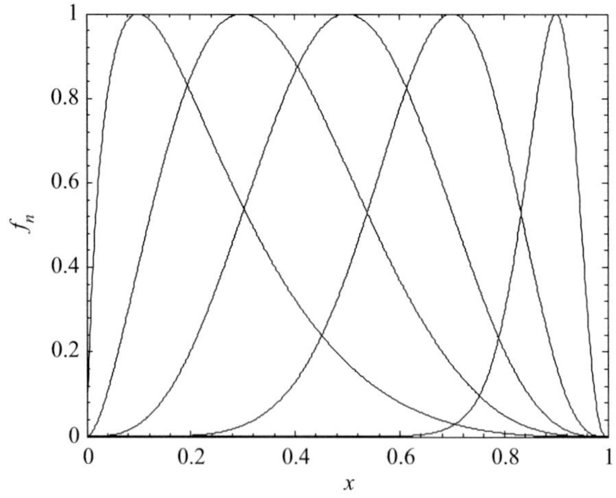

Fig. 10.9 Hicks-Henne bump functions with maxima at $x_n = 0.1$, 0.3, 0.5 0.7, and 0.9, $b = 3.0$, and $\delta = 1.0$

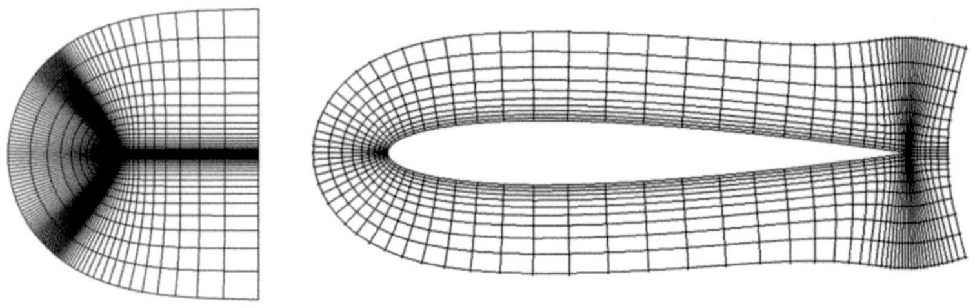

Fig. 10.10 A hyperbolic C-grid showing a far-field view (*left*) and a view close to the surface (*right*)

(SU2) (Palacios et al. 2013) computer code for fluid flow simulations. The high-fidelity model grids are constructed using the hyperbolic C-mesh generator of Kinsey and Barth (1984). The mesh has roughly 163,000 cells. A typical evaluation time of the high-fidelity model is around 27 min. An example grid is shown in Fig. 10.10. The low-fidelity model is based on the same CFD model as the high-fidelity one but reduced levels of discretization as well as relaxed flow solver convergence criteria. Typical evaluation time is 1 min. The average high-to-low simulation time ratio is approximately 25.

The initial and optimized shapes are shown in Fig. 10.11a. It can be seen that both direct optimization with adjoints and the adjoint-enhanced space mapping (AESM) algorithm obtain comparable airfoil shapes. Figure 10.11b shows the surface pressure responses of the initial and optimized designs. Table 10.4 shows the optimization results. The direct approach requires 49 high-fidelity model evaluations, whereas the AESM needs 77 low-fidelity model evaluations and 7 high-fidelity ones. This is less than 11 equivalent high-fidelity model evaluations in total.

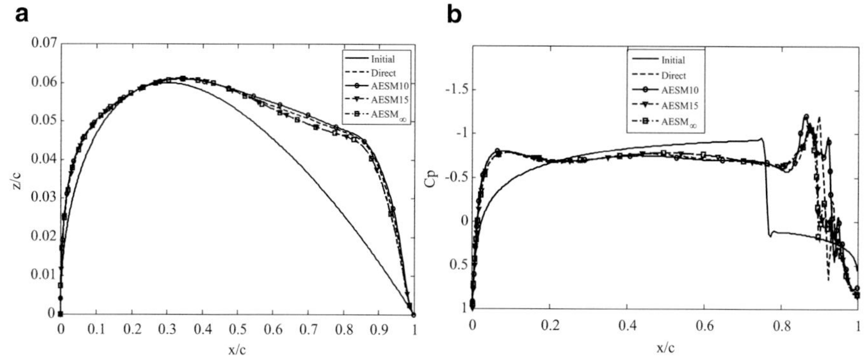

Fig. 10.11 Optimization results (Tesfahunegn et al. 2015a, b): (**a**) the initial and optimized airfoil shapes, and (**b**) pressure coefficient distributions for initial and optimized designs

Table 10.4 Transonic airfoils: SM optimization results (Tesfahunegn et al. 2015a, b)

	Initial	Gradient based with adjoint sensitivity	SM with adjoint sensitivity
C_l	0.0000	0.0092	0.0408
C_d	0.0471	0.0139	0.0144
A	0.0817	0.0989	0.0998
N_c	–	–	77
N_f	–	49	7
Cost[a]	–	49	<11

[a]The total optimization cost is expressed in terms of the equivalent number of high-fidelity model evaluations. The ratio of high-to-coarse simulation time is 25

10.3 Manifold Mapping with Adjoints

Enhancing manifold mapping (MM) with adjoint sensitivity is described. The MM algorithm is applied to antenna and filter design problems.

10.3.1 MM Surrogate Modeling and Optimization

The MM technique (Echeverria and Hemker 2005) can be exploited to construct a surrogate model using an auxiliary low-fidelity model $\boldsymbol{c}$, and the derivatives of both $\boldsymbol{c}$ and $\boldsymbol{f}$ obtained using adjoint sensitivity. The MM surrogate model is defined as (Echeverria and Hemker 2005)

$$s^{(i)}(x) = f\left(x^{(i)}\right) + S^{(i)}\left(c(x) - c\left(x^{(i)}\right)\right), \tag{10.28}$$

where c is a low-fidelity model of the structure under consideration (obtained from coarse-discretization simulations), whereas $S^{(i)}$ is a $m \times m$ correction matrix defined as

$$S^{(i)} = \Delta F \cdot \Delta C^{\dagger}. \tag{10.29}$$

The matrices ΔF and ΔC can be constructed using evaluations of f and c accumulated during the optimization run. Here, we exploit adjoint sensitivity data for both the high- and low-fidelity model so that $\Delta F = J_f$ and $\Delta C = J_c$, where J_f and J_c denote the Jacobian of the respective model. Thus, we have

$$S^{(i)} = J_f\left(x^{(i)}\right) \cdot J_c\left(x^{(i)}\right)^{\dagger}. \tag{10.30}$$

The pseudoinverse, denoted by †, is defined as

$$J_c^{\dagger} = V_{J_c} \Sigma_{J_c}^{\dagger} U_{J_c}^{T}, \tag{10.31}$$

where U_{Jc}, $\sum_{Jc}$, and V_{Jc} are factors in a singular value decomposition of J_c. The matrix $\sum_{Jc}{}^{\dagger}$ is the result of inverting the nonzero entries in $\sum_{Jc}$, leaving the zeroes invariant (Echeverria and Hemker 2005).

By using the sensitivity data to construct the correction matrix (10.29) we ensure a first-order consistency (Alexandrov and Lewis 2001) between the surrogate (10.28) and the high-fidelity model, i.e., $s^{(i)}(x^{(i)}) = f(x^{(i)})$ and $J[s^{(i)}(x^{(i)})] = J[f(x^{(i)})]$.

Convergence using the MM surrogate to a local f optimum is ensured by using the SBO algorithm (10.1) embedded within the trust region framework and the first-order consistency relations. However, in practice, the optimum cannot be reached exactly because neither the high-fidelity nor the surrogate model are smooth (both contain numerical noise).

Input space mapping (ISM) (Bandler et al. 2004a) can be used to precondition the low-fidelity model, i.e., reduce its misalignment with respect to the high-fidelity model. The ISM preconditioning is realized by introducing a parameter shift with the shift vector p obtained as

$$p^{*} = \arg\min_{p} \left\| f\left(x^{(0)}\right) - c\left(x^{(0)} + p\right) \right\|. \tag{10.32}$$

This procedure allows us to initially improve the alignment between the low- and high-fidelity models. The preconditioned low-fidelity model is then $c(x + p^{*})$ and it is used in (10.1) and (10.28) instead of $c(x)$. As mentioned before, the low-fidelity model is obtained from coarse-discretization simulations, so that it is relatively expensive.

To speed up the extraction process (10.32), adjoint sensitivities can be exploited similarly as in (10.7). Using a simple trust region-based (Conn et al. 2000) algorithm where the approximate solution $\boldsymbol{p}^{(k+1)}$ of $\boldsymbol{p}^*$ is found (k is the iteration index for the extraction process) the process is as follows:

$$\boldsymbol{p}^{(k+1)} = \arg \min_{\left\|\boldsymbol{p}-\boldsymbol{p}^{(k)}\right\| \le \delta_{PE}^{(k)}} \left\| \boldsymbol{f}\left(\boldsymbol{x}^{(0)}\right) - \boldsymbol{L}_{c.p}{}^{(k)}(\boldsymbol{p}) \right\|, \tag{10.33}$$

where $\boldsymbol{L}_{c.p}{}^{(k)}(\boldsymbol{p}) = \boldsymbol{c}(\boldsymbol{x}^{(0)} + \boldsymbol{p}^{(k)}) + \boldsymbol{J}_c(\boldsymbol{x}^{(0)} + \boldsymbol{p}^{(k)}) \cdot (\boldsymbol{p} - \boldsymbol{p}^{(k)})$ is a linear approximation of $\boldsymbol{c}(\boldsymbol{x}^{(0)} + \boldsymbol{p})$ at $\boldsymbol{p}^{(k)}$. The TR radius $\delta_{PE}{}^{(k)}$ is updated according to the standard rules (Koziel et al. 2010a).

Surrogate optimization with the MM surrogate and adjoint sensitivities can be realized using a TR-based algorithm that produces a sequence of approximations $\boldsymbol{x}^{(i+1.k)}$ of the solution $\boldsymbol{x}^{(i+1)}$ to (3.1) as follows (k is the iteration index for surrogate model optimization process; cf. (10.8)):

$$\boldsymbol{x}^{(i+1.k+1)} = \arg \min_{\left\|\boldsymbol{x}-\boldsymbol{x}^{(i+1.k)}\right\| \le \delta_{SO}^{(k)}} U\left(\boldsymbol{L}_{c.x}{}^{(i.k)}(\boldsymbol{x})\right), \tag{10.34}$$

where $\boldsymbol{L}_{c.x}{}^{(i.k)}(\boldsymbol{x}) = \boldsymbol{s}^{(i)}(\boldsymbol{x}^{(i+1.k)}) + \boldsymbol{J}_{s(i)}(\boldsymbol{x}^{(i+1.k)}) \cdot (\boldsymbol{x} - \boldsymbol{x}^{(i+1.k)})$ is a linear approximation of $\boldsymbol{s}^{(i)}(\boldsymbol{x})$ at $\boldsymbol{x}^{(i+1.k)}$. The TR radius $\delta_{SO}{}^{(k)}$ is updated according to the standard rules (Koziel et al. 2010a). Typically, because of adjoint sensitivities, the surrogate model optimization requires only a few evaluations of the low-fidelity model $\boldsymbol{c}$. Note that sensitivity of the surrogate model (10.28) can be easily calculated using the Jacobian of $\boldsymbol{c}$ as $\boldsymbol{J}_{s(i)}(\boldsymbol{x}) = \boldsymbol{S}^{(i)} \cdot \boldsymbol{J}_c(\boldsymbol{x} + \boldsymbol{p}^*)$.

10.3.2 Design Example: UWB Monopole Antenna

The problem of Sect. 10.2.2 is repeated using adjoint-enhanced MM of Sect. 10.3.1. Figure 10.12a shows the responses of $\boldsymbol{f}$ and $\boldsymbol{c}$ at $\boldsymbol{x}^{init}$, and Fig. 10.12b shows the response of the low-fidelity model at the final design $\boldsymbol{x}^{(2)} = [20.22\ 19.48\ 2.43\ 0.128]^T$ obtained after only two MM iterations, i.e., only four evaluations of the high-fidelity model. The number of function evaluations is larger than the number of MM iterations because some designs can be rejected by the trust region mechanism.

The total optimization cost is summarized in Table 10.5. Figure 10.13 shows the value of the minimax specification error versus the iteration index. It should be emphasized that the use of adjoint sensitivities allowed us to also keep the number of both low-fidelity model evaluations quite low in the design process (the total number of these is only 31). In particular, the average cost of surrogate model optimization is only about six $\boldsymbol{c}$ evaluations.

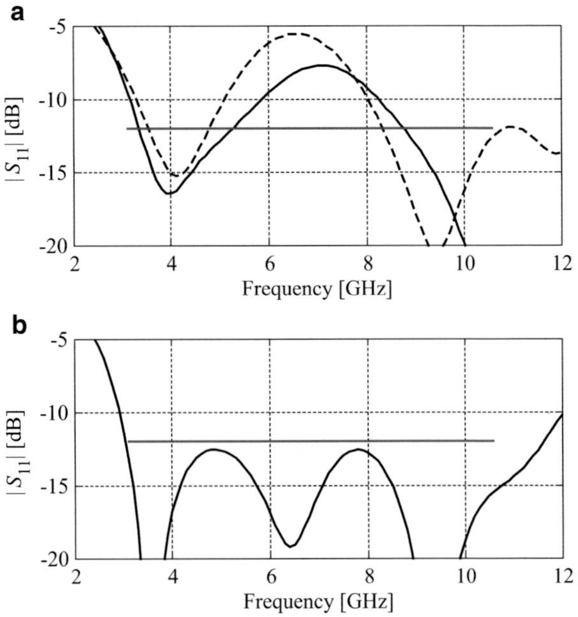

Fig. 10.12 $|S_{11}|$ responses of the UWB monopole (Koziel et al. 2013f): (**a**) high-fidelity model $\boldsymbol{f}$ (—) and low-fidelity model $\boldsymbol{c}$ (- - -) at the initial design $\boldsymbol{x}^{init}$; (**b**) high-fidelity model $\boldsymbol{f}$ (—) at the final designKoziel, Leifsson, and Ogurtsov (2013d) has been changed to Koziel et al. (2013f) as per the reference list. Please check if okay.OK

Table 10.5 UWB monopole antenna: MM optimization results (Koziel et al. 2013f)

Algorithm component	Number of model evaluations[a]	CPU time Absolute (min)	CPU time Relative to $\boldsymbol{f}$
Evaluation of $\boldsymbol{c}$	31	31	0.8
Evaluation of $\boldsymbol{f}$	4	120	4.0
Total cost[a]	N/A	151	**4.8**

[a]Includes $\boldsymbol{f}$ evaluation at the initial design

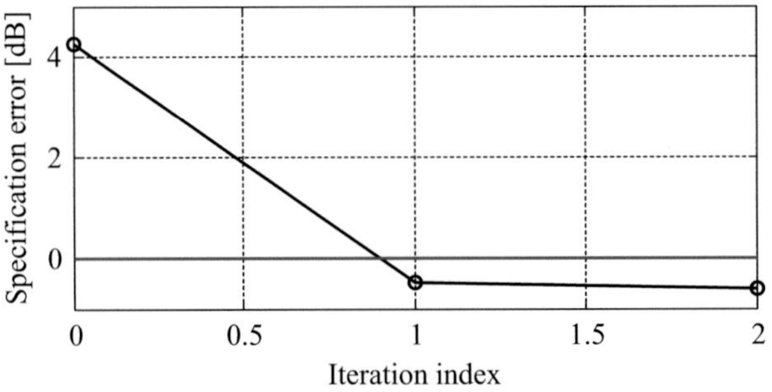

Fig. 10.13 UWB monopole (Koziel et al. 2013f): minimax specification error versus manifold mapping algorithm iteration index

10.3.3 Design Example: Third-Order Chebyshev Band-Pass Filter

Consider the third-order Chebyshev band-pass filter (Kuo et al. 2003), Fig. 10.14. The design variables are $\boldsymbol{x} = [L_1 L_2 S_1 S_2]^T$ mm. Other parameters are $W_1 = W_2 = 0.4$ mm. Both high-fidelity (396,550 mesh cells, evaluation time 45 min) and low-fidelity (82,350 mesh cells, evaluation time 1 min) models are evaluated in CST MWS (2011).

The filter is designed assuming the following specifications: $|S_{21}| \geq -3$ dB for 1.8 GHz $\leq \omega \leq$ 2.2 GHz, and $|S_{21}| \leq -20$ dB for 1.0 GHz $\leq \omega \leq$ 1.55GHz and 2.45 GHz $\leq \omega \leq$ 3.0 GHz. The initial design is the coarse model optimal solution $\boldsymbol{x}^{init} = [16\ 16\ 1\ 1]^T$ mm.

Optimization results are shown in Figs. 10.15 and 10.16, as well as in Table 10.6. The final design $\boldsymbol{x}^{(2)} = [14.88\ 14.62\ 0.472\ 0.976]^T$ is obtained after two manifold mapping iterations. As in the previous example, the optimization cost is very low, which is partially because of using cheap sensitivity information. Also, in this

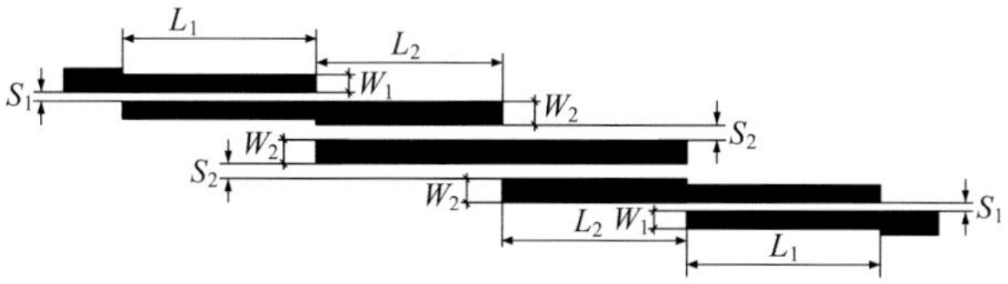

Fig. 10.14 Geometry of the third-order Chebyshev band-pass filter (Kuo et al. 2003)

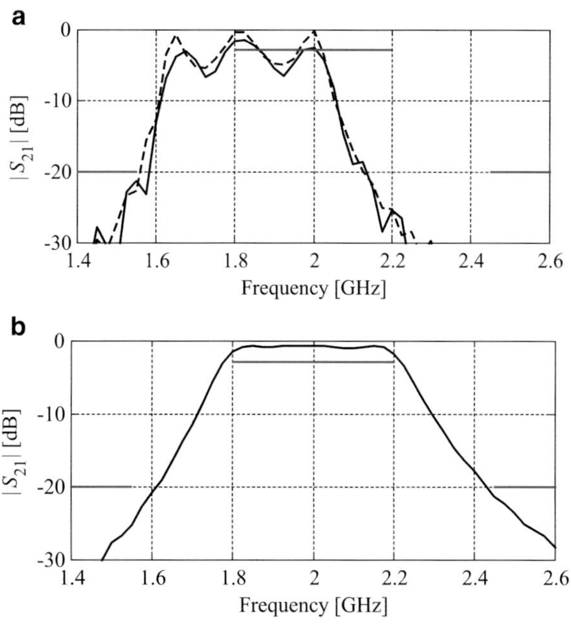

Fig. 10.15 $|S_{21}|$ responses of the third-order Chebyshev filter (Koziel et al. 2013f): (**a**) high-fidelity model $\boldsymbol{f}$ (—) and low-fidelity model $\boldsymbol{c}$ (- - -) at the initial design $\boldsymbol{x}^{init}$; (**b**) high-fidelity model $\boldsymbol{f}$ (—) at the final design

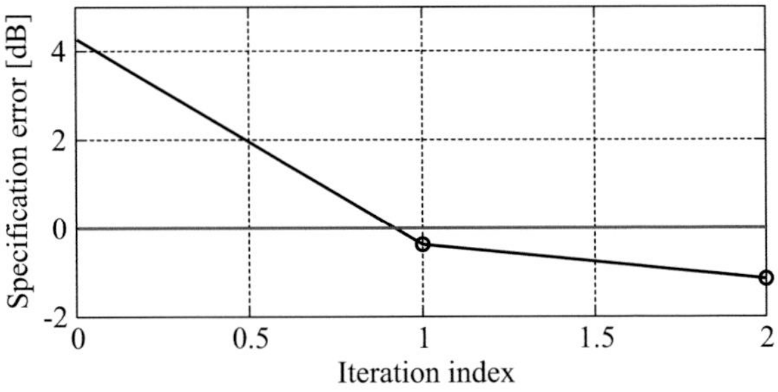

Fig. 10.16 Third-order Chebyshev filter (Koziel et al. 2013f): minimax specification error versus manifold mapping iteration index

Table 10.6 Third-order Chebyshev filter: MM optimization results (Koziel et al. 2013d)

Algorithm component	Number of model evaluations[a]	CPU time	
		Absolute (min)	Relative to $\boldsymbol{f}$
Evaluation of $\boldsymbol{c}$	36	36	0.8
Evaluation of $\boldsymbol{f}$	4	135	4.0
Total cost[a]	N/A	171	**4.8**

[a]Includes $\boldsymbol{f}$ evaluation at the initial design

example the number of high-fidelity model evaluations is larger than the number of iterations because of the trust region scheme (some designs may be rejected if they do not bring the improvement of the objective function value).

It should be emphasized that the resolution with which the high-fidelity model optimum can be located is limited by the numerical noise contained in the models (both high and low fidelity). In practice, this corresponds to the termination condition at the level of $||\boldsymbol{x}^{(i)} - \boldsymbol{x}^{(i-1)}|| < 10^{-2}$ or so, which is sufficient for most applications.

10.4 Shape-Preserving Response Prediction with Adjoints

The last technique we describe in this chapter is the shape-preserving response prediction (SPRP) enhanced with adjoint sensitivities. The approach is demonstrated through the design of an antenna and a microwave filter.

10.4.1 SPRP Surrogate Modeling and Optimization

In SPRP (Koziel 2010a), the surrogate is constructed assuming that the change of $\boldsymbol{f}$ due to the adjustment of the design variables (from $\boldsymbol{x}^{(i)}$ to $\boldsymbol{x}$) can be predicted using the corresponding changes of the low-fidelity model response $\boldsymbol{c}$. The latter is described by translation vectors connecting the characteristic points of the responses (from $\boldsymbol{x}^{(i)}$ to $\boldsymbol{x}$). The translation vectors are subsequently used to predict the response change of the high-fidelity model, whereas the response of $\boldsymbol{f}$ at the current iteration point $\boldsymbol{x}^{(i)}$ is treated as a reference. The concept of SPRP is explained in detail in Chap. 7.

Operator form is a convenient way to describe the SPRP surrogate (Koziel et al. 2014). Define an operation $\boldsymbol{C}(\cdot)$ that extracts k characteristic points (represented by k two-dimensional vectors) from a frequency sweep. Characteristic points are the points describing the characteristics of a frequency sweep such as level points, end points, local maximum points, and local minimum points. Furthermore, define an operation $\boldsymbol{F}(\cdot)$ that carries out the inverse process (i.e., it converts the characteristic points back to a frequency sweep). The SPRP surrogate in Koziel (2010a) can be simply written as follows (Koziel et al. 2014):

$$\boldsymbol{s}^{(i)}(\boldsymbol{x}) = \boldsymbol{F}\left\{\boldsymbol{C}\left[\boldsymbol{f}\left(\boldsymbol{x}^{(i)}\right), \boldsymbol{w}\right] + \boldsymbol{C}[\boldsymbol{c}(\boldsymbol{x}), \boldsymbol{w}] - \boldsymbol{C}\left[\boldsymbol{c}\left(\boldsymbol{x}^{(i)}\right), \boldsymbol{w}\right]\right\}, \tag{10.35}$$

where $\boldsymbol{w}$ represents swept frequencies (assuming identical frequency sweep for all the designs and all the models).

The SPRP surrogate can be enhanced using sensitivity information in the following way (Koziel et al. 2014):

$$\boldsymbol{s}_{sens}^{(i)}(\boldsymbol{x}) = \boldsymbol{s}^{(i)}(\boldsymbol{x}) + \left[\boldsymbol{J}_f\left(\boldsymbol{x}^{(i)}\right) - \boldsymbol{J}_{s^{(i)}}\left(\boldsymbol{x}^{(i)}\right)\right] \cdot \left(\boldsymbol{x} - \boldsymbol{x}^{(i)}\right), \tag{10.36}$$

where $\boldsymbol{s}^{(i)}$ is the original SPRP surrogate (Koziel 2010a) at the ith iteration, i.e., (10.35), whereas $\boldsymbol{J}_{s(i)}$ and $\boldsymbol{J}_f$ are Jacobians of the SPRP and high-fidelity models, respectively. The Jacobian of the high-fidelity model is obtained directly from an EM solver. The Jacobian of $\boldsymbol{s}^{(i)}$ is obtained indirectly using the Jacobian of the low-fidelity model and finite differentiation of the SPRP characteristic points. Using a first-order Taylor approximation $\bar{\mathbf{c}}$ instead of the actual low-fidelity model $\boldsymbol{c}$ saves computational effort. The following formula approximates the perturbed SPRP surrogate (Koziel et al. 2014):

$$\boldsymbol{s}^{(i)}(\boldsymbol{x} + \Delta \boldsymbol{x}) = \boldsymbol{F}\left\{\boldsymbol{C}\left[\boldsymbol{f}\left(\boldsymbol{x}^{(i)}\right), \boldsymbol{w}\right] + \boldsymbol{C}[\bar{\boldsymbol{c}}(\boldsymbol{x} + \Delta \boldsymbol{x}), \boldsymbol{w}] - \boldsymbol{C}\left[\boldsymbol{c}\left(\boldsymbol{x}^{(i)}\right), \boldsymbol{w}\right]\right\}. \tag{10.37}$$

$\boldsymbol{J}_{s(i)}$ can be estimated using (10.35) and (10.37). The practical feasibility of the method relies on the fact that the Jacobian information can be obtained without considerable computational effort, i.e., using adjoint sensitivity.

The sensitivity-enhanced surrogate (10.36) does not have a zero-order correction term since, by its definition, the original SPRP surrogate (Koziel 2010a) already ensures a zero-order consistency condition of $\boldsymbol{s}^{(i)}(\boldsymbol{x}^{(i)}) = \boldsymbol{f}(\boldsymbol{x}^{(i)})$. The enhanced surrogate (10.36) ensures first-order consistency, i.e., $\boldsymbol{J}_{ssens(i)}(\boldsymbol{x}^{(i)}) = \boldsymbol{J}_f(\boldsymbol{x}^{(i)})$, which improves the local search capability of the SPRP optimization process.

The search process is embedded in the trust-region framework (Conn et al. 2000) so that (10.1) is replaced by

$$x^{(i+1)} = \arg \min_{\left\|x - x^{(i)}\right\| \le \delta^{(i)}} U\left(s_{sens}^{(i)}(x)\right), \tag{10.38}$$

where the search radius $\delta^{(i)}$ is updated using the classical rules (Conn et al. 2000). This greatly improves convergence of the SPRP algorithm. It should be emphasized that zero- and first-order consistency together with some standard assumptions concerning the smoothness of the models guarantees global convergence of the algorithm (10.38) to the local minimum of the high-fidelity model (Alexandrov and Lewis 2001). The key point is that for a sufficiently small trust-region radius $\delta^{(i)}$, the optimum prediction generated through (10.38), along with a surrogate model satisfying the above consistency conditions, will lead to an improvement of the design at the high-fidelity model level. This is not guaranteed when only zero-order consistency is satisfied (as, e.g., in the standard SPRP model). While in practice EM-simulated responses are prone to the numerical noise, the proposed technique greatly improves the SPRP performance.

Similarly as for the SM algorithm (cf. (10.8)), sensitivity information can also be utilized to speed up the SPRP optimization step (10.38) with the TR-based algorithm producing a sequence of approximations $x^{(i+1.k)}$ of solution $x^{(i+1)}$ to (10.38) as follows (k is the iteration index for the surrogate optimization process):

$$x^{(i+1.k+1)} = \arg \min_{\left\|x - x^{(i+1.k)}\right\| \le \delta_{SO}^{(k)}} U\left(L_s^{(i.k)}(x)\right), \tag{10.39}$$

where $L_s^{(i.k)}(x) = s_{sens}^{(i)}(x^{(i+1.k)}) + J_{ssens(i)}(x^{(i+1.k)})\cdot(x - x^{(i+1.k)})$ is a linear approximation of $s_{sens}^{(i)}(x)$ at $x^{(i+1.k)}$. The TR radius $\delta_{SO}^{(k)}$ is updated according to the standard rules. In practice, the surrogate optimization process (10.39) requires a few evaluations of the underlying low-fidelity model c. Sensitivity of $s_{sens}^{(i)}(x)$ can be calculated using the sensitivities of both f and s as: $J_{s(i)}(x) + [J_f(x^{(i)}) - J_{s(i)}(x^{(i)})]$.

We summarize adjoint-enhanced SPRP algorithm here (Koziel et al. 2014):

Step 0: Find $x^{(0)}$ through optimization of $c(x)$ and evaluate $f(x^{(0)})$; let $i = 0$.
Step 1: Build a sensitivity-enhanced SPRP surrogate using $f(x^{(i)})$, $c(x^{(i)})$, $c(x)$, and derivatives of $f(x^{(i)})$ and $c(x^{(i)})$ with the method described in Koziel (2010a) [also see (10.35), (10.36), and (10.37)].
Step 2: Obtain a new design $x^{(i+1)}$ by optimizing the sensitivity-enhanced SPRP surrogate w.r.t. x using (10.36) and (10.39).
Step 3: Evaluate $f(x^{(i+1)})$.
Step 4: Stop if termination criteria are satisfied.
Step 5: $i = i+1$ and go to Step 1.

Note that a parameter extraction process is not required in this algorithm since the response residual between the high-fidelity model and the low-fidelity model is represented by the SPRP surrogate using both the zero-order and the first-order consistency conditions. This is analogous to the response residual space mapping technique (Bandler et al. 2004c) and output space mapping (Koziel et al. 2006a, b) where the residual between the coarse (low-fidelity) and the fine (high-fidelity) model is calculated and "embedded" in a surrogate. The SPRP method should perform better than the response residual space mapping or output space mapping since the knowledge (of similarity) in the response shapes has been exploited. Moreover, the adjoint-enhanced SPRP algorithm should be more efficient than space mapping exploiting sensitivity (Koziel et al. 2013d) because of the removal of the parameter extraction step.

10.4.2 Design Example: UWB Monopole Antenna

Let us repeat the problem of Sect. 10.2.2 using SPRP with adjoint sensitivities (final design $x^* = [20.14 \;\; 19.02 \;\; 2.30 \;\; 0.0396]^T$ mm, specification error –1.3 dB, Fig. (10.17), original SPRP (specification error –0.1 dB), SM with adjoint

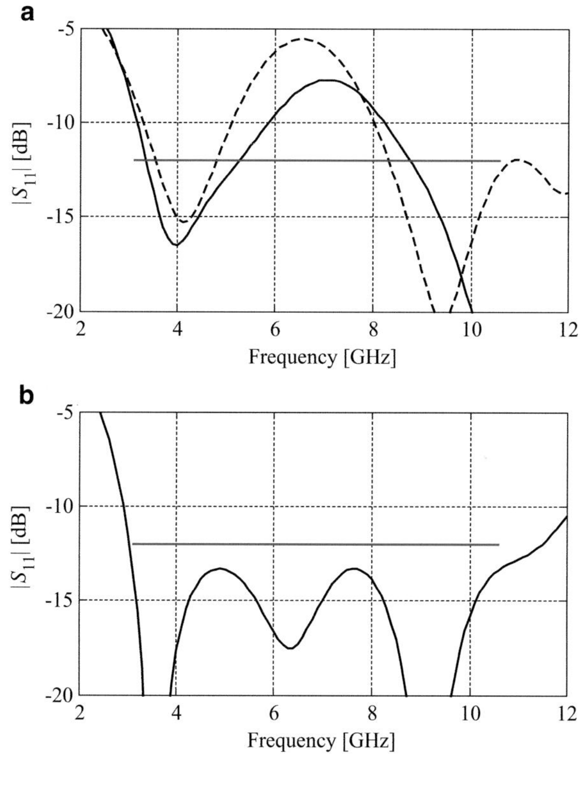

Fig. 10.17 UWB monopole (Koziel et al. 2014): (**a**) responses of $\boldsymbol{f}$ (—) and $\boldsymbol{c}$ (- - -) at the initial design $\boldsymbol{x}^{init}$; (**b**) response of $\boldsymbol{f}$ (—) at the final design

Table 10.7 UWB monopole: SPRP optimization results (Koziel et al. 2014)

Algorithm	Algorithm component	Number of model evaluations[a]	CPU time Absolute (min)	CPU time Relative to f	Specification error
SPRP with sensitivities	Evaluation of $\boldsymbol{c}$	34	34	0.8	
	Evaluation of $\boldsymbol{f}$	5	200	5.0	–1.3 dB
	Total cost[a]	N/A	234	5.8	
Original SPRP (no sensitivity)	Evaluation of $\boldsymbol{c}$	88	88	2.2	
	Evaluation of $\boldsymbol{f}$	3	120	3.0	–0.1 dB
	Total cost[a]	N/A	208	5.2	
Space mapping with sensitivities	Evaluation of $\boldsymbol{c}$	45	45	1.1	
	Evaluation of $\boldsymbol{f}$	4	160	4.0	–0.8 dB
	Total cost[a]	N/A	205	5.1	
Matlab's fminimax	Total cost	$62 \times \boldsymbol{f}$	2480	62.0	–0.8 dB

[a]Excludes $\boldsymbol{f}$ evaluation at the initial design

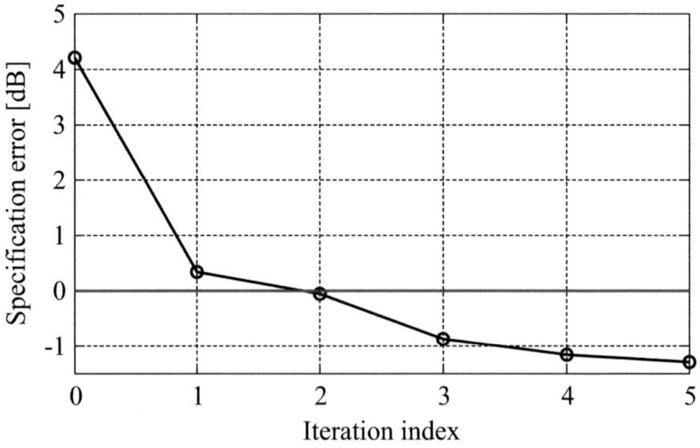

Fig. 10.18 Minimax specification error versus SPRP iteration index for the UWB monopole antenna (Koziel et al. 2014)

sensitivities (specification error –0.8 dB), and gradient-based algorithm with adjoint sensitivities (here, *fminimax*, specification error –0.8 dB).

The computational costs of both SPRP versions and SM are similar (around $5 \times \boldsymbol{f}$, Table 10.7, Fig. 10.18); however, the adjoint-enhanced SPRP algorithm produces the best design. It can be seen that adjoint sensitivity enhancement is crucial here: the original SPRP algorithms stop after two iterations and it is not able to yield a design as good as the methods exploiting derivative data.

10.4.3 Design Example: Dielectric Resonator Filter

Let us reconsider the dielectric resonator filter design problem described in Sect. 10.2.3 using the SPRP algorithm enhanced with adjoint sensitivities. The final design, found in three iterations, is $\boldsymbol{x}^* = [8.00\ 11.96\ 24.6\ 23.1\ -12.43]^T$ mm

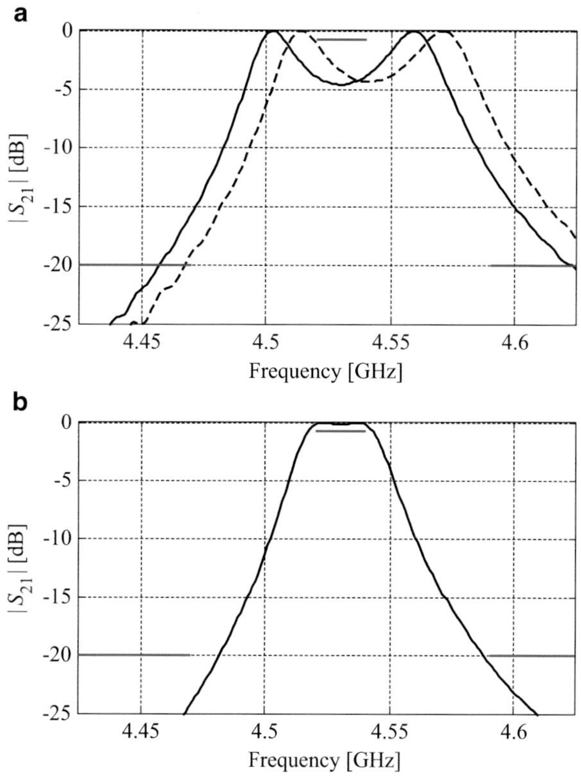

Fig. 10.19 Dielectric resonator filter (Koziel et al. 2014): (**a**) responses of $\boldsymbol{f}$ (—) and $\boldsymbol{c}$ (- - -) at the initial design $\boldsymbol{x}^{init}$; (**b**) response of $\boldsymbol{f}$ (—) at the final design

(Fig. 10.19), with specification error of −0.56 dB. The overall design cost corresponds to less than five evaluations of $\boldsymbol{f}$ (Table 10.8, Fig. 10.20).

One of the important features of SPRP is its very good generalization capability. At the beginning of the optimization process the norm of the SPRP surrogate Jacobian $||\boldsymbol{J}_{s(0)}(\boldsymbol{x}^{(0)})||$ is about five times larger than $||\boldsymbol{J}_{s(0)}(\boldsymbol{x}^{(0)}) - \boldsymbol{J}_f(\boldsymbol{x}^{(0)})||$, which means that the contribution of the linear term in (10.36) is very small. This, in contrast to the first-order Taylor approximation used by conventional gradient-based search, means that the linear term in (10.36) does not degrade the prediction capability of the SPRP surrogate while getting away from $\boldsymbol{x}^{(0)}$. On the other hand, towards the end of the optimization run, the contribution of the linear term increases, improving the local search capability of the algorithm.

10.5 Summary and Discussion

In this chapter, we have described how adjoint sensitivities (i.e., gradients obtained at a low computational cost) can be used to enhance response correction techniques. In particular, adjoint sensitivities can be comprehensively utilized in all stages of

Table 10.8 Dielectric resonator filter: SPRP optimization results (Koziel et al. 2014)

Algorithm	Algorithm component	Number of model evaluations[a]	CPU time Absolute (min)	CPU time Relative to ***f***	Specification error
SPRP with sensitivities	Evaluation of ***c***	34	34	1.2	
	Evaluation of ***f***	3	84	3.0	−0.56 dB
	Total cost[a]	N/A	118	4.2	
Original SPRP (no sensitivity)	Evaluation of ***c***	205	205	7.3	
	Evaluation of ***f***	5	140	5.0	–0.56 dB
	Total cost[a]	N/A	345	12.3	
Space mapping with sensitivities	Evaluation of ***c***	80	80	2.9	
	Evaluation of ***f***	11	308	11.0	–0.47 dB
	Total cost[a]	N/A	398	13.9	
Matlab's fminimax	Total cost	67 × ***f***	1867 min	67.0	+0.06 dB

[a]Excludes ***f*** evaluation at the initial design

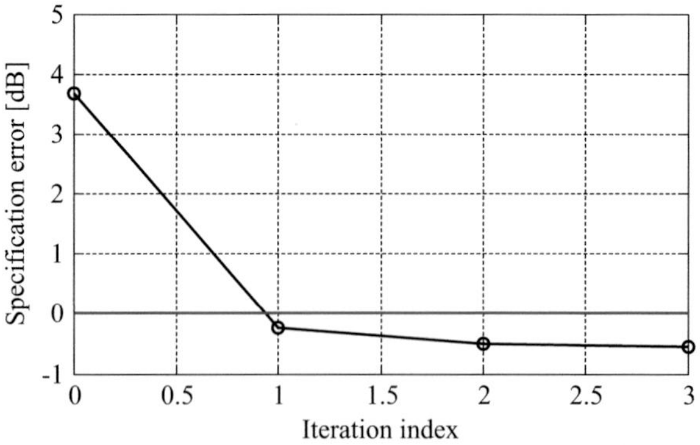

Fig. 10.20 Minimax specification error versus SPRP iteration index for the dielectric resonator filter (Koziel et al. 2014)

the surrogate-based optimization process to (1) align the surrogate with the high-fidelity model, (2) speed up parameter extraction, and (3) reduce the cost of the surrogate model optimization process. These enhancements of the surrogate optimization process, combined with embedding the algorithm in the trust region framework, ensure convergence. Moreover, design constraints can be conveniently handled. The resulting algorithms are capable of carrying out the design optimization process at a low cost in terms of the number of both high- and low-fidelity model simulations.

It should be noted that adjoint sensitivities are not always available. In fact, they have only recently been made available in many commercial solvers in various fields, such as microwave engineering and aerospace engineering. However, obtaining the adjoints can be limited to certain physical situations. It is believed that adjoint sensitivity information will be more and more important since the systems under consideration typically grow in complexity and scope. Therefore, providing adjoint sensitivity information will become a requirement in future computer codes.

Chapter 11
Multi-objective Optimization Using Variable-Fidelity Models and Response Correction

Vast majority of practical optimization problems are of multi-objective nature. In many cases, especially if the designer's priorities are known beforehand, the problem can be turned into a single-objective one by selecting the primary goals and handling the remaining objectives through appropriately defined constraints. Also, it is possible to aggregate the objectives into a scalar cost function using a weighted sum approach or penalty functions. In some situations, however, it is important to obtain more comprehensive information about the system at hand, in particular, to identify the best possible trade-offs between conflicting criteria. In such cases, defaulting to genuine multi-objective optimization is a necessity, which further increases the complexity of the optimization task. Perhaps the most popular multi-objective optimization approaches are population-based metaheuristics (Deb 2001). These techniques are capable of yielding the entire representation of the so-called Pareto front (Fonseca 1995) in one algorithm run; however, the computational cost of evolutionary optimization may be very high: thousands or even tens of thousands of objective function evaluations. Consequently, direct multi-objective optimization of expensive simulation models is normally prohibitive. In this chapter, we discuss multi-objective optimization of expensive models using surrogate-based optimization, particularly response correction techniques. Our considerations are illustrated using examples from the areas of microwave and antenna design as well as aerospace engineering.

11.1 Multi-objective Optimization Problem Formulation

As before, we denote by $\boldsymbol{f}(\boldsymbol{x})$ the high-fidelity simulation model of the system at hand, where $\boldsymbol{x}$ is a vector of designable parameters. We consider N_{obj} design objectives, $F_k(\boldsymbol{x})$, $k = 1, \ldots, N_{obj}$. In most cases, these objectives are conflicting with each other, so that improvement of one leads to degradation of the others.

S. Koziel, L. Leifsson, *Simulation-Driven Design by Knowledge-Based Response Correction Techniques*, DOI 10.1007/978-3-319-30115-0_11

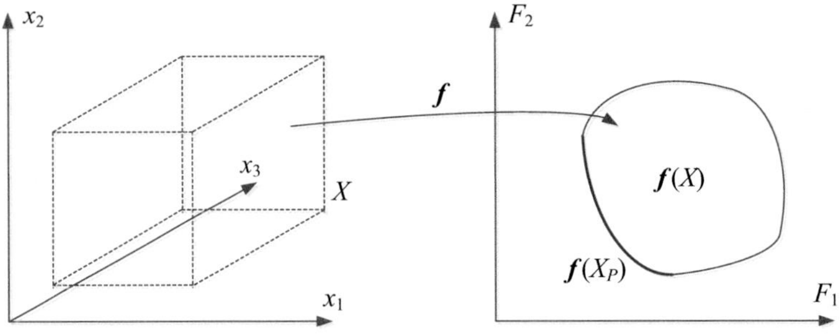

Fig. 11.1 Conceptual illustration of the Pareto front for two-objective design problem with three-dimensional design space: design space X (*left*), its image $\boldsymbol{f}(X)$ (*right*), and the image of the Pareto front $\boldsymbol{f}(X_P)$ marked using a thick line

If $N_{obj} > 1$ then any two designs $\boldsymbol{x}^{(1)}$ and $\boldsymbol{x}^{(2)}$ for which $F_k(\boldsymbol{x}^{(1)}) < F_k(\boldsymbol{x}^{(2)})$ and $F_l(\boldsymbol{x}^{(2)}) < F_l(\boldsymbol{x}^{(1)})$ for at least one pair $k \neq l$ are not commensurable; that is, none is better than the other in the multi-objective sense. We define Pareto dominance relation $<$ (Fonseca 1995) saying that for the two designs $\boldsymbol{x}$ and $\boldsymbol{y}$, we have $\boldsymbol{x} < \boldsymbol{y}$ ($\boldsymbol{x}$ dominates $\boldsymbol{y}$) if $F_k(\boldsymbol{x}) \leq F_k(\boldsymbol{y})$ for all $k = 1, \ldots, N_{obj}$, and $F_k(\boldsymbol{x}) < F_k(\boldsymbol{y})$ for at least one k. The goal of the multi-objective optimization if to find a representation of a so-called Pareto front (of Pareto optimal set) $\boldsymbol{f}(X_P)$ of the design space image through $\boldsymbol{f}(X)$, such that for any $\boldsymbol{x} \in X_P$, there is no $\boldsymbol{y} \in X$ for which $\boldsymbol{y} < \boldsymbol{x}$ (Fonseca 1995).

Figure 11.1 explains the concept of the Pareto front for the two-objective design problem. The part of the boundary of the design space image through $\boldsymbol{f}$, $\boldsymbol{f}(X)$, marked with the thick line, is the Pareto front; that is, there are no designs in X that would dominate any design in X_P. In general, the Pareto front is at most $N_{obj} - 1$ dimensional manifold in the feature space (e.g., a surface for $N_{obj} = 3$).

11.2 Multi-objective Optimization Using Variable-Fidelity Models and Pareto Front Refinement

In this section, a multi-objective design optimization procedure is discussed exploiting variable-fidelity simulation models, response surface approximation surrogates, as well as response correction techniques for Pareto front refinement (Koziel and Ogurtsov 2013c). The essential components of the technique are described in Sects. 11.2.1 through 11.2.3. The overall design flow is presented in Sect. 11.2.4. A numerical example concerning an ultra-wideband antenna is discussed in Sect. 11.2.5.

11.2.1 Design Space Reduction

The optimization methodology presented in this section relies on a fast response surface model that is subjected to multi-objective evolutionary optimization in order to create the initial approximation of the Pareto set. The first step of the optimization procedure is therefore to reduce the design space so that the RSA model is only constructed in the region containing the Pareto optimal designs. This step is crucial particularly for higher dimensional spaces where the number of training points necessary to ensure sufficient model accuracy may easily exceed any available computational budget when the model is set up in the original space. In particular, the cost of gathering the training data increases exponentially with the number of design variables, which makes utilization of such a model questionable if the number of antenna geometry parameters is larger than 5 or 6 (Bekasiewicz et al. 2014a; Koziel and Ogurtsov 2013c).

The Pareto optimal set is usually located in a small region of the initially defined design space: normally, the lower/upper bounds for each geometry parameter of the antenna are defined rather wide to ensure that the desired solutions are located within these prescribed limits. Figure 11.2 shows a typical example of such a situation, here, for a UWB monopole antenna (Bekasiewicz et al. 2014b). Nonetheless, setting up the RSA model in such a large solution space is virtually impractical. However, the variable bounds can be conveniently reduced using single-objective optimizations with respect to each design goal. Consider $\boldsymbol{l}$ and $\boldsymbol{u}$ as initial lower/upper bounds for the design variables. Let

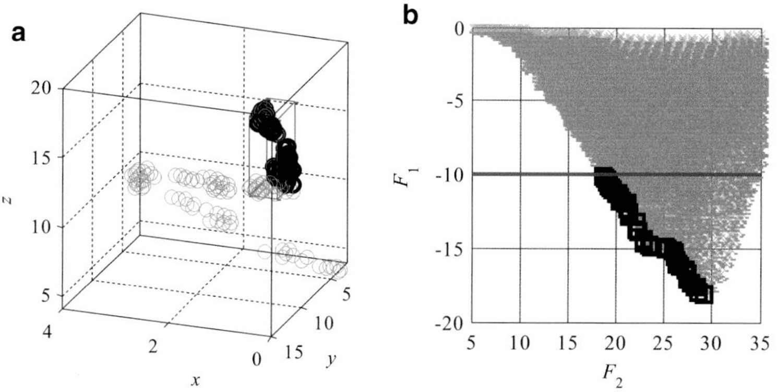

Fig. 11.2 Feature space and Pareto set: (**a**) visualization of the Pareto optimal set (*open circle*) for 3-dimensional solution space (data for UWB monopole antenna, Bekasiewicz et al. 2014b). The portion of the design space that contains the part of the Pareto set we are interested in (*red* cuboid, where $F_1 \leq -10$) is only a small fraction of the initial space. (**b**) The Pareto set of interest (*open square*) versus the entire design space mapped to the feature space (×)

$$\boldsymbol{x}_c^{*(k)} = \arg \min_{\boldsymbol{l} \leq \boldsymbol{x} \leq \boldsymbol{u}} F_k(\boldsymbol{c}(\boldsymbol{x})), \tag{11.1}$$

where $k = 1, \ldots, N_{obj}$, is an optimal design of the low fidelity-model $\boldsymbol{c}$ with respect to the kth objective.

These extreme (or corner) points of the Pareto optimal set are denoted by $\boldsymbol{x}_c^{*(k)}$. The boundaries of the reduced design space can then be defined as follows:

$$\boldsymbol{l}^* = \min\left\{\boldsymbol{x}_c^{*(1)}, \ldots, \boldsymbol{x}_c^{*(Nobj)}\right\}, \tag{11.2}$$

$$\boldsymbol{u}^* = \max\left\{\boldsymbol{x}_c^{*(1)}, \ldots, \boldsymbol{x}_c^{*(Nobj)}\right\}. \tag{11.3}$$

The reduced solution space is usually orders of magnitude (volume-wise) smaller than the initial one, which makes the generation of an accurate RSA model possible at a low computational expense. Note that some of the Pareto optimal solutions may fall outside the reduced design space; however, the majority of them are usually accounted for.

11.2.2 Surrogate Model Construction and Initial Pareto Set Determination

The next step of the procedure is to construct—in the reduced design space—a Kriging interpolation model $\boldsymbol{s}_{KR}$. The model is obtained using sampled low-fidelity simulation data. The design of experiments approach utilized here is Latin Hypercube Sampling (Beachkofski and Grandhi 2002).

Having $\boldsymbol{s}_{KR}$, we apply a multi-objective evolutionary algorithm (MOEA) to find a set of designs representing Pareto-optimal solutions with respect to the objectives F_k of interest. A standard multi-objective evolutionary algorithm with fitness sharing, Pareto-dominance tournament selection, and mating restrictions (Fonseca 1995) is utilized at this stage. Note that the high-fidelity model $\boldsymbol{f}$ is not evaluated in the above procedure.

11.2.3 Pareto Set Refinement

The design solutions obtained by optimizing the Kriging surrogate with MOEA represent the initial approximation of the Pareto optimal set. In the next stage, we select K designs from the initial solution set: $\boldsymbol{x}_s^{(k)}$, $k = 1, \ldots, K$. The chosen designs are subsequently refined using SBO to find their corresponding high-fidelity model $\boldsymbol{f}$ responses. The description of the SBO scheme given below assumes two design objectives: F_1 and F_2; however, the procedure can be generalized to any number of objectives. For each $\boldsymbol{x}_s^{(k)}$, the corresponding high-fidelity model solution $\boldsymbol{x}_f^{(k)}$ is found using the output space mapping (OSM) algorithm of the form (Koziel and Ogurtsov 2013c):

$$\boldsymbol{x}_f^{(k.i+1)} = \arg \min_{\boldsymbol{x},\ F_2(\boldsymbol{x}) \le F_2\left(\boldsymbol{x}_s^{(k.i)}\right)} F_1\left(\boldsymbol{s}_{KR}(\boldsymbol{x}) + \left[\boldsymbol{f}\left(\boldsymbol{x}_s^{(k.i)}\right) - \boldsymbol{s}_{KR}\left(\boldsymbol{x}_s^{(k.i)}\right)\right]\right), \quad (11.4)$$

where $\boldsymbol{x}_f^{(k.i)}$ is the ith approximation of $\boldsymbol{x}_f^{(k)}$. The objective of design refinement is to minimize F_1 for each $\boldsymbol{x}_f^{(k)}$ without degrading F_2. The utilization of OSM ensures perfect alignment of the surrogate model $\boldsymbol{s}_{KR}$ with the high-fidelity model at the beginning of each iteration of (11.4). Typically, 2–3 iterations are required to find the desired high-fidelity model solutions $\boldsymbol{x}_f^{(k)}$. The OSM-driven refinement procedure is repeated for all K chosen samples. One should emphasize that the evaluation of the high-fidelity model $\boldsymbol{f}$ is performed only during the refinement step. The construction of the Kriging interpolation model is realized by means of the DACE toolbox (Lophaven et al. 2002). More detailed explanation of the optimization algorithm can be found in Koziel and Ogurtsov (2013c).

11.2.4 Design Optimization Flow

The overall design flow can be summarized as follows:

1. Perform design space reduction.
2. Sample the design space and acquire the low-fidelity model data.
3. Construct the Kriging interpolation model $\boldsymbol{s}_{KR}$.
4. Obtain the Pareto front by optimizing $\boldsymbol{s}_{KR}$ using MOEA.
5. Refine selected elements of the Pareto front, $\boldsymbol{x}_s^{(k)}$, to obtain corresponding high-fidelity model designs $\boldsymbol{x}_f^{(k)}$.

Note that the high-fidelity model $\boldsymbol{f}$ is not evaluated until the refinement stage (Step 5). Furthermore, the cost of finding the high-fidelity model Pareto optimal set is only about three evaluations of the high-fidelity model per design. In case of considerable discrepancy between $\boldsymbol{s}_{KR}$ and $\boldsymbol{f}$ it is possible to enhance (precondition) the Kriging model by aligning it with the high-fidelity model at certain (usually small) number of designs using space mapping. Typically, output space mapping and frequency scaling are preferred (Koziel and Ogurtsov 2013c).

11.2.5 Case Study: Optimization of UWB Dipole Antenna

For the sake of illustration, consider a planar, single-layer dipole antenna (Fig. 11.3b) consisting of the main radiator element and two parasitic strips (Spence

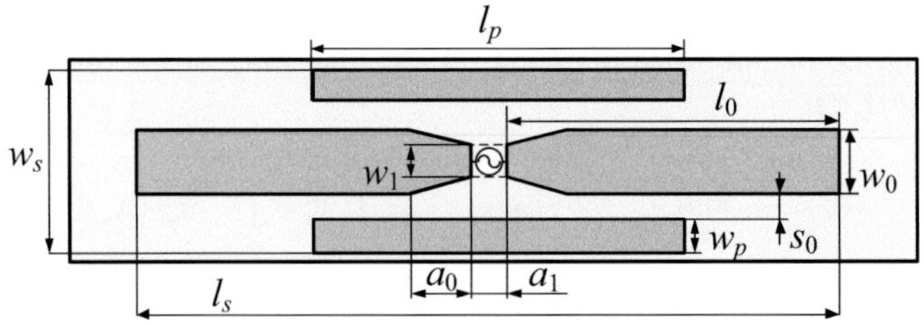

Fig. 11.3 Geometry of the UWB dipole antenna (Spence and Werner 2006)

and Werner 2006). The design variables are $\boldsymbol{x} = [l_0 w_0 a_0 l_p w_p s_0]^T$, while $a_1 = 0.5$ mm and $w_1 = 0.5$ mm remain fixed. Substrate is a 1.58 mm thick Rogers RT5880. Both, the low-fidelity model $\boldsymbol{c}$ (~167,900 mesh cells, 30-s evaluation time) and the high-fidelity model $\boldsymbol{f}$ (~12,510,000 mesh cells, 20-min evaluation time) are simulated in CST Studio (CST 2013). The following objectives are considered: (1) minimization of antenna reflection within 3.1–10.6 GHz band, and (2) reduction of the area defined by a rectangle $w_s \times l_s$, where $w_s = 2w_p + 2s_0 + w_0$ and $l_s = 2l_0 + 2a_0 + a_1$.

The initial design space is defined by the following lower/upper bounds: $\boldsymbol{l} = [10\ 5\ 0.5\ 2\ 1\ 0.1]^T$ and $\boldsymbol{u} = [30\ 20\ 5\ 20\ 10\ 5]^T$. A methodology of Sect. 11.2.1 is used for the determination of refined lower/upper bounds: $\boldsymbol{l} = [18\ 7.96\ 0.5\ 12.8\ 4.01\ 1.08]^T$ and $\boldsymbol{u} = [18.7\ 12.98\ 0.53\ 13.72\ 8.45\ 1.54]^T$, resulting in reduction by a factor of 10^6 (volume-wise). The total cost of the corresponding two single-objective optimization runs is 250 evaluations of the coarse-mesh EM model $\boldsymbol{c}$. A Kriging interpolation model $\boldsymbol{s}_{KR}$ has been constructed within a reduced DS using 500 low-fidelity model samples (423 samples obtained from Latin hypercube sampling, supplemented with 64 design space corners and 13 based on star distribution). The methodology of Sect. 11.2.3 has been utilized to fine the initial Pareto set and carry out the refinement procedure (cf. (11.4)) using 12 designs $\boldsymbol{x}_s^{(k)}$. Figure 11.4 shows the initial and refined Pareto set (see also Table 11.1 for antenna dimensions).

The total aggregated cost of the optimization process corresponds to 61 evaluations of the high-fidelity model (750 $\boldsymbol{c}$ samples, and 42 $\boldsymbol{f}$ evaluations required by the refinement step), which is negligible in comparison to direct multi-objective optimization, here, a few tens of thousands (estimated on the basis of $\boldsymbol{s}_{KR}$ evaluations during MOEA optimization).

It should be noted that the initial design space reduction is crucial for the generation of a reliable RSA model from small amount of $\boldsymbol{c}$ samples. The average generalization error of the RSA model constructed within the initial design space is almost 1 dB with maximum over 4 dB (too inaccurate to be used in the optimization process) and only 0.2 dB for the RSA model in the reduced DS (see Fig. 11.5).

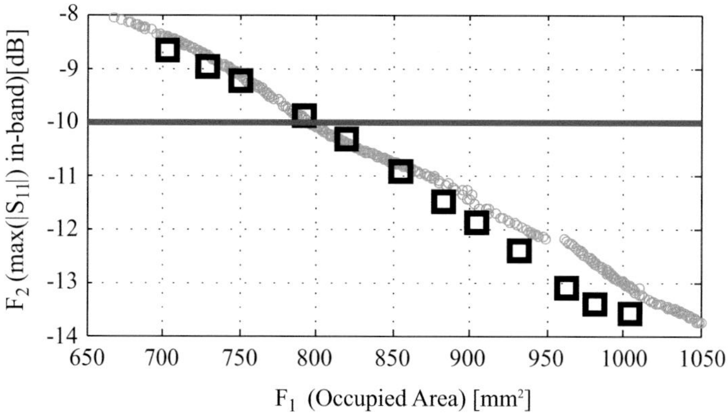

Fig. 11.4 UWB dipole antenna: initial Pareto set representation obtained by multi-objective optimization of the RSA model (*open circle*) and final (refined) Pareto set (*open square*)

Table 11.1 UWB dipole optimization results

Size (mm)^2	Design variables (mm)						max(/S_{11}/)
	l_0	w_0	a_0	l_p	w_p	s_0	
703	18.1	8.60	0.51	12.8	4.28	1.15	−8.7
730	18.0	8.74	0.52	12.8	4.58	1.16	−9.0
750	18.0	9.02	0.52	12.8	4.74	1.15	−9.2
792	18.0	9.44	0.52	12.8	5.11	1.17	−9.9
820	18.0	9.76	0.52	12.8	5.31	1.18	−10.3
856	18.2	10.2	0.50	12.8	5.60	1.08	−10.9
883	18.1	10.3	0.50	12.8	5.98	1.08	−11.5
905	18.0	10.6	0.50	12.8	6.15	1.10	−11.9
932	18.0	11.0	0.51	12.8	6.31	1.14	−12.4
964	18.1	12.6	0.50	12.8	5.52	1.47	−13.1
982	18.1	12.6	0.50	12.9	5.80	1.45	−13.4
1004	18.2	12.4	0.51	12.8	6.25	1.36	−13.6

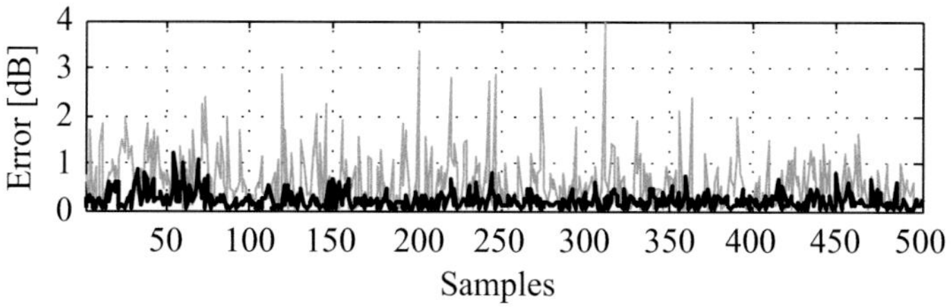

Fig. 11.5 Generalization error at 500 samples for RSA for model set up in the reduced design space (*black line*) and model set up in the initial design space (*gray line*)

11.3 Multi-objective Optimization Using Pareto Front Exploration

In this section, we describe a technique for multi-objective optimization of expensive simulation models that exploits Pareto front exploration and space mapping. We start from introducing the design case study (Sect. 11.3.1). The optimization method is outlined in Sects. 11.3.2 and 11.3.3. The numerical and experimental results are provided in Sect. 11.3.4.

11.3.1 Design Case Study: Compact Rat-Race Coupler

One of the most popular microwave circuits designed for space-limited applications is a rat-race coupler (RRC). The multi-objective design procedure considered in this section is illustrated using a folded equal-split RRC example; however it is valid for other microwave passives that can be represented by an equivalent circuit model. By folding each 70.7-Ω section of a conventional circuit, a significant RRC size reduction can be expected. We choose Taconic RF-35 dielectric substrate ($\varepsilon_r = 3.5$, $h = 0.762$ mm, $\tan\delta = 0.018$), 50-Ω port impedance, and 1-GHz operating frequency for the prototype design. A parameterized layout of the novel RRC is shown in Fig. 11.6a. Designable parameters are given by $\boldsymbol{x} = [l_1\ l_2\ l_3 d\ w]^T$, with $w_0 = 1.7$, $l_0 = 15$ fixed (all dimensions in mm). The low- and high-fidelity models of the structure are implemented in Agilent ADS (Agilent 2011) and CST Microwave Studio (CST 2013) (~220,000 mesh cells, ~15-min simulation time per design), respectively. The initial design is $\boldsymbol{x} = [5\ 14\ 21\ 0.7\ 0.9]^T$. Lower/upper bounds $\boldsymbol{l}/\boldsymbol{u}$ of the solution space are given by $\boldsymbol{l} = [2\ 10\ 17\ 0.2\ 0.5]^T$ and $\boldsymbol{u} = [8\ 18\ 25\ 1.2\ 1.3]^T$. The initial design has been chosen so that the lengths and widths of respective sections of the folded RRC correspond to their counterparts of the conventional coupler. Next, the solution space has been defined based on the initial design solution and some arbitrary perturbations. One should note that the range of each parameter should be sufficiently large for the Pareto set to be located inside of the solution

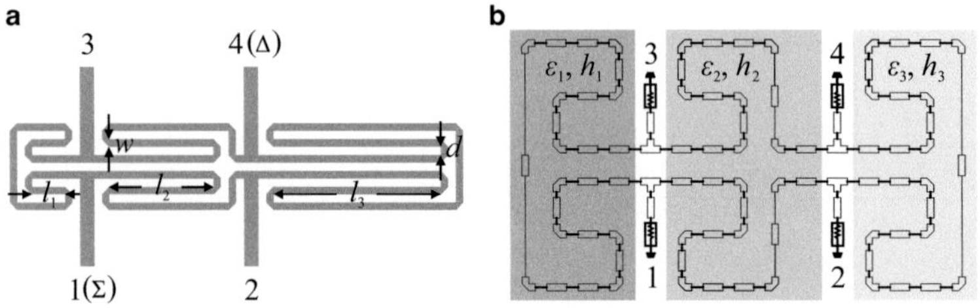

Fig. 11.6 Folded RRC (Koziel et al. 2015): (**a**) layout; (**b**) equivalent circuit model (Agilent ADS). Highlighted regions correspond to different sets of implicit SM parameters $\boldsymbol{p}$

space. The following design objectives are considered: F_1 – bandwidth (BW) maximization (defined as intersection of $|S_{11}|$ and $|S_{41}|$, both below –20 dB) and F_2 – RRC size minimization (layout area). Equal power split between the output ports at the operating frequency is ensured—in the design process—by means of a suitably defined penalty function.

11.3.2 Space Mapping Surrogate

The optimization approach presented here exploits a low-fidelity model of the structure of interest corrected using space mapping. The low-fidelity model is faster than the high-fidelity one but of limited accuracy. Its corrected version, a surrogate model $\boldsymbol{s}$, will be utilized in the optimization process. Here, the space mapping technology (Koziel et al. 2015) is used for model correction. As explained in Chap. 4, various space mapping methods are available. Here, we use implicit and frequency space mapping; however, other types of space mapping (in particular, the output SM) can be used as well if suitable. The surrogate is defined as

$$\boldsymbol{s}(\boldsymbol{x}) = \boldsymbol{c}_F(\boldsymbol{x}; \boldsymbol{F}, \boldsymbol{p}), \tag{11.5}$$

where $\boldsymbol{c}_{\mathrm{F}}$ is a frequency-scaled low-fidelity model, whereas $\boldsymbol{F}$ and $\boldsymbol{p}$ are frequency SM and implicit SM parameters, respectively. Let $\boldsymbol{c}(\boldsymbol{x}) = [c(\boldsymbol{x},\omega_1)\ c(\boldsymbol{x},\omega_2)\ \ldots\ c(\boldsymbol{x},\omega_m)]^T$, where $c(\boldsymbol{x},\omega_j)$ is evaluation of the circuit model at a frequency ω_j. Then, $\boldsymbol{c}_c(\boldsymbol{x};\boldsymbol{F},\boldsymbol{p}) = [c(\boldsymbol{x},f_0 + \omega_1 \cdot f_1,\boldsymbol{p})\ \ldots\ c(\boldsymbol{x}, f_0 + \omega_m \cdot f_1,\boldsymbol{p})]^T$, with f_0 and f_1 being frequency scaling parameters. Here, implicit SM parameters $\boldsymbol{p}$ are dielectric permittivity and substrate thickness of the microstrip components of the circuit corresponding to $\boldsymbol{p} = [\varepsilon_1\varepsilon_2\varepsilon_3h_1h_2h_3]^T$. SM parameters are extracted to minimize misalignment between $\boldsymbol{s}$ and $\boldsymbol{f}$ as follows:

$$\left[\boldsymbol{F}^*, \boldsymbol{p}^*\right] = \arg\min_{\boldsymbol{F},\boldsymbol{p}} \left|\left|\boldsymbol{f}(\boldsymbol{x}) - \boldsymbol{c}_F(\boldsymbol{x}; \boldsymbol{F}, \boldsymbol{p})\right|\right|. \tag{11.6}$$

Figure 11.7 shows the responses of the high- and low-fidelity model at a selected design $\boldsymbol{x}$, as well as the response of the surrogate model $\boldsymbol{s}$ at the same design. It can be observed that the model alignment is greatly improved; however, generalization capability of the surrogate is limited (cf. Fig. 11.7b). In particular, it is not possible to find a single set of SM parameters that would ensure surrogate model accuracy across the entire design space. As a consequence, in order to lead towards a satisfactory design, the surrogate has to be iteratively refined during the optimization process.

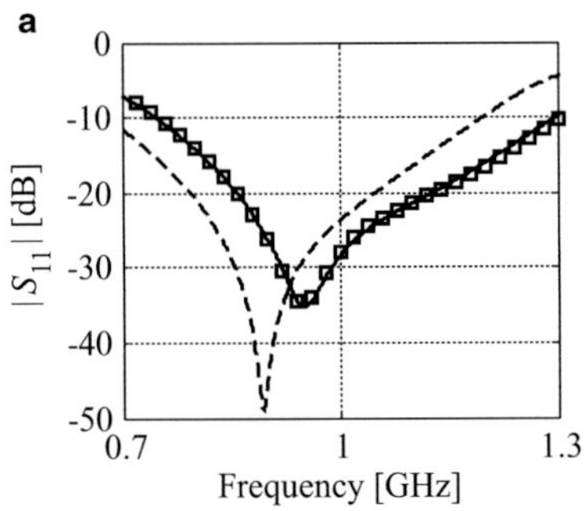

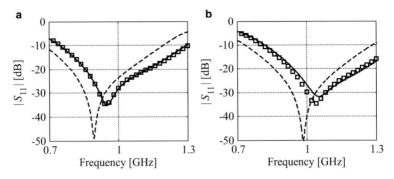

Fig. 11.7 Responses of the high- (—) and low-fidelity (– –) coupler models as well as the SM surrogate (*open square*): (**a**) at certain design $\boldsymbol{x}$ (at which the surrogate is extracted); (**b**) at some other design. Plot (**b**) indicates limited generalization capability of the surrogate (Koziel et al. 2015)

11.3.3 Multi-objective Optimization Algorithm

The optimization procedure works as follows. In the first step, the coupler structure is optimized taking into account the first objective only (here, electrical parameters). The obtained value $F_1(\boldsymbol{f}(\boldsymbol{x}_p^{(1)}))$ at the optimum design $\boldsymbol{x}_p^{(1)}$ determines, together with the corresponding value of the second objective (here, layout area), $F_2^{(1)} = F_2(\boldsymbol{f}(\boldsymbol{x}_p^{(1)}))$, the extreme points of the Pareto set:

$$\boldsymbol{x}_p^{(j)} = \arg \min_{\boldsymbol{x},\ F_2(\boldsymbol{f}(\boldsymbol{x})) \le F_2^{(j)}} F_1(\boldsymbol{f}(\boldsymbol{x})). \tag{11.7}$$

In the subsequent steps, we set the threshold values for the second objective $F_2^{(j)}$, and optimize the structure with respect to the first objective so that the above threshold value is preserved:

Here, $\boldsymbol{x}_p^{(j)}$ is the jth element of the Pareto set. The process is continued until $F_1(\boldsymbol{f}(\boldsymbol{x}_p^{(j)}))$ is still satisfactory from the point of view of given design specifications. Problem (11.7) is solved using the SM surrogate model and it is itself realized as an iterative process:

$$\boldsymbol{x}_p^{(j.k)} = \arg \min_{\boldsymbol{x},\ F_2(\boldsymbol{s}^{(j.k)}(\boldsymbol{x})) \le F_2^{(j)}} F_1\left(\boldsymbol{s}^{(j.k)}(\boldsymbol{x})\right), \tag{11.8}$$

where

$$\boldsymbol{s}^{(j.k)}(\boldsymbol{x}) = \boldsymbol{c}_F\left(\boldsymbol{x}; \boldsymbol{F}^{(j.k)}, \boldsymbol{p}^{(j.k)}\right) \tag{11.9}$$

and

$$\left[F^{(j.k)}, p^{(j.k)}\right] = \arg\min_{F,p} \left\| f\left(x^{(j.k)}\right) - c_F\left(x^{(j.k)}; F, p\right) \right\|. \tag{11.10}$$

The starting point for the algorithm (11.8) is $x_p^{(j-1)}$ (the previously obtained Pareto set point). Typically, two iterations of (11.8) are sufficient to obtain $x_p^{(j)}$, which is because the starting point is already a good approximation of the optimum. In practice, the thresholds $F_2^{(j)}$ can be obtained as $F_2^{(j)} = \alpha \cdot F_2^{(j-1)}$ with $\alpha < 1$ (e.g., $\alpha = 0.95$), or $F_2^{(j)} = F_2^{(j-1)} - \beta$ with $\beta > 0$ (e.g., $\beta = 0.05 \cdot F_2^{(1)}$). Equal power split between the output ports at the operating frequency is secured by adding to F_1 in (11.8) the terms proportional to $(|S_{21}| + 3)^2$ and $(|S_{31}| + 3)^2$ (at 1 GHz), which penalize the designs violating the aforementioned requirement.

The computational cost of the entire multi-objective design process using the proposed methodology can be estimated (in terms of the number of EM simulations of the structure) as $N \cdot K$, where N is the number of point in the Pareto set, and K is the average number of iterations (11.8) necessary to obtain the next point. In practice, $K \leq 3$.

11.3.4 Results

The RRC of Sect. 11.3.1 has been designed using the methodology of Sect. 11.3.3. The first design obtained with (11.8) without any area constraints resulted in bandwidth (BW) of 281 MHz and the layout area of 570 mm^2. Nine other designs have been obtained by setting up $F_2^{(j)}$ to 540, 500, 475, 450, 425, 400, 375, 350, and 325 mm^2, respectively.

Figure 11.8a shows the obtained representation of the Pareto front. For the layout area of 300 mm^2, it was impossible to obtain a design with positive value of BW as defined in Sect. 11.3.1. Table 11.2 and Figure 11.8b–d show the numerical data as well as simulated and measured frequency characteristics for the selected designs (Fig. 11.9 depicts the fabricated prototypes). The range of sizes of designs in Fig. 11.8a is from 570 mm^2 down to 325 mm^2, the latter corresponding to 40 % reduction w.r.t. the former (that exhibits the best possible electrical performance). At the same time, the Pareto optimal designs offer around 91 to 94 % size reduction w.r.t. the conventional RRC.

The total cost of the design process corresponds to less than 30 high-fidelity model evaluations (~7.5 h of CPU time), including the overhead related to multiple evaluations of the circuit model $\boldsymbol{c}$ (the latter does not exceed 20 % of the overall EM simulation cost).

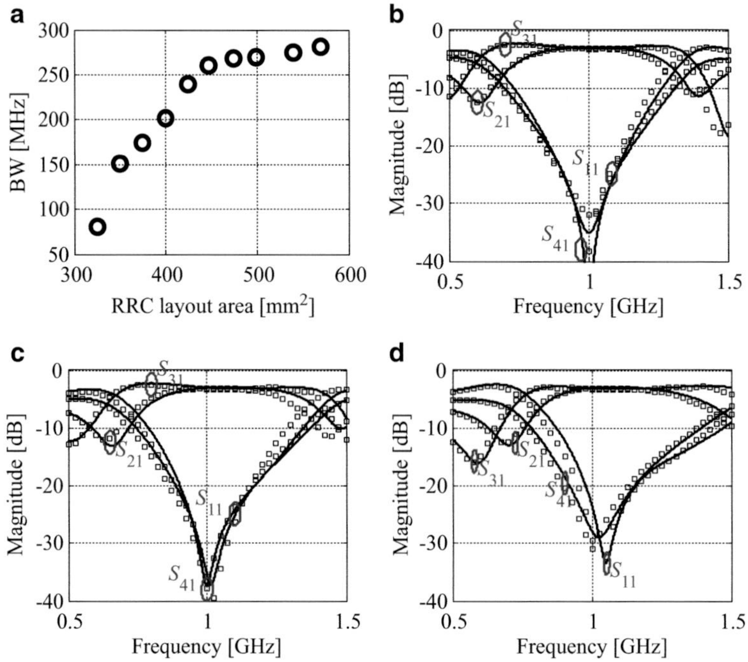

Fig. 11.8 Multi-objective optimization of compact RRC (Koziel et al. 2015): (**a**) A Pareto set representing the best possible RRC design trade-offs, obtained using the proposed multi-objective optimization methodology. (**b**)–(**d**) Simulated (—) and measured (*open square*) frequency characteristics for selected designs, corresponding to the layout areas of (**b**) 570 mm^2; (**c**) 448 mm^2; (**d**) 375 mm^2

Table 11.2 Multi-objective RRC optimization: selected results

Design variables (mm)					Objectives		
l_1	l_2	l_3	d	W	BW (MHz)	Layout area (mm^2)	Miniaturization[a] (%)
4.18	13.20	20.68	0.994	0.865	281	570	90.8
3.83	11.76	20.44	0.825	0.877	270	500	91.9
4.10	13.78	21.14	0.581	0.887	260	450	92.7
4.25	12.17	22.12	0.400	0.923	202	400	93.5
3.95	10.87	21.71	0.350	0.936	174	375	93.9
4.37	12.33	22.52	0.350	0.820	151	350	94.3

[a]With respect to conventional RRC (radius = 44.39 mm, size = 6190 mm^2)

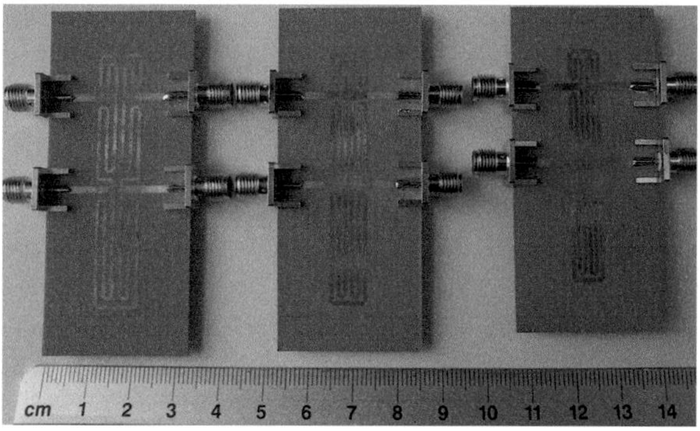

Fig. 11.9 A photograph of the fabricated circuits. From the left: 570, 448, and 375 mm^2

11.4 Multi-objective Optimization Using Multipoint Response Correction

The multi-objective optimization procedure presented in Sect. 11.2 is based on sequential low-fidelity model enhancement by means of response correction techniques and using the high-fidelity model data sampled along the current representation of the Pareto front. This section illustrates the approach on an example of low-speed airfoils (Leifsson et al. 2015).

11.4.1 Optimization Approach

The multi-objective optimization methodology presented here is similar to that of Sect. 11.2; however, a correction of the low-fidelity model is executed before construction of the Kriging interpolation surrogate. Initially, it is done in the entire design space, in further iterations only in the vicinity of the current approximation of the Pareto front using the high-fidelity model data points sampled along the front. The main optimization engine is a multi-objective evolutionary algorithm (MOEA), cf. Sect. 11.2.2.

The design optimization flow can be summarized as follows:

1. Correct the low-fidelity model $\boldsymbol{c}$ using multipoint output space mapping as described in Sect. 6.2.
2. Sample the design space and acquire the data from the corrected low-fidelity model $\boldsymbol{s}_0$.
3. Construct the Kriging interpolation surrogate model $\boldsymbol{s}$.
4. Obtain the Pareto front by optimizing $\boldsymbol{s}$ using MOEA.
5. Evaluate the high-fidelity model at the geometries selected from the Pareto front.
6. If the termination condition is not satisfied, add the new high-fidelity data set to the existing one, and go to 1.
7. END.

The initial representation of the Pareto front is obtained in the first four steps of the procedure. Upon concluding the verification stage (Step 5), the surrogate model is updated by executing multipoint space mapping procedure using the entire high-fidelity model data set (the original one and the Pareto front representation). The improved surrogate model is then re-optimized. It is expected that after a few iterations a much better agreement between the surrogate and the high-fidelity Pareto front will be obtained. The computational cost of each iteration of the above procedure is only due to the evaluation of the high-fidelity model at the geometries picked up from the Pareto front (in practice, a few points are sufficient). Additionally, the design space in the refinement iterations can be restricted to only the part that contains the Pareto set. The termination condition is based on the distance (e.g., in the norm sense) between the Pareto set generated by MOEA and the high-fidelity verification samples.

11.4.2 Case Study

The approach is illustrated using multi-objective optimization of a low-speed airfoil, specifically, seeking the best possible trade-offs between its aerodynamic and aeroacoustic performances. We consider only the clean wing, trailing-edge (TE) noise, and we aim at minimizing the section drag coefficient ($C_{d.f}$) for a given section lift coefficient ($C_{l.f}$), and at the same time minimize the TE noise (given by a noise metric NM_f explained later). Therefore, we have two objectives, $F_1(\boldsymbol{x}) = C_{d.f}$ and $F_2(\boldsymbol{x}) = NM_f$, both obtained by a high-fidelity simulation.

The specific design case considered is optimization for a target lift coefficient of $C_l = 1.5$. The operating conditions are a free-stream Mach number of $M_\infty = 0.208$ and a Reynolds number of $Re_c = 0.665$ million. The angle of attack α is a dependent variable utilized to obtain, for any given airfoil geometry, the target lift coefficient. We use again the NACA four-digit parameterization (Abbott et al. 1959) with the following parameter bounds: $0.0 \leq m \leq 0.1$, $0.3 \leq p \leq 0.6$, and $0.08 \leq t \leq 0.14$. No other constraints are considered.

The Kriging surrogate model constructed for the purpose of evolutionary-based multi-objective optimization was obtained using 343 low-fidelity model samples allocated on a uniform rectangular $7 \times 7 \times 7$ grid. The surrogate was further corrected using the multipoint output space mapping, and utilizing nine high-fidelity model samples: seven samples allocated according to the star-distribution factorial design of experiments, and two additional random samples necessary to ensure that the regression problem has a unique solution.

The noise metric model was developed by Hosder et al. (2010), to give an accurate relative noise measure suitable for design studies. The noise metric therefore does not give an accurate prediction of the absolute noise level. However, it does give an accurate measure of the change in noise due to changes in the wing shape.

The CFD simulation models are set up as follows. The flow is assumed to be steady, compressible, and viscous. The steady Reynolds-averaged Navier-Stokes equations are taken as the governing fluid flow equations with the k-ω SST turbulence model (Menter 1994). The solution domain boundaries are placed at 25 chord lengths in front of the airfoil, 50 chord lengths behind it, and 25 chord lengths above and below it. The computational meshes are of structured curvilinear body-fitted C-topology with elements clustering around the airfoil and growing in size with distance from the airfoil surface. The grids are generated using the hyperbolic C-mesh (Kinsey and Barth 1984). The high-fidelity model grid has around 400,000 mesh cells. Numerical fluid flow simulations are performed using the computer code FLUENT (2015). The flow solver is of implicit density-based formulation and the fluxes are calculated by an upwind-biased second-order spatially accurate Roe flux scheme. Asymptotic convergence to a steady-state solution is obtained for each case. The solution convergence criterion for the high-fidelity model is the one that occurs first of the following: a reduction of the residuals by six orders of magnitude, or a maximum number of iterations of 4,000.

The low-fidelity CFD model is constructed in the same way as the high-fidelity model, but with a coarser computational mesh and relaxed convergence criteria. The low-fidelity mesh has around 30,000 mesh cells. Although the flow equation residuals are not converged, the lift and drag coefficients and the noise metric typically converge within 1,200 iterations. Therefore, the maximum number of iterations is set to 1,200.

11.4.3 Numerical Results

Figure 11.10 shows the distribution of the solutions in the feature space at the first iteration of the evolutionary algorithm. The population size used by the algorithm was $N = 500$. Random initialization with uniform probability distribution is utilized. One can observe a strong correlation between the drag coefficient and the noise metric, indicating that the two objectively are weakly conflicting.

The Pareto set obtained after optimizing the surrogate model is shown in Fig. 11.11, together with the allocation of the solution in the design space. Note that all the Pareto optimal solutions correspond to the thinnest possible airfoil shapes (here, $t = 0.08$). Figure 11.12a shows the high-fidelity model verification samples, indicating certain discrepancies between the drag/noise figures predicted by the surrogate model and actual values. The distribution of the Pareto optimal designs on the drag and noise metric landscape is shown in Fig. 11.13. It can be observed that some of the verification samples indicate that the Pareto front is allocated lower (in terms of the objective function values) than predicted by the surrogate model.

The Pareto front refinement has been subsequently executed in the refined design space of $0.045 \leq m \leq 0.075$, $0.3 \leq p \leq 0.6$, and $0.08 \leq t \leq 0.82$. The verification samples obtained in the previous step have been utilized to reset the surrogate

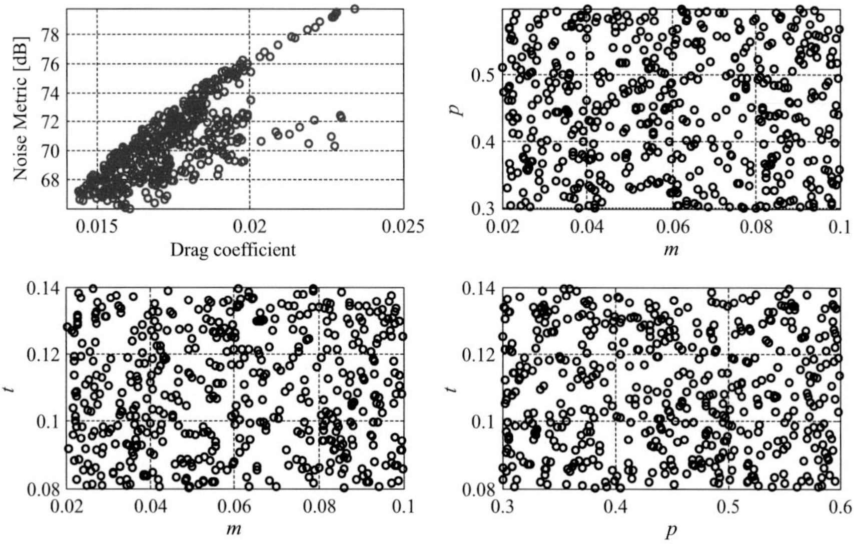

Fig. 11.10 Multi-objective optimization of the surrogate model (Leifsson et al. 2015): Distribution of the initial population in the feature and design spaces

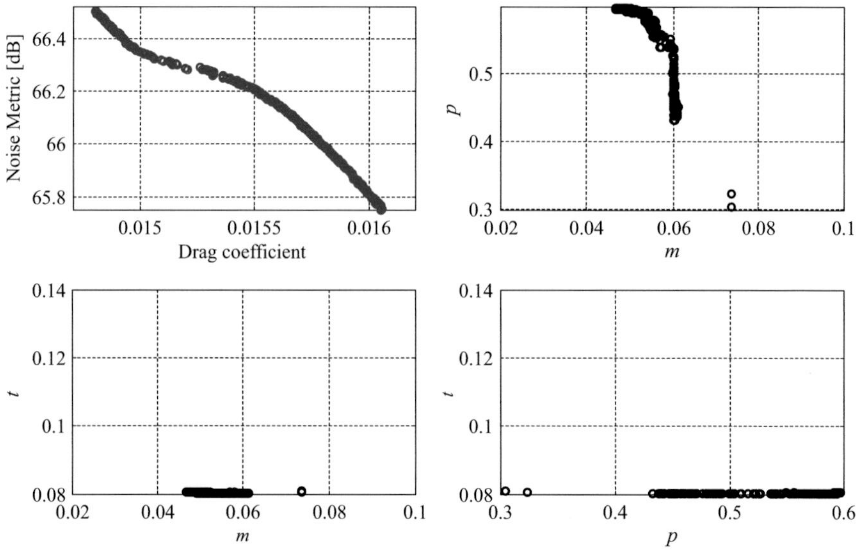

Fig. 11.11 Multi-objective optimization of the surrogate model (Leifsson et al. 2015): Pareto set found by optimizing the surrogate model and the corresponding allocation of Pareto optimal solutions in the design space

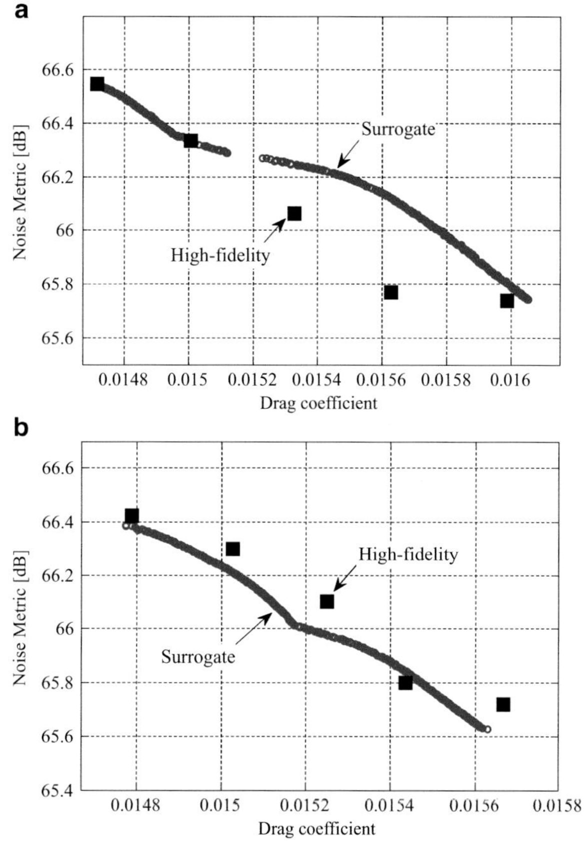

Fig. 11.12 Pareto front representation obtained by optimizing the surrogate (Leifsson et al. 2015): surrogate model points (*circles*), and selected high-fidelity model points (*squares*): (**a**) initial Pareto set; (**b**) final Pareto set

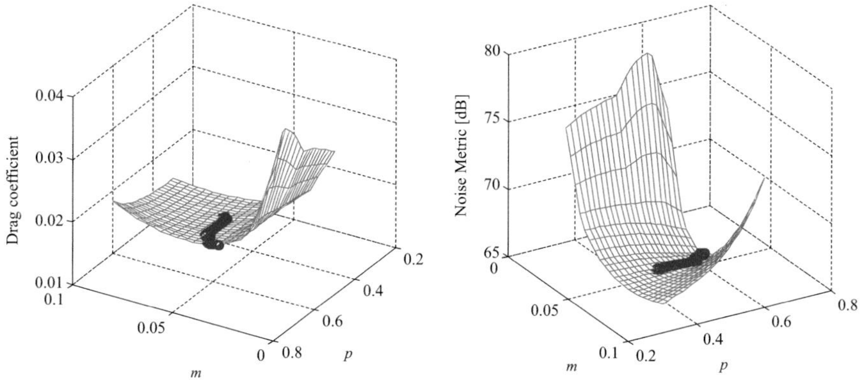

Fig. 11.13 Visualization of the surrogate model for $t = 0.08$ and the allocation of the initial Pareto set representation obtained by optimizing the surrogate model (*circles*) (Leifsson et al. 2015)

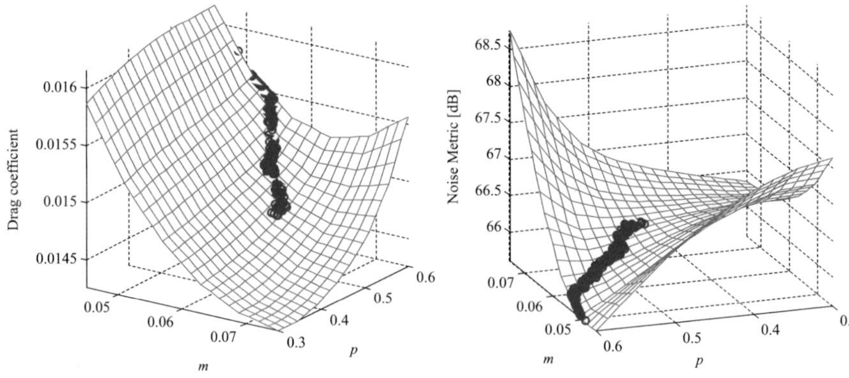

Fig. 11.14 Visualization of the surrogate model for $t = 0.08$ and the allocation of the final Pareto set representation obtained by optimizing the surrogate model (*circles*) (Leifsson et al. 2015)

model. The results of the refinement iteration are shown in Figs. 11.12b and 11.14. The overall optimization procedure is terminated at this point because the assumed accuracy of <1 drag count and <0.1 dB with respect to the noise metric is met. The final Pareto shows that the range of the drag coefficients for the trade-off solutions is from 0.0148 to 0.0156 with the corresponding noise metric from 66.4 to 65.6 dB. Thus, improvement of the noise performance by 0.8 dB can be obtained by increasing the drag by eight counts.

Furthermore, comparing Fig. 11.12a, b it can be concluded that the refinement iteration brings meaningful improvement of the Pareto front representation by moving it down in the noise metric (by about 0.2 dB), and by four drag counts at the right-hand side end.

It should be mentioned that the overall size of the Pareto front for this particular problem setup is rather small; that is, it occupies a very small fraction of the feature space (compare top-left plot of Figs. 11.10 versus 11.12).

11.5 Summary

In this chapter, we highlighted several variable-fidelity methods for multi-objective optimization of computationally expensive simulation models. As demonstrated, response correction techniques play an important role in linking the fast replacement model of the system at hand (either the physics-based low-fidelity model itself or its response surface approximations) to the high-fidelity system representation. This can be done either locally, at the Pareto set refinement stage (cf. Sect. 11.2) or Pareto front exploration (cf. Sect. 11.3), or globally (cf. Sect. 11.4). In any case, the computational cost of the multi-objective optimization process may be greatly reduced compared to direct handling of the high-fidelity model.

Chapter 12
Physics-Based Surrogate Modeling Using Response Correction

The surrogate modeling and response correction techniques considered in this book were mostly discussed in the context of design optimization. In such a setup, the primary purpose of the surrogate is to ensure good local alignment with the high-fidelity model, whereas global accuracy of the model is not of a major concern. In a more general setting, i.e., global or quasi-global modeling, the surrogate is to be valid within a larger portion of the design space. This is important for creating multiple-use library models and applications such as statistical analysis, uncertainty quantification, or global optimization. This chapter describes approaches to quasi-global surrogate modeling using physics-based surrogates and response correction techniques. Formulation of the modeling problem is followed by a discussion of global modeling using space mapping, and space mapping enhanced by function approximation layers, as well as surrogate modeling with the shape-preserving response prediction. Finally, feature-based modeling for statistical design is described. The chapter ends with summary and discussion.

12.1 Formulation of the Modeling Problem

Let $\boldsymbol{f}$: $X_f \rightarrow R^m$, $X_f \subseteq R^n$ denote the high-fidelity model response vector of the structure of interest. The task is to build a surrogate model $\boldsymbol{s}$ of $\boldsymbol{f}$ so that the match between the two models is as good as possible in the region of interest $X_R \subseteq X_f$. Typically, X_R is an n-dimensional interval in R^n with center at the reference point $\boldsymbol{x}^0 = [x_{0.1} \ldots x_{0.n}]^T \in R^n$ and size $\boldsymbol{\delta} = [\delta_1 \ldots \delta_n]^T$ (Koziel et al. 2005).

The quality of the surrogate can be assessed using a suitable error measure, for example, the relative error measure $||\boldsymbol{f}(\boldsymbol{x}) - \boldsymbol{s}(\boldsymbol{x})||/||\boldsymbol{f}(\boldsymbol{x})||$ expressed in percent, where $\boldsymbol{f}(\boldsymbol{x})$ and $\boldsymbol{s}(\boldsymbol{x})$ denote the high-fidelity and the respective surrogate model response at a given test point (design) $\boldsymbol{x}$. The figures of merit are the average and maximum

S. Koziel, L. Leifsson, *Simulation-Driven Design by Knowledge-Based Response Correction Techniques*, DOI 10.1007/978-3-319-30115-0_12

errors over a set of test points. Typically, the test points are allocated in X_R using a uniform space-filling design of experiments such as Latin hypercube sampling (LHS) (Beachkofski and Grandhi 2002).

12.2 Global Modeling Using Multipoint Space Mapping

Let $\boldsymbol{c}$: $X_c \rightarrow R^m$, $X_c \subseteq R^n$ denote the response vector of a computationally cheap representation of the high-fidelity model $\boldsymbol{f}$ evaluation of the system of interest. It is assumed that the base set $X_B = \{\boldsymbol{x}^1, \boldsymbol{x}^2, \ldots, \boldsymbol{x}^N\} \subset X_R$ is available such that the high-fidelity model response is known at all points $\boldsymbol{x}^j$, $j = 1, 2, \ldots, N$. In general, we do not assume any particular location of these base points. The goal is to enhance the low-fidelity model $\boldsymbol{c}$ and create a space-mapping surrogate model $\boldsymbol{s}$ using auxiliary mappings with parameters determined so that $\boldsymbol{s}$ matches the high-fidelity model as well as possible at all base points. Because the low-fidelity model is assumed to be physics based (meaning that it describes the same physical phenomenon as the high-fidelity model but with a different level of fidelity), we hope that the resulting surrogate model will retain a good match with the high-fidelity model over the whole region of interest.

A standard SM model (SM-Standard) is defined as (see Sect. 4.5.1)

$$s_{SM}(\boldsymbol{x}) = \bar{s}_{SM}(\boldsymbol{x}, \boldsymbol{p}), \tag{12.1}$$

where the space-mapping parameters $\boldsymbol{p}$ are obtained using the parameter extraction process

$$\boldsymbol{p} = \arg\min_{\boldsymbol{r}} \sum\nolimits_{k=1}^{N} \left|\left|\boldsymbol{f}(\boldsymbol{x}^k) - s_{SM}(\boldsymbol{x}^k, \boldsymbol{r})\right|\right|, \tag{12.2}$$

while $\bar{s}$ is a generic space-mapping model, i.e., the low-fidelity model composed with some suitable mappings.

A model often used in practice has the form

$$\bar{s}_{SM}(\boldsymbol{x}, \boldsymbol{p}) = \bar{s}_{SM}(\boldsymbol{x}, \boldsymbol{A}, \boldsymbol{B}, \boldsymbol{c}, \boldsymbol{d}) = \boldsymbol{A} \cdot \boldsymbol{c}(\boldsymbol{B} \cdot \boldsymbol{x} + \boldsymbol{q}) + \boldsymbol{d}, \tag{12.3}$$

where $\boldsymbol{A} = diag\{a_1, \ldots, a_m\}$, $\boldsymbol{B}$ is an $n \times n$ matrix, $\boldsymbol{q}$ is an $n \times 1$ vector, and $\boldsymbol{d}$ is an $m \times 1$ vector.

The flexibility of the model represented by (12.1)–(12.3) can be enhanced in many ways, e.g., by exploiting so-called pre-assigned or implicit parameters. These parameters (e.g., dielectric constants, substrate height) are fixed in the high-fidelity model but can be freely modified in the low-fidelity model in order to allow better alignment between the high-fidelity model and the surrogate (Koziel et al. 2006a; Bandler et al. 2004b; Cheng et al. 2008b).

The standard space-mapping surrogate model is simple and fast because once the SM parameters are established, the model evaluation cost is roughly the same as the evaluation cost of the low-fidelity model (which is assumed to be much cheaper than the high-fidelity model). However, linear mappings such as (12.3) may not be able to provide sufficient accuracy. Also, (12.3) may only provide a limited modification of the range of the low-fidelity model, and this modification is basically independent of the design variables. Finally, because of the finite number of parameters, which are extracted in one shot for the entire region of interest, the surrogate is, in fact, a regression model. Consequently, the modeling error might not decrease below certain, problem-dependent, nonzero limits even if the number of base points becomes unlimited (cf. Koziel et al. 2006b).

In Koziel and Bandler (2006), an approach with variable weight coefficients was proposed which provides better accuracy than the standard method, however, at the expense of a significant increase in the evaluation time, which is due to a separate parameter extraction required for each evaluation of the surrogate model. This limits the application range of the method.

12.3 Mixed Modeling: Space Mapping with a Function Approximation Layer

The limitations of the standard space-mapping technique can be alleviated, without compromising computational cost, by using a function approximation layer on top of the standard model. The approach can be done as follows. We define an enhanced space-mapping surrogate model as

$$\boldsymbol{s}(\boldsymbol{x}) = \boldsymbol{s}_{SM}(\boldsymbol{x}) + \widetilde{\boldsymbol{s}}(\boldsymbol{x}), \tag{12.4}$$

where $\boldsymbol{s}_{SM}$ is the standard space-mapping surrogate model (12.1)–(12.3), while $\widetilde{\boldsymbol{s}}$ is a function approximation model. One can consider $\boldsymbol{s}_{SM}$ as a trend function and $\widetilde{\boldsymbol{s}}$ as an output space-mapping term that models the residuals between the high-fidelity model and $\boldsymbol{s}_{SM}$ at all base points (Koziel and Bandler 2007c). This has certain advantages:

1. A relatively good modeling accuracy can be obtained with limited high-fidelity model data due to the space-mapping approach implemented with an underlying physics-based low-fidelity model.
2. The resulting surrogate is computationally as cheap as the low-fidelity model because the function approximation layer typically exploits analytical formulas.
3. It is possible to take advantage of any amount of available high-fidelity model data, so that modeling accuracy can be as good as required provided that the base set is sufficiently "dense."

Several approaches exploiting (12.4) have been proposed, including $\tilde{s}$ implemented through radial basis function interpolation (Koziel and Bandler 2007c), fuzzy systems (Koziel and Bandler 2007d), and Kriging interpolation (Koziel and Bandler 2012). It has been demonstrated that the modeling accuracy of the model (12.4) is better than the accuracy of the standard space-mapping surrogate, and, at the same time, better than the accuracy of the function approximation model used alone, provided that in each case the same amount of high-fidelity model data was used to set up the model. This is because of features (1) and (3) mentioned above.

In the next two sections, we demonstrate the mixed modeling approach on microwave filter examples where $\tilde{s}$ is realized using radial-basis functions (described in Sect. 4.2.3.2), Kriging interpolation (described in Sect. 4.2.3.3), and fuzzy systems (described in Sect. 4.2.3.6).

12.3.1 Example: Fourth-Order Ring Resonator Band-Pass Filter

Consider the fourth-order ring resonator band-pass filter (Salleh et al. 2008) shown in Fig. 12.1a. The design parameters are $\boldsymbol{x} = [L_1 L_2 L_3 S_1 S_2]^T$ mm; the other parameters are $W_1 = 1.2$ mm and $W_2 = 0.8$ mm. The high-fidelity model $\boldsymbol{f}$ is simulated in FEKO (2008). The low-fidelity model $\boldsymbol{c}$, Fig. 12.2b, is implemented in Agilent ADS (2011). The region of interest is defined by the reference point $\boldsymbol{x}^0 = [24.0\ 20.0\ 25.0\ 0.2\ 0.2]^T$ mm, and the region size $\boldsymbol{\delta} = [2.0\ 2.0\ 2.0\ 0.1\ 0.1]^T$ mm. The base set contains 100 points allocated in the region of interest according to the modified LHS algorithm (Beachkofski and Grandhi 2002). The standard SM surrogate $\boldsymbol{s}_{SM}$ (SM-Standard) is the model (12.1)–(12.3) enhanced by implicit SM (Koziel et al. 2006a, b) with six preassigned parameters: the dielectric constant (initial value 4.32) and the substrate height (initial value 1.52 mm) for the microstrip segments corresponding to L_1, L_2, and L_3.

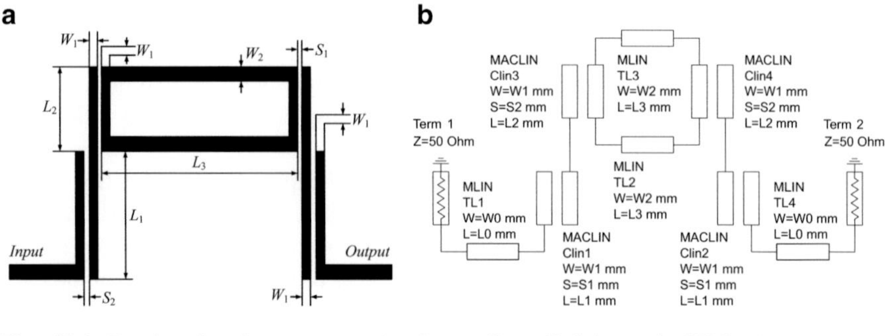

Fig. 12.1 Fourth-order ring resonator band-pass filter (Salleh et al. 2008): (**a**) geometry, (**b**) low-fidelity model (Agilent ADS)

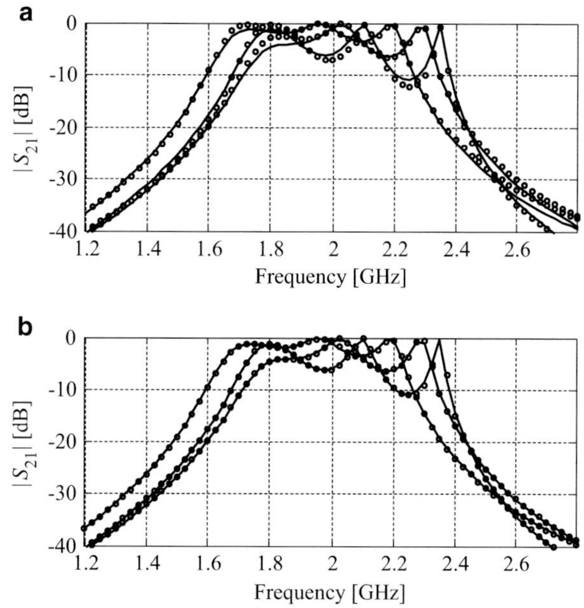

Fig. 12.2 Fourth-order ring resonator filter (Koziel and Bandler 2012): high-fidelity model (*solid line*) and surrogate model (*circles*) responses at the three selected test points for (**a**) SM-Standard, and (**b**) SM-Kriging

Table 12.1 Results for fourth-order ring resonator filter (Koziel and Bandler 2012)

Model type	Model name	Average error (%)	Maximum error (%)
Space mapping	SM-standard	2.9	5.6
Enhanced	SM-RBF	1.8	3.8
Space mapping	SM-Fuzzy	1.3	2.7
	SM-Kriging	0.8	2.2
Function	RBF	4.6	16.9
Approximation	Fuzzy	7.4	17.6
	Kriging	3.9	11.8

Table 12.1 shows the average and maximum modeling error for all considered models. Here, we use the relative error measure defined in Sect. 12.1. It follows that the proposed combination of SM and Kriging outperforms all the other techniques. Relatively poor performance of non-SM models indicates that using only function approximation is not recommended for a limited number of training points. Figure 12.2 shows the high-fidelity and the surrogate model responses at the three selected test points for both SM-Standard and SM-Kriging.

As an application example, we use the SM-Kriging model to optimize the filter with respect to the following design specifications: $|S_{21}| \geq -1$ dB for 1.75 GHz $\leq \omega \leq$ 2.25 GHz, and $|S_{21}| \leq -20$ dB for 1.0 GHz $\leq \omega \leq$ 1.5 GHz and 2.5 GHz $\leq \omega \leq$ 3.0 GHz using $\boldsymbol{x}^0$ as a starting point (specification error +4.5 dB). The optimized surrogate model design is $\boldsymbol{x}^* = [22.98\ 19.78\ 26.80\ 0.183\ 0.053]^T$ mm (corresponding high-fidelity model specification error –0.32 dB).

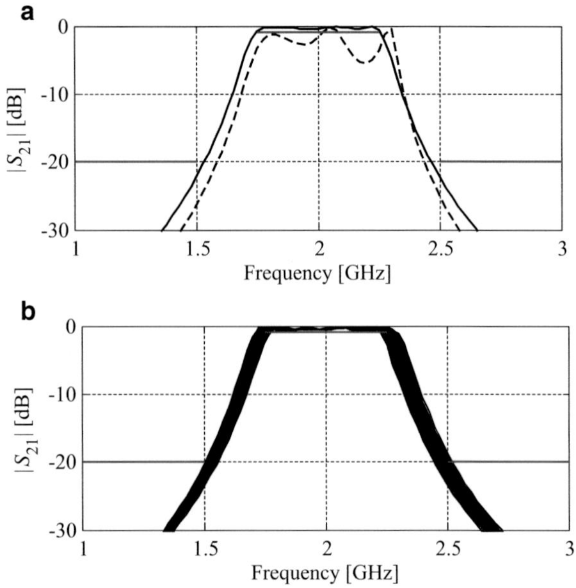

Fig. 12.3 Fourth-order ring resonator filter (Koziel and Bandler 2012): (**a**) high-fidelity model responses at the reference point $\boldsymbol{x}^0$ (*dashed line*) and at the optimal solution $\boldsymbol{x}^*$ of the SM-Kriging surrogate model (*solid line*); (**b**) statistical analysis at $\boldsymbol{x}^*$ using SM-Kriging (200 random samples and 2 % deviation for all design variables)

Figure 12.3a shows the high-fidelity model response at $\boldsymbol{x}^0$ and at $\boldsymbol{x}^*$. The SM-Kriging model was subsequently used to estimate yield at $\boldsymbol{x}^*$ using 200 random samples and assuming 2 % deviation for all design variables. The estimated yield is 38 % (Fig. 12.3b). This value is close to the estimation obtained directly with the high-fidelity model (34 %). The yield estimation obtained using the SM-Standard model (24 %) is less accurate.

12.3.2 Example: Microstrip Band-Pass Filter

Consider the microstrip band-pass filter with two transmission zeros shown in Fig. 12.4a (Hsieh and Chang 2003). The design parameter is $\boldsymbol{x} = [L\ g\ s\ d]^T$. The high-fidelity model is simulated in FEKO. The region of interest is defined by the reference point $\boldsymbol{x}^0 = [7.0\ 0.1\ 0.5\ 1.5]^T$ mm, and the region size $\boldsymbol{\delta} = [0.25\ 0.05\ 0.1\ 0.5]^T$ mm. The low-fidelity model (Fig. 12.4b) is implemented in Agilent ADS. The base set contains 100 points allocated using the LHS algorithm. The standard SM surrogate $\boldsymbol{s}_{SM}$ (SM-Standard) is the model (12.1)–(12.3) enhanced by implicit SM with two preassigned parameters: dielectric constant (initial value 10.2) and the substrate height (initial value 0.635 mm).

The average and maximum modeling error for the models is given in Table 12.2. Figure 12.5 shows the high-fidelity and the surrogate model responses at the three selected test points for both SM-Standard and SM-Kriging. We use the SM-Kriging model to optimize the filter with respect to the following design specifications:

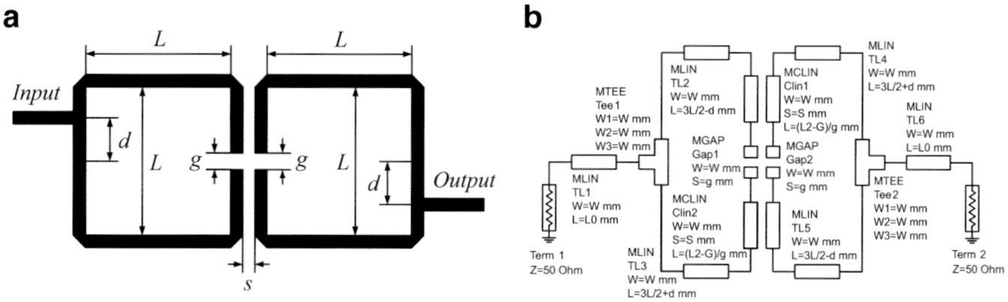

Fig. 12.4 Band-pass filter with two transmission zeros: geometry (Hsieh and Chang 2003): (**a**) geometry, (**b**) low-fidelity model (Agilent ADS)

Table 12.2 Modeling results for band-pass filter with two transmission zeros (Koziel and Bandler 2012)

Model type	Model name	Average error (%)	Maximum error (%)
Space mapping	SM-Standard	3.8	6.8
Enhanced space mapping	SM-RBF	1.8	5.9
	SM-Fuzzy	3.4	6.7
	SM-Kriging	1.5	4.1
Function approximation	RBF	6.9	24.8
	Fuzzy	9.5	22.0
	Kriging	6.1	20.1

Fig. 12.5 Band-pass filter with two transmission zeros (Koziel and Bandler 2012): high-fidelity model (*solid line*) and surrogate model (*circles*) responses at the three selected test points for (**a**) SM-Standard, and (**b**) SM-Kriging

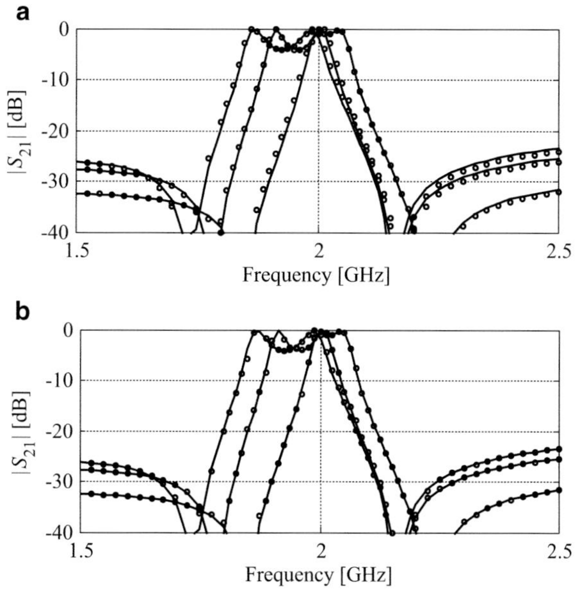

$|S_{21}| \geq -1$ dB for 1.75 GHz $\leq \omega \leq$ 2.25 GHz, and $|S_{21}| \leq -20$ dB for 1.0 GHz $\leq \omega \leq$ 1.5GHz and 2.5 GHz $\leq \omega \leq$ 3.0 GHz using $\boldsymbol{x}^0$ as a starting point (specification error +9.5 dB). The optimized surrogate model design is $\boldsymbol{x}^* = [22.98\ 19.78\ 26.80\ 0.183\ 0.053]^T$ mm (high-fidelity model specification error –0.34 dB). Figure 12.6a shows the high-fidelity model response at $\boldsymbol{x}^0$ and at $\boldsymbol{x}^*$. The SM-Kriging model was also used to estimate yield at $\boldsymbol{x}^*$ (200 random samples, 1 % deviation). The obtained value, 75 % (Fig. 12.6b), is close to the estimation obtained directly with the high-fidelity model (73 %). As in the first example, the estimate obtained with the SM-Standard model, 92 %, is less accurate.

12.4 Surrogate Modeling Using Multipoint Output Space Mapping

The multipoint space-mapping (SM) technique discussed in Sect. 12.2 is classified as input space mapping (multipoint ISM). In this section, we describe the use of multipoint output space mapping (OSM) for modeling purposes, and give a numerical example involving robust design of transonic airfoil shapes.

12.4.1 Model Formulation

The multipoint OSM applies correction terms directly to the low-fidelity model response components. Here, two low-fidelity response components are assumed, $C_{d.c}(\boldsymbol{x})$ and $C_{l.c}(\boldsymbol{x})$, i.e., the drag and lift coefficients of an airfoil, respectively. The OSM surrogate model is defined as (Leifsson and Koziel 2014)

$$\begin{aligned} \boldsymbol{s}(\boldsymbol{x}) &= \boldsymbol{A}(\boldsymbol{x}) \circ \boldsymbol{c}(\boldsymbol{x}) + \boldsymbol{D}(\boldsymbol{x}) \\ &= [a_d(\boldsymbol{x}) C_{d.c}(\boldsymbol{x}) + d_d(\boldsymbol{x}) \quad a_l(\boldsymbol{x}) C_{l.c}(\boldsymbol{x}) + d_l(\boldsymbol{x})]^T, \end{aligned} \tag{12.5}$$

where $^\circ$ denotes component-wise multiplication. Both the multiplicative and additive correction terms are design-variable dependent and take the form

$$\boldsymbol{A}(\boldsymbol{x}) = \left[a_{d.0} + [a_{d.1}\ a_{d.2}\ \ldots\ a_{d.n}] \cdot (\boldsymbol{x} - \boldsymbol{x}^0) \quad a_{l.0} + [a_{l.1}\ a_{l.2}\ \ldots\ a_{l.n}] \cdot (\boldsymbol{x} - \boldsymbol{x}^0)\right]^T, \tag{12.6}$$

$$\boldsymbol{D}(\boldsymbol{x}) = \left[d_{d.0} + [d_{d.1}\ d_{d.2}\ \ldots\ d_{d.n}] \cdot (\boldsymbol{x} - \boldsymbol{x}^0) \quad d_{l.0} + [d_{l.1}\ d_{l.2}\ \ldots\ d_{l.n}] \cdot (\boldsymbol{x} - \boldsymbol{x}^0)\right]^T, \tag{12.7}$$

where $\boldsymbol{x}^0$ is the center of the design space. Response correction parameters $\boldsymbol{A}$ and $\boldsymbol{D}$ are obtained as

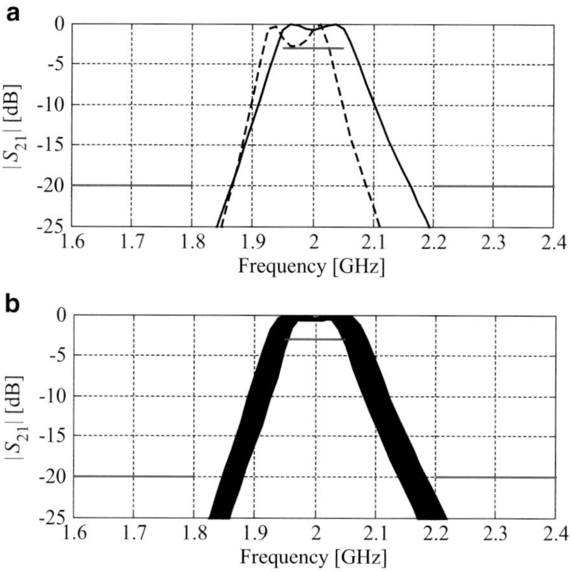

Fig. 12.6 Band-pass filter with two transmission zeros (Koziel and Bandler 2012): (**a**) high-fidelity model responses at the reference point $\boldsymbol{x}^0$ (*dashed line*) and at the optimal solution $\boldsymbol{x}^*$ of the SM-Kriging surrogate model (*solid line*); (**b**) statistical analysis at $\boldsymbol{x}^*$ using SM-Kriging (200 random samples and 1 % deviation for all design variables) – estimated yield 75 %

$$[\boldsymbol{A}, \boldsymbol{D}] = \arg \min_{[\overline{\boldsymbol{A}}, \overline{\boldsymbol{D}}]} \sum\nolimits_{k=1}^{N} \left|\left|\boldsymbol{f}(\boldsymbol{x}^k) - \left(\overline{\boldsymbol{A}}(\boldsymbol{x}^k) \circ \boldsymbol{c}(\boldsymbol{x}^k) + \overline{\boldsymbol{D}}(\boldsymbol{x}^k)\right)\right|\right|^2, \qquad (12.8)$$

i.e., the response scaling is supposed to globally improve the matching for all training points $\boldsymbol{x}^k$, $k = 1, \ldots, N$. The problem (12.8) is equivalent to the linear regression problems $[a_{d.0} a_{d.1} \ldots a_{d.n} d_{d.0} d_{d.1} \ldots d_{d.n}]^T \mathbf{C}_d = \mathbf{F}_d$ and $[a_{l.0} a_{l.1} \ldots a_{l.n} d_{l.0} d_{l.1} \ldots d_{l.n}]^T \mathbf{C}_l = \mathbf{F}_l$. The problems have an analytical solution. More specifically, the correction parameters $\boldsymbol{A}$ and $\boldsymbol{D}$ can be found as (Leifsson and Koziel 2015a, b, c)

$$\begin{bmatrix} a_{d.0} \\ a_{d.1} \\ \vdots \\ a_{d.n} \\ d_{d.0} \\ \vdots \\ d_{d.n} \end{bmatrix} = \left(\mathbf{C}_d^T \mathbf{C}_d\right)^{-1} \mathbf{C}_d^T \mathbf{F}_d \quad \begin{bmatrix} a_{l.0} \\ a_{l.1} \\ \vdots \\ a_{l.n} \\ d_{l.0} \\ \vdots \\ d_{l.n} \end{bmatrix} = \left(\mathbf{C}_l^T \mathbf{C}_l\right)^{-1} \mathbf{C}_l^T \mathbf{F}_l \qquad (12.9)$$

where

$$\boldsymbol{C}_d = \begin{bmatrix} C_{d.c}(\boldsymbol{x}^1) & C_{d.c}(\boldsymbol{x}^1)\cdot\left(x_1^1 - x_1^0\right) & \cdots & C_{d.c}(\boldsymbol{x}^1)\cdot\left(x_n^1 - x_n^0\right) & 1 & \left(x_1^1 - x_1^0\right) & \cdots & \left(x_n^1 - x_n^0\right) \\ C_{d.c}(\boldsymbol{x}^2) & C_{d.c}(\boldsymbol{x}^2)\cdot\left(x_1^2 - x_1^0\right) & \cdots & C_{d.c}(\boldsymbol{x}^2)\cdot\left(x_n^2 - x_n^0\right) & 1 & \left(x_1^1 - x_1^0\right) & \cdots & \left(x_n^1 - x_n^0\right) \\ \vdots & \vdots & \ddots & \vdots & \vdots & \vdots & \vdots & \vdots \\ C_{d.c}(\boldsymbol{x}^N) & C_{d.c}(\boldsymbol{x}^N)\cdot\left(x_1^N - x_1^0\right) & \cdots & C_{d.c}(\boldsymbol{x}^N)\cdot\left(x_n^N - x_n^0\right) & 1 & \left(x_1^1 - x_1^0\right) & \cdots & \left(x_n^1 - x_n^0\right) \end{bmatrix} \tag{12.10}$$

$$\boldsymbol{F}_d = \left[C_{d.f}\left(\boldsymbol{x}^1\right) \quad C_{d.f}\left(\boldsymbol{x}^2\right) \quad \ldots \quad C_{d.f}\left(\boldsymbol{x}^N\right)\right]^T \tag{12.11}$$

$$\boldsymbol{C}_l = \begin{bmatrix} C_{l.c}(\boldsymbol{x}^1) & C_{l.c}(\boldsymbol{x}^1)\cdot\left(x_1^1 - x_1^0\right) & \cdots & C_{l.c}(\boldsymbol{x}^1)\cdot\left(x_n^1 - x_n^0\right) & 1 & \left(x_1^1 - x_1^0\right) & \cdots & \left(x_n^1 - x_n^0\right) \\ C_{l.c}(\boldsymbol{x}^2) & C_{l.c}(\boldsymbol{x}^2)\cdot\left(x_1^2 - x_1^0\right) & \cdots & C_{l.c}(\boldsymbol{x}^2)\cdot\left(x_n^2 - x_n^0\right) & 1 & \left(x_1^1 - x_1^0\right) & \cdots & \left(x_n^1 - x_n^0\right) \\ \vdots & \vdots & \ddots & \vdots & \vdots & \vdots & \vdots & \vdots \\ C_{l.c}(\boldsymbol{x}^N) & C_{l.c}(\boldsymbol{x}^N)\cdot\left(x_1^N - x_1^0\right) & \cdots & C_{l.c}(\boldsymbol{x}^N)\cdot\left(x_n^N - x_n^0\right) & 1 & \left(x_1^1 - x_1^0\right) & \cdots & \left(x_n^1 - x_n^0\right) \end{bmatrix} \tag{12.12}$$

$$\boldsymbol{F}_l = \left[C_{l.f}\left(\boldsymbol{x}^1\right) \quad C_{l.f}\left(\boldsymbol{x}^2\right) \quad \ldots \quad C_{l.f}\left(\boldsymbol{x}^N\right)\right]^T. \tag{12.13}$$

Here, $C_{d.f}(\boldsymbol{x})$ and $C_{l.f}(\boldsymbol{x})$ are the high-fidelity model drag and lift coefficients, respectively. Note that the matrices $\mathbf{C}_l^T\mathbf{C}_l$ and $\mathbf{C}_d^T\mathbf{C}_d$ are non-singular for $N > n+1$ assuming that all training points are distinct. Choosing a star distribution training set satisfies this condition and is sufficient in many modeling cases. The star distribution training set consists of $N = 2n + 1$ points allocated at the center of the design space $\boldsymbol{x}^0 = (\boldsymbol{l} + \boldsymbol{u})/2$ ($\boldsymbol{l}$ and $\boldsymbol{u}$ being the lower and upper bound for the design variables, respectively), and the centers of its faces, i.e., points with all coordinates but one equal to those of $\boldsymbol{x}^0$, and the remaining one equal to the corresponding component of $\boldsymbol{l}$ or $\boldsymbol{u}$.

12.4.2 Example: Low-Cost Modeling for Robust Design of Transonic Airfoils

Multipoint OSM is illustrated in an example involving the robust design of transonic airfoil shapes (Leifsson et al. 2013). The goal of robust design in this case is to find an airfoil shape with minimum drag and a given lift coefficient that is least sensitive to changes in the operating conditions. The problem is formulated such that the operating condition (in this case it is only the free-stream Mach number) is treated as an input uncertainty and is represented as a uniform random variable with bounds. The objective function is to reduce the mean drag coefficient (μ_{Cd}) and the standard deviation of the drag coefficient (σ_{Cd}) subject to a given minimum mean lift coefficient (μ_{Cl}). The airfoil shape parameters are taken as the deterministic design variables.

The high-fidelity model solves steady, two-dimensional, compressible Reynolds-averaged Navier-Stokes (RANS) equations and the Spalart-Allmaras turbulence model on a structured C-grid (Tannehill et al. 1997). The model has around 400,000 mesh cells and the simulation time is around 2 h. The low-fidelity model is constructed in the same way as the high-fidelity one, but with a coarser mesh (~32,000 cells) and is around 80 times faster than the high-fidelity one.

The airfoil shape is parameterized using NACA 4-digit airfoils using three parameters. In order to satisfy the lift constraint the angle of attack is taken as a design variable. Thus, in total, there are four deterministic design variables. The free-stream Mach number (M_∞) is the only uncertain variable and it is bounded as follows: $0.7 \leq M_\infty \leq 0.8$. The minimum lift coefficient is set as $C_l^* = 0.5$.

The statistical properties are calculated based on stochastic expansions derived from the non-intrusive polynomial chaos (NIPC) technique (see, e.g., Hosder 2012; Zhang et al. 2012). In NIPC, a stochastic response surface approximation (RSA) model is created based on high-fidelity data. Our case requires at least 42 high-fidelity model evaluations to set up the stochastic RSA model. To reduce the cost, the multipoint OSM, described in Sect. 12.5.1, is used to create a globally accurate surrogate model which is used in place of the high-fidelity one to construct the stochastic RSA. Given the problem size, 53 low-fidelity and 11 high-fidelity model evaluations are needed, corresponding to less than 12 equivalent high-fidelity model evaluations in total.

Figure 12.7a shows the optimized airfoil shapes when using the high-fidelity model, and the surrogate as well as the low-fidelity model. The shapes produced by using the high-fidelity model and the surrogate have the same thickness but different camber. As a result, the angle of attack is different to attain the prescribed lift coefficient. The shape produced by using the low-fidelity model is different than the others. Figure 12.7b shows the variation of the drag coefficients of the shapes with respect to the Mach number. The comparator shape, NACA 2412, has a significant drag-rise over this Mach number range, whereas the optimized shapes maintain lower drag coefficient values. Furthermore, the variation of the drag coefficient values of the shapes obtained by the high-fidelity and the OSM-based method is very similar. The variation of the airfoil obtained using the low-fidelity model is significantly higher than the shapes obtained by the other two methods.

12.5 Surrogate Modeling with Shape-Preserving Response Prediction

This section illustrates the use of the shape-preserving response prediction (SPRP) technique, described in Sect. 7.3, for general surrogate modeling. As opposed to its application in optimization (Sects. 7.3.2–7.3.4), where the SPRP model is only supposed to be valid in the vicinity of the current design, generic surrogate modeling requires that the surrogate is valid (or accurate in any given sense) in a larger region, usually defined by certain lower/upper bounds for the problem parameters.

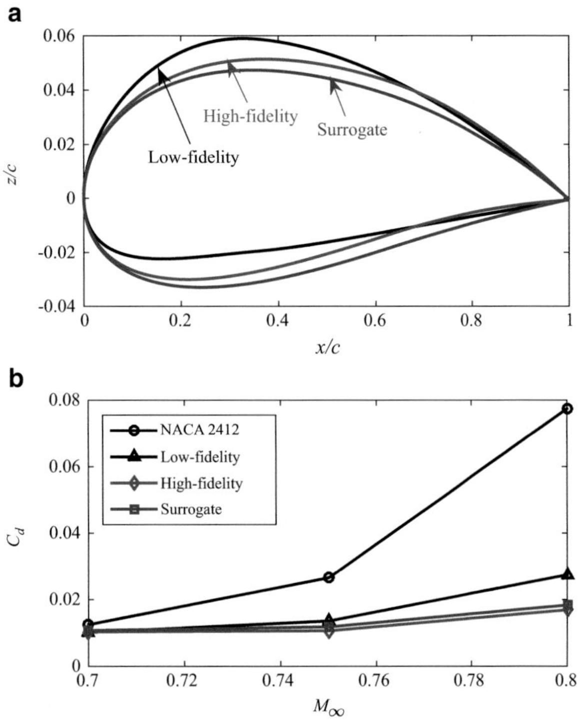

Fig. 12.7 Characteristics of the initial and optimized airfoils (Leifsson et al. 2013): (**a**) optimized shapes, (**b**) variation of the drag coefficient with Mach number at a lift coefficient of $C_l = 0.5$

Three versions of SPRP surrogates are considered: the basic one, the modified implementation that exploits multiple training points, and the generalized formulation that allows for arbitrary allocation of the training data samples. Further discussion on recent developments of SPRP models can be found in (Koziel and Leifsson 2012d).

12.5.1 Basic SPRP

Basic SPRP surrogate modeling (Koziel 2010a) exploits the SPRP concept in a straightforward fashion. Let $X_R \subseteq X$ be the region of interest where we want the surrogate model to be valid. Typically, X_R is an n-dimensional interval in R^n with center at reference point $\boldsymbol{x}^0 = [x_{0.1} \ \ldots \ x_{0.n}]^T \in R^n$ and size $\boldsymbol{\delta} = [\delta_1 \ \ldots \ \delta_n]^T$. Let $X_B = \{\boldsymbol{x}^1, \boldsymbol{x}^2, \ldots, \boldsymbol{x}^N\} \subset X_R$ be the base set, such that the high-fidelity model response is known at all points $\boldsymbol{x}^j$, $j = 1, 2, \ldots, N$. Here, the base points are allocated using a star distribution (Bandler et al. 2006), which is a design of experiments traditionally used by space mapping.

The SPRP surrogate model is defined as

$$s(\boldsymbol{x}) = \boldsymbol{S}(\boldsymbol{x}, \boldsymbol{x}^r), \tag{12.14}$$

where $\boldsymbol{x}^r$ is the base point that is the closest to $\boldsymbol{x}$, i.e.,

$$\boldsymbol{x}^r = \arg \min_{\boldsymbol{y} \in X_B} ||\boldsymbol{x} - \boldsymbol{y}||, \tag{12.15}$$

whereas $\boldsymbol{S}(\boldsymbol{x},\boldsymbol{x}^r)$ is the SPRP model created with $\boldsymbol{x}^r$ used as a reference design (cf. Sect. 7.3.1).

Although, as demonstrated in (Koziel 2010a), this simple modeling approach proves to be more accurate than SM, it has some drawbacks. The model (12.14), (12.15) utilizes only one base point at a time. As shown in Fig. 12.8a, the region of interest is divided into regions of "attractions" of particular base points. For all evaluation points $\boldsymbol{x}$ located in a given region of "attraction," the surrogate model (12.14) is determined using the same single base point as a reference design. Due to this, the surrogate does not utilize all available $\boldsymbol{f}$-model data at a time. Also, the surrogate model is discontinuous at the borders of the areas of "attraction" because the solution to (12.15) is not unique at these points. This may cause some problems while using the surrogate for design optimization.

12.5.2 Modified SPRP

The modified SPRP modeling technique (Koziel and Szczepanski 2011) utilizes multiple reference designs and solves the discontinuity problem described in the previous section. Again, the base set is assumed to be allocated using star distribution (Bandler et al. 2004a); however, the model can also be formulated for more general setups.

The concept of the SPRP model exploiting multiple reference designs is explained in Fig. 12.8b. For an evaluation point $\boldsymbol{x}$, we find a subset X_S of the base set X_B that defines a rectangular area (hypercube) of the region of interest containing $\boldsymbol{x}$. The surrogate model is set up using all points from X_S. The star distribution base set contains $N = 2n + 1$ points. Without loss of generality, we can assume that $X_S = \{\boldsymbol{x}^0, \boldsymbol{x}^1, \ldots, \boldsymbol{x}^n\}$. We have

$$\boldsymbol{x} = \boldsymbol{x}^0 + \beta_1 \boldsymbol{v}_1 + \beta_2 \boldsymbol{v}_2 + \ldots + \beta_n \boldsymbol{v}_n, \tag{12.16}$$

where $\beta_1, \ldots, \beta_n$, determines a unique representation of $\boldsymbol{x} - \boldsymbol{x}_0$ using vectors $\boldsymbol{v}_i = \boldsymbol{x}_i - \boldsymbol{x}_0$, $i = 1, \ldots, n$. Coefficients β_i can be found as

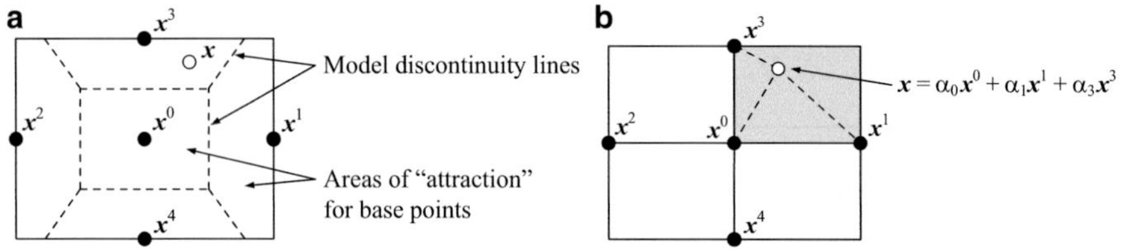

Fig. 12.8 SPRP modeling ($n = 2$): (**a**) Original: Star-distributed base points are denoted using *black circles*. The region of interest is divided into areas of "attraction" of particular base points, determined by the Euclidean distance. An example evaluation design $\boldsymbol{x}$ is close to the base design $\boldsymbol{x}^3$, and this point becomes a reference design for SPRP model. (**b**) Modified: Base points are denoted using *black circles*. A shaded area denotes a hypercube defined by a subset X_S of base points being the closest to an example evaluation design $\boldsymbol{x}$. The surrogate at $\boldsymbol{x}$ is defined as a linear combination of SPRP models using all base points from X_S as reference designs. Coefficients of this linear combination are calculated by representing $\boldsymbol{x}$ through all points from X_S

$$\begin{bmatrix} \beta_1 \\ \beta_2 \\ \vdots \\ \beta_n \end{bmatrix} = \begin{bmatrix} \boldsymbol{v}_1 & \boldsymbol{v}_2 & \dots & \boldsymbol{v}_n \end{bmatrix}^{-1} \cdot \left(\boldsymbol{x} - \boldsymbol{x}^0\right), \tag{12.17}$$

The vector $\boldsymbol{x}$ can be uniquely represented as

$$\boldsymbol{x} = \alpha_0 \boldsymbol{x}^0 + \alpha_1 \boldsymbol{x}^1 + \alpha_2 \boldsymbol{x}^2 + \ldots + \alpha_n \boldsymbol{x}^n, \tag{12.18}$$

where $\alpha_0 = 1 - (\alpha_1 + \ldots + \alpha_n)$, and $\alpha_i = \beta_i$, $i = 1, \ldots, n$. The modified SPRP surrogate model is then defined as

$$\hat{\boldsymbol{s}}(\boldsymbol{x}) = \alpha_0 \boldsymbol{S}\left(\boldsymbol{x}, \boldsymbol{x}^0\right) + \alpha_1 \boldsymbol{S}\left(\boldsymbol{x}, \boldsymbol{x}^1\right) + \ldots + \alpha_n \boldsymbol{S}(\boldsymbol{x}, \boldsymbol{x}^n) \tag{12.19}$$

with $\boldsymbol{S}(\boldsymbol{x}, \boldsymbol{x}^i)$, $i = 0, 1, \ldots, n$, being the SPRP models (12.14) determined using respective reference designs. It can be verified that the model (12.19) is continuous with respect to $\boldsymbol{x}$ provided that both $\boldsymbol{f}$ and $\boldsymbol{c}$ are continuous functions of $\boldsymbol{x}$. Also, it is expected to be more accurate than the model (12.14), (12.15) because it exploits the available high-fidelity model data in a more comprehensive way.

12.5.3 Generalized SPRP

The generalized SPRP modeling procedure (Koziel and Leifsson 2012d) assumes that the number and the location of the training points, $\boldsymbol{x}^k$, $k = 1, \ldots, N$, are arbitrary. The training points can be allocated uniformly in the region of interest using, e.g., the LHS algorithm. The entire training data set will contribute to the surrogate model evaluation at any given design variable vector.

For the sake of subsequent considerations we will denote the training set as $X_T = \{\boldsymbol{x}^1, \boldsymbol{x}^2, \ldots, \boldsymbol{x}^N\} \subset X_R$. The high-fidelity model responses at the base designs, $\boldsymbol{f}(\boldsymbol{x}^k)$, are assumed to be known. The following notation is used: $\boldsymbol{f}(\boldsymbol{x}) = [f(\boldsymbol{x},\omega_1) \ldots f(\boldsymbol{x},\omega_m)]^T$, where ω_j, $j = 1, \ldots, m$, is the frequency sweep.

The surrogate is constructed following the generalized SPRP (GSPRP), which is based on the SPRP formulation (Koziel 2011b); however, no restrictions on the training set are imposed. As explained in Sect. 7.3, the SPRP technique is based on processing the sets of so-called characteristic points that describe the response of the structure under consideration. In particular, these characteristic points can correspond to specific levels of the response (e.g., $|S_{21}|$ equal to –3 dB or –20 dB, cf. Fig. 7.18), local response minima and/or maxima, etc. The main assumption here is that the characteristic points of the low-fidelity and high-fidelity model responses are in one-to-one correspondence (see (Koziel 2010a) for details and (Koziel 2010e) for generalizations). The SPRP model created using the high-fidelity model response at a specific reference design as well as its characteristic points and similar data from the low-fidelity model allows us to predict the high-fidelity model response at other designs of interest (Koziel 2010e).

Figure 12.9 shows the response of the high-fidelity model at a certain number of training points (only a few points are shown for clarity). A set of characteristic points is distinguished on each of the plots, here, corresponding to $|S_{21}| = -6$ dB, –15 dB, as well as local $|S_{21}|$ minima within the pass band. A discussion on selecting a set of characteristic points for a given design case can be found in Koziel (2010e). The notation $\boldsymbol{p}_k^j = [\omega_k^j r_k^j]^T$, $j = 1, \ldots, K$, is used to denote the characteristic points of $\boldsymbol{f}(\boldsymbol{x}^k)$. Here, ω_k^j and r_k^j denote the frequency and magnitude components of $\boldsymbol{p}_k^j$, respectively.

GSPRP predicts the response of $\boldsymbol{f}$ at any design variable vector $\boldsymbol{x}$ using the information contained in all training points. The model is initialized by constructing auxiliary models $\boldsymbol{s}_{\omega.j}(\boldsymbol{x})$ and $\boldsymbol{s}_{r.j}(\boldsymbol{x})$, $j = 1, \ldots, K$, of the sets of corresponding characteristic points for all training points, $\{\boldsymbol{p}_1^j, \boldsymbol{p}_2^j, \ldots, \boldsymbol{p}_N^j\}$, $j = 1, \ldots, K$. Here, we use Kriging interpolation (Lophaven et al. 2002). The initialization process is shown graphically in Fig. 12.10a.

Evaluation of the GSPRP model at any vector $\boldsymbol{x}$ is a three-step process (shown in Fig. 12.10b). In the first step, the characteristic points corresponding to the vector $\boldsymbol{x}$ are obtained as

$$\boldsymbol{p}^j(\boldsymbol{x}) = \left[\boldsymbol{s}_{\omega.j}(\boldsymbol{x}) \quad \boldsymbol{s}_{r.j}(\boldsymbol{x})\right]^T, \tag{12.20}$$

where $j = 1, \ldots, K$. In the second step, the index $k_{\min}(\boldsymbol{x})$ of the training point is the one identified being the closest to $\boldsymbol{x}$, i.e.,

$$k_{\min}(\boldsymbol{x}) = \text{argmin}\{k \in \{1, \ldots, N\} \ : \ ||\boldsymbol{x}^k - \boldsymbol{x}||\}, \tag{12.21}$$

and calculate the translation vectors defined as

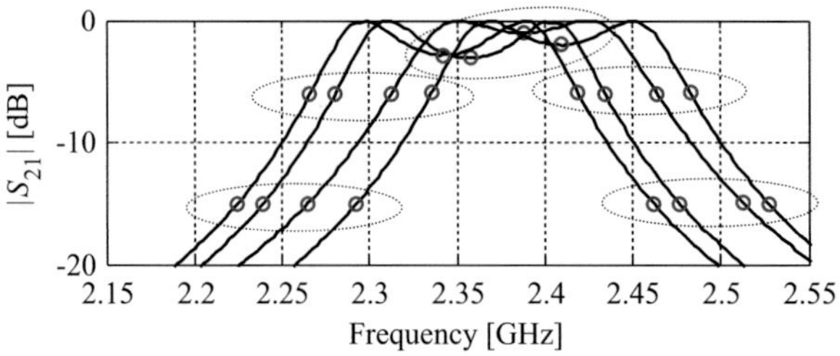

Fig. 12.9 Example response of the high-fidelity model $\boldsymbol{f}$ at several training points (*solid lines*) (Leifsson and Koziel 2015c). *Circles* indicate the characteristic points of the responses. Ellipses indicate the groups of corresponding characteristic points

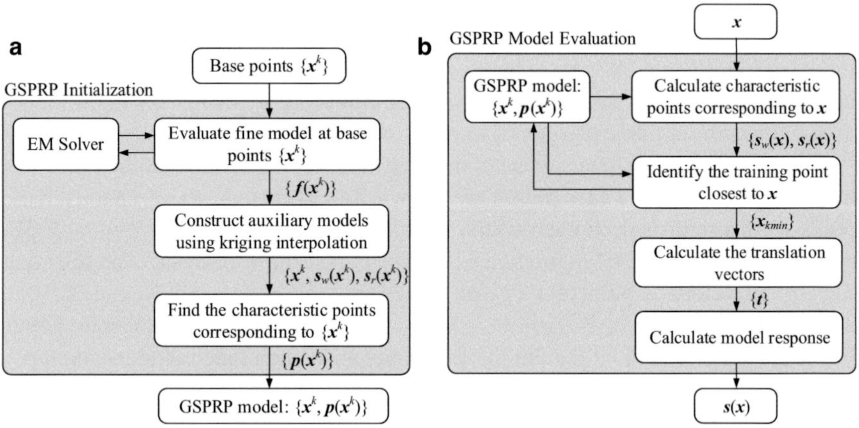

Fig. 12.10 Flowchart of the GSPRP-based surrogate modeling methodology (Leifsson and Koziel 2015c); (**a**) model initialization: EM solver is used to evaluate the high-fidelity model response at the training points, auxiliary models are constructed using Kriging interpolation, and the characteristic points of the training points are found; (**b**) model evaluation at a design $\boldsymbol{x}$: the characteristic points corresponding to $\boldsymbol{x}$ are obtained using (12.20), then the index $k_{min}(\boldsymbol{x})$ of the training point closest to $\boldsymbol{x}$ is identified as in (12.21),the corresponding translation vectors are calculated through (12.22), and finally the surrogate model response $\boldsymbol{s}(\boldsymbol{x})$ is calculated using (12.23)–(12.25)

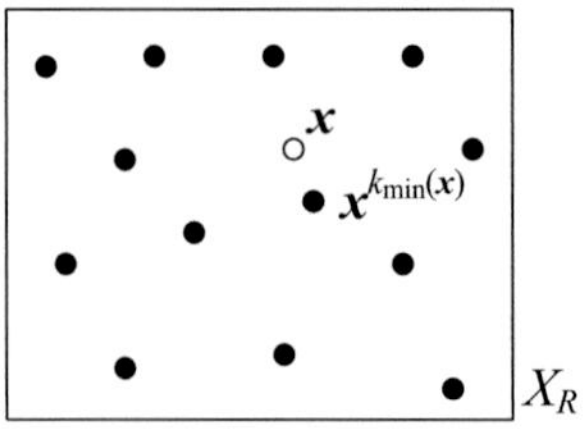

Fig. 12.11 Illustration of the region of interest X_R, the training points (*filled circles*), as well as an example vector of interest $\boldsymbol{x}$ (evaluation point) (Leifsson and Koziel 2015c). The training point that is the closest to $\boldsymbol{x}$ is denoted as $\boldsymbol{x}^{k_{\min}(\boldsymbol{x})}$

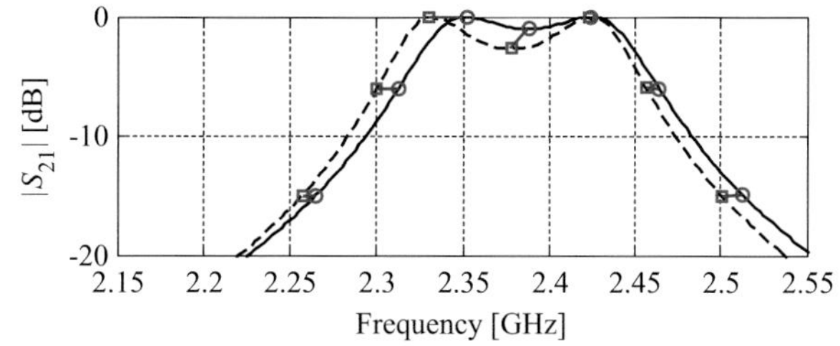

Fig. 12.12 The high-fidelity model $\boldsymbol{f}$ at $\boldsymbol{x}^{k_{\min}(\boldsymbol{x})}$ (the training point closest to the evaluation point $\boldsymbol{x}$), $\boldsymbol{f}(\boldsymbol{x}^{k_{\min}(\boldsymbol{x})})$, (—), the characteristic points of $\boldsymbol{f}(\boldsymbol{x}^{k_{\min}(\boldsymbol{x})})$ (*open circle*), the characteristic points $\boldsymbol{p}^j(\boldsymbol{x})$ corresponding to $\boldsymbol{x}$ obtained using (12.20) (*open square*), as well as the translation vectors $\boldsymbol{t}^j$ (12.22) (short line segments) (Leifsson and Koziel 2015c). The GSPRP-predicted high-fidelity model response at $\boldsymbol{x}$, $\boldsymbol{s}(\boldsymbol{x})$ (- - -) is obtained using $\boldsymbol{p}^j(\boldsymbol{x})$, $\boldsymbol{t}^j$, and $\boldsymbol{f}(\boldsymbol{x}^{k_{\min}(\boldsymbol{x})})$ from (12.20)–(12.25)

$$\boldsymbol{t}^j = \begin{bmatrix} \omega_t^j & \boldsymbol{r}_t^j \end{bmatrix}^T = \begin{bmatrix} \boldsymbol{s}_{\omega.j}(\boldsymbol{x}) - \omega_{k_{\min}(\mathbf{x})}^j & \boldsymbol{s}_{r.j}(\boldsymbol{x}) - \boldsymbol{r}_{k_{\min}(\mathbf{x})}^j \end{bmatrix}^T, \tag{12.22}$$

$j = 1, \ldots, K$. These vectors indicate the change of the characteristic points of the $\boldsymbol{f}$ response while moving from $\boldsymbol{x}^{k_{\min}(\boldsymbol{x})}$) to $\boldsymbol{x}$. Figure 12.11 shows a conceptual illustration of a training set, an example vector $\boldsymbol{x}$, as well as a corresponding vector $\boldsymbol{x}^{k_{\min}(\boldsymbol{x})}$. Figure 12.12 shows the high-fidelity model responses at $\boldsymbol{x}^{k_{\min}(\boldsymbol{x})}$ and $\boldsymbol{x}$, the translation vectors $\boldsymbol{t}^j$, as well as the GSPRP model response at $\boldsymbol{x}$.

Using the translation vectors $\boldsymbol{t}^j$, defined in (12.22), we can define the GSPRP surrogate model $\boldsymbol{s}$ of $\boldsymbol{f}$ as

$$\boldsymbol{s}(\boldsymbol{x}) - [\boldsymbol{s}(\boldsymbol{x}, \omega_1) \quad \ldots \quad \boldsymbol{s}(\boldsymbol{x}, \omega_m)]^T, \tag{12.23}$$

where $\boldsymbol{s}$ is defined at frequencies $\omega_{k_{\min}(\mathbf{x})}{}^j + \omega_t^j$, $j = 0, 1, \ldots, K, K+1$, as (here, $\omega k_{\min}(\boldsymbol{x})^0 = \omega_1$, $\omega k_{\min}(\boldsymbol{x})^{K+1} = \omega_m$, and $\omega_t{}^0 = \omega_t{}^{K+1} = 0$)

$$\boldsymbol{s}\left(\boldsymbol{x}, \omega_{k_{\min}(\mathbf{x})}^j + \omega_t^j\right) = \bar{\boldsymbol{f}}\left(\boldsymbol{x}^{k_{\min}(\mathbf{x})}, \omega_{k_{\min}(\mathbf{x})}^j\right) + r_t^j \tag{12.24}$$

for $j = 1, \ldots, m$. For other frequencies, the model $\boldsymbol{s}$ is obtained through linear interpolation

$$\boldsymbol{s}(\boldsymbol{x}, \omega) = \bar{\boldsymbol{f}}\left(\boldsymbol{x}^0, (1-\alpha)\omega_{k_{\min}(\mathbf{x})}^j + \alpha\omega_{k_{\min}(\mathbf{x})}^{j+1}\right) + \left[(1-\alpha)r_t^j + \alpha r_t^{j+1}\right], \tag{12.25}$$

where $\omega_{k_{\min}(\boldsymbol{x})}{}^j + \omega_t^j \leq \omega \leq \omega_{k_{\min}(\boldsymbol{x})}{}^{j+1} + \omega_t^{j+1}$ and $\alpha = [\omega - (\omega_{k_{\min}(\boldsymbol{x})}{}^j + \omega_t^j)]/[(\omega_{k_{\min}(\boldsymbol{x})}{}^{j+1} + \omega_t^{j+1}) - (\omega_{k_{\min}(\boldsymbol{x})}{}^j + \omega_t^j)]$. $\bar{\boldsymbol{f}}\left(\boldsymbol{x}^{k_{\min}(\mathbf{x})}, \omega\right)$ is an interpolation of $\{\boldsymbol{f}(\boldsymbol{x}^{k_{\min}(\boldsymbol{x})}, \omega_1), \ldots, \boldsymbol{f}(\boldsymbol{x}^{k_{\min}(\boldsymbol{x})}, \omega_m)\}$ onto the frequency interval $[\omega_1, \omega_m]$. This interpolation is necessary because the original frequency sweep is a discrete set.

12.5.4 Example: Microwave Filter Modeling

Here, the SPRP technique is applied to the modeling of a microstrip filter. A comparison of both the basic and the modified SPRP with surrogate modeling using standard space mapping (Bandler et al. 2006) is provided. The standard SM model is quite involved because it is using input and output SM of the form $\boldsymbol{A}\cdot\boldsymbol{c}$ $(\boldsymbol{B}\cdot\boldsymbol{x}+\boldsymbol{c})$, enhanced by the implicit and frequency space mapping (Bandler et al. 2006). All surrogate models are set up using the same base set consisting of $N=2n+1$ points allocated according to the star distribution (Bandler et al. 2006). The quality of the models is assessed using a relative error measure $||\boldsymbol{f}(\boldsymbol{x})-\boldsymbol{s}(\boldsymbol{x})||/||\boldsymbol{f}(\boldsymbol{x})||$ expressed in percent.

Consider the fourth-order ring resonator band-pass filter considered in Sect. 12.3.1 (Fig. 12.1a). The design parameters are $\boldsymbol{x}=[L_1L_2L_3S_1S_2]^T$ mm. The high-fidelity model $\boldsymbol{f}$ is simulated in FEKO (2008). The low-fidelity model, Fig. 12.1b, is implemented in Agilent ADS (2011). The region of interest is defined by the reference point $\boldsymbol{x}^0=[24.0\ 21.0\ 26.0\ 0.2\ 0.1]^T$ mm, and the region size $\boldsymbol{\delta}=[2.0\ 2.0\ 2.0\ 0.1\ 0.05]^T$ mm. The SPRP characteristic points were set to capture the filter passband, i.e., points at both slopes (at levels of –20 dB, –18 dB, up to –8 dB) and 30 points in between the –8 dB points.

The modeling accuracy has been verified using 50 random test points. The results shown in Table 12.3 and in Fig. 12.13 indicate that the modified SPRP model ensures better accuracy than the standard SM model and the original version of SPRP (Koziel 2010e).

As an application example, the modified SPRP surrogate was utilized to optimize the filter with respect to the following design specifications: $|S_{21}|\geq-1$ dB for 1.75 GHz $\leq\omega\leq$ 2.25 GHz, and $|S_{21}|\leq-20$ dB for 1.0 GHz $\leq\omega\leq$ 1.5GHz and 2.5 GHz $\leq\omega\leq$ 3.0 GHz. The initial design was $\boldsymbol{x}^0=[24.0\ 21.0\ 26.0\ 0.2\ 0.1]^T$ mm. Figure 12.14a shows the high-fidelity model response of the filter at the initial design and at the design $\boldsymbol{x}^*=[22.61\ 20.11\ 26.626\ 0.156\ 0.040]^T$ mm obtained by optimizing the surrogate. The specification error at the optimized design is –0.45 dB.

The SPRP model was also used to estimate yield at the optimized design, assuming 0.2 mm deviation for length parameters (L_1 to L_2) and 0.02 mm for spacing parameters (S_1 and S_2). The yield estimation based on 200 random samples is 68 % (Fig. 12.14b). This value is very close to the yield estimated directly using the high-fidelity model (70 %). The estimation performed with the SM model is less accurate (50 %). Note that the total computational cost of building the surrogate model, design closure, and statistical analysis is only 11 full-wave simulations of the filter structure.

To illustrate the operation and performance of GSPRP, two examples of microstrip filters are considered: the stacked slotted resonator band-pass filter (Huang et al. 2008) shown in Fig. 12.15a (Filter 1), and the microstrip band-pass filter with open stub inverter (Lee et al. 2000), Fig. 12.15b (Filter 2). Design variables are $\boldsymbol{x}=[L_1L_2W_1S_1S_2d]^T$ (Filter 1), and $\boldsymbol{x}=[L_1L_2L_3S_1S_2W_1]^T$ (Filter 2). Filter 1 is simulated in Sonnet ***em*** (2010) using a grid of 0.05 mm × 0.05 mm. Filter

Table 12.3 Fourth-order ring resonator filter: modeling results (Leifsson and Koziel 2015c)

Model	Average error (%)	Maximum error (%)
SM	1.8	4.5
Basic SPRP	1.1	2.7
Modified SPRP	0.3	0.6

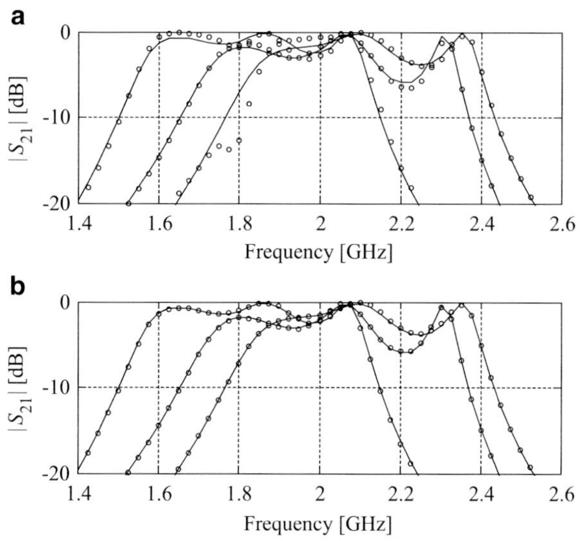

Fig. 12.13 Fourth-order ring resonator band-pass filter: high-fidelity model (*solid line*) and surrogate model (*circles*) responses at three selected test points for (**a**) standard SM model, (**b**) modified SPRP surrogate model (Leifsson and Koziel 2015c)

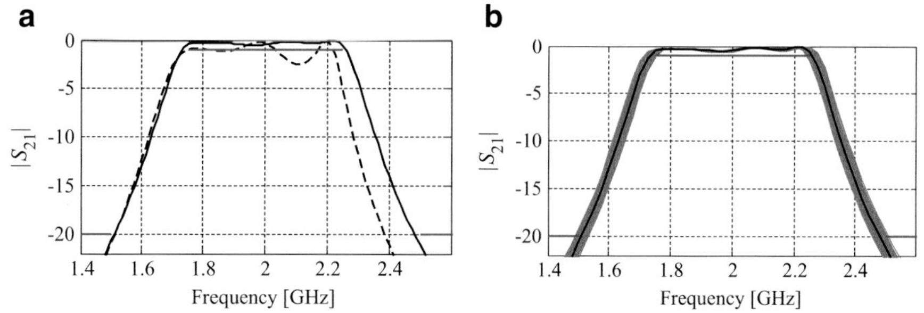

Fig. 12.14 Fourth-order ring resonator band-pass filter optimization results (Leifsson and Koziel 2015c): (**a**) fine model responses at the reference point $\boldsymbol{x}^0$ (*dashed line*) and at the optimal solution $\boldsymbol{x}^*$ of the modified shape-preserving response prediction surrogate model (*solid line*); (**b**) statistical analysis at $\boldsymbol{x}^*$ using the modified shape-preserving response prediction model. Estimated yield is 68 %. Thick black solid line denotes the fine model response at optimal design $\boldsymbol{x}^*$

2 is evaluated in FEKO (2008) with the total mesh number of 432. The region of interest for Filter 1 is defined by the reference point $\boldsymbol{x}^0 = [6\ 9.6\ 1\ 1\ 2\ 2]^T$ mm, and the region size $\boldsymbol{\delta} = [0.8\ 0.8\ 0.2\ 0.2\ 0.4\ 0.4]^T$ mm. For Filter 4, we have $\boldsymbol{x}^0 = [24\ 10\ 2\ 0.6\ 0.2\ 0.5]^T$ mm, and $\boldsymbol{\delta} = [2\ 2\ 1\ 0.4\ 0.1\ 0.4]^T$ mm.

Table 12.4 shows the average modeling error for the GSPRP model (12.11)–(12.16), as well as Kriging interpolation (Forrester and Keane 2009; Salleh

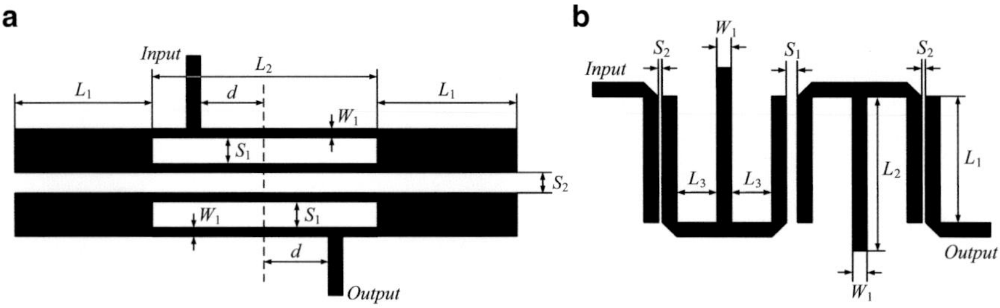

Fig. 12.15 Filter structures for GSPRP modeling: (**a**) stacked slotted resonator filter (Huang et al. 2008), (**b**) band-pass filter with open stub inverter (Lee et al. 2000)

Table 12.4 GSPRP modeling results for Filters 1 and 2 (Leifsson and Koziel 2015c)

	Average error (Filter 1)		Average error (Filter 2)	
Number of training points[a]	GSPRP (%)	Kriging interpolation (%)	GSPRP (%)	Kriging interpolation (%)
20	3.9	15.6	7.7	11.6
50	1.8	11.9	2.7	8.8
100	0.9	11.0	1.8	6.1
200	0.6	9.3	1.4	4.8
400	0.4	6.7	1.2	3.1

[a]Training points allocated using Latin hypercube sampling (Beachkofski and Grandhi 2002)

et al. 2008), with different number of training points from 20 to 400. For both filters, the accuracy of the GSPRP model is better than the accuracy of the Kriging surrogate for the corresponding number of training points.

Figure 12.16 shows the high-fidelity and GSPRP model responses at selected test points for Filters 1 and 2. A comparison with Kriging interpolation of high-fidelity model date reveals that the comparable error level can be obtained for the training set that is several times smaller than for Kriging interpolation. The results are consistent for all considered examples.

Compared to the original SPRP technique (Koziel 2010e) and the modified SPRP of Koziel (2011b), the GSPRP is better at utilizing available high-fidelity model data. This results not only in further reduction of the modeling error but also in the surrogate model being continuous with respect to the design variables across the entire design space, which was not possible in that of Koziel (2010e).

12.5.5 Example: Fluid Flow Through a Converging-Diverging Nozzle

The use of SPRP and GSPRP is demonstrated for the modeling of fluid flow (Leifsson et al. 2014b). One-dimensional, steady, viscous, turbulent, non-heat-conducting compressible flow through a converging-diverging nozzle is

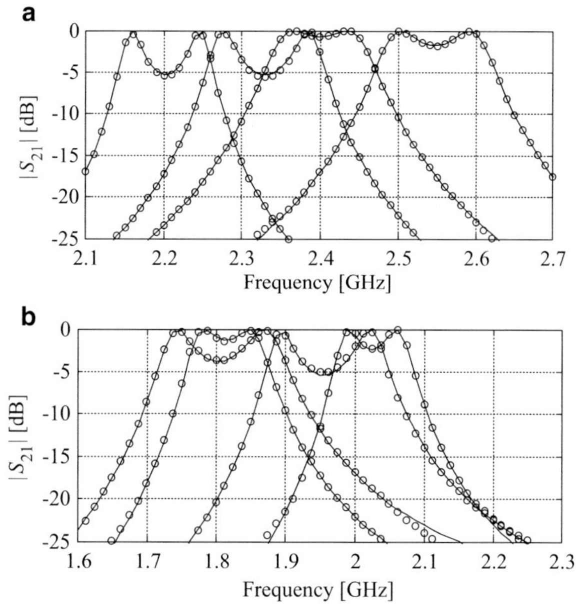

Fig. 12.16 High-fidelity (—) and GSPRP surrogate model (12.11)–(12.16) responses (*open circle*) obtained for 100 base points at the selected test designs: (**a**) Filter 1, (**b**) Filter 2 (Leifsson and Koziel 2015c)

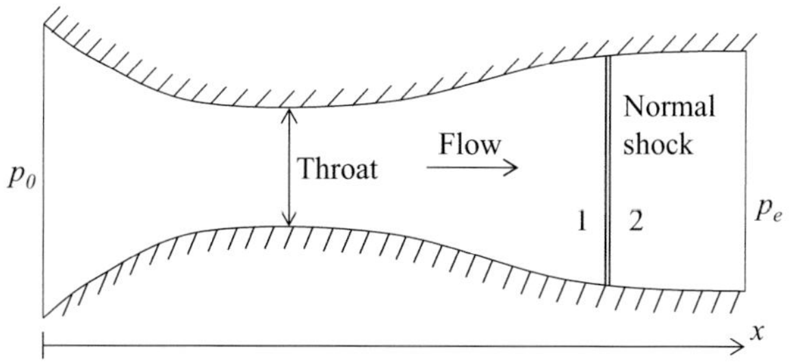

Fig. 12.17 A sketch of a convergent-divergent nozzle with the relevant inlet and outlet flow boundary conditions. The geometry considered in this example is presented and discussed in Liou (1987)

considered. A sketch of a typical converging-diverging nozzle is shown in Fig. 12.17. The operating parameters of the nozzle flow are the stagnation pressure (p_0) at the inlet and the static pressure at the exit (p_e). Depending on these parameters, as well as the nozzle geometry, the flow is characterized by subsonic and/or supersonic flow, and even with normal shocks occurring after behind the throat.

The high-fidelity model is based on a computational fluid dynamics (CFD) simulation solving the steady-state Reynolds-averaged Navier-Stokes equations with the one-equation Spalart-Allmaras turbulence model (see, for example, Tannehill et al. (1997)). The CFD calculations were performed using Fluent (2015) and the grids are generated using ICEM CFD (2012). The low-fidelity model is analytical and based on the isentropic equations and the normal shock equations (the model details can be found in Anderson (2007)).

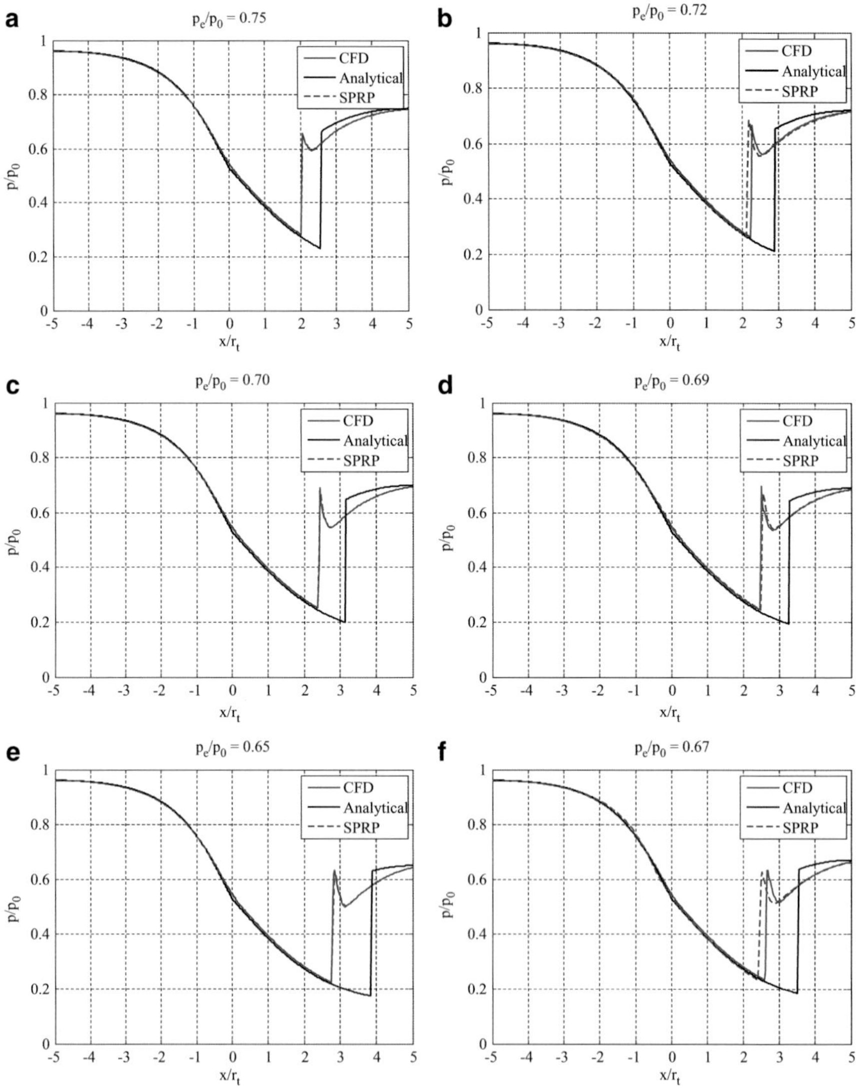

Fig. 12.18 Centerline pressure ratio distributions obtained by the CFD, analytical, and SPRP models (Leifsson and Koziel 2015c)

Figure 12.18 shows the centerline pressure distributions for several pressure ratios p_e/p_0, obtained for the analytical model, CFD simulations, as well as the basic SPRP surrogate. The basic SPRP model was set up using the following set of characteristic points: the 2 end points ($x/r = -5, 5$), the 2 points corresponding to the pressure shock (the lowest and the highest value), as well as the 20 additional points equally distributed along the x/r axis, between the four aforementioned points.

The training set consists of four locations, corresponding to $p_e/p_0 = 0.65$, 0.70, 0.75, and 0.8. It should be noted that the SPRP predictions coincide with the training point responses (Fig. 12.18a–c), because SPRP is an interpolative model. The predictions for other values of the pressure ratio (Fig. 12.18d–f) are also very good. It should be emphasized that SPRP allows for excellent prediction of the CFD model pressure shock shape even though the shock shape is different for the underlying analytical model.

Figure 12.19 shows the pressure distributions for the generalized SPRP for the same values of the pressure ratio as for the SPRP model. GSPRP surrogate was established using the same sets of characteristic points and training samples as for the SPRP model. Note that the agreement between the GSPRP response and the CFD model response is excellent and better than for the SPRP surrogate. However, in general, the SPRP model tends to be more accurate than GSPRP in situations where the training set is sparse in the design space and the variability of the model responses is large. In such cases, the knowledge embedded in the low-fidelity model is essential to obtain good predictive power of the surrogate.

12.6 Feature-Based Modeling for Statistical Design

Reliable design of microwave components and circuits has to account for manufacturing tolerances and uncertainties. In many cases, the objective is a robust design, i.e., maximization of the probability that the fabricated structure satisfies given performance specifications under assumed deviations from the nominal values of geometry and/or material parameters (for yield-driven design or design centering see, e.g., Bandler et al. 1976a, b; Abdel-Malek and Bandler 1978; Styblinski and Oplaski 1986; Abdel-Malek et al. 1999; Scotti et al. 2005). In this context, statistical analysis and yield estimation are indispensable steps in the design process (Bandler and Chen 1988; Biernacki et al. 2012; Swidzinski and Chang 2012).

In this section, rapid and reliable statistical analysis and yield estimation of EM-simulated microwave structures are described (Koziel and Bandler 2014). The essence of the approach is approximation-based modeling of suitably selected features of the filter response. The features are chosen so that they can be used to uniquely determine whether or not the structure satisfies given performance requirements. Utilization of response features for design optimization has been discussed in Chap. 9. The approximation model is constructed using few training designs (and, consequently, only a few corresponding EM simulations of the structure are necessary for its setup), which grows only linearly with the dimensionality of the design space. We describe the technique in detail and then demonstrate it on a band-pass filter to estimate the yield (Sect. 12.6.1). The technique is also demonstrated on yield-driven design optimization (Sect. 12.6.2).

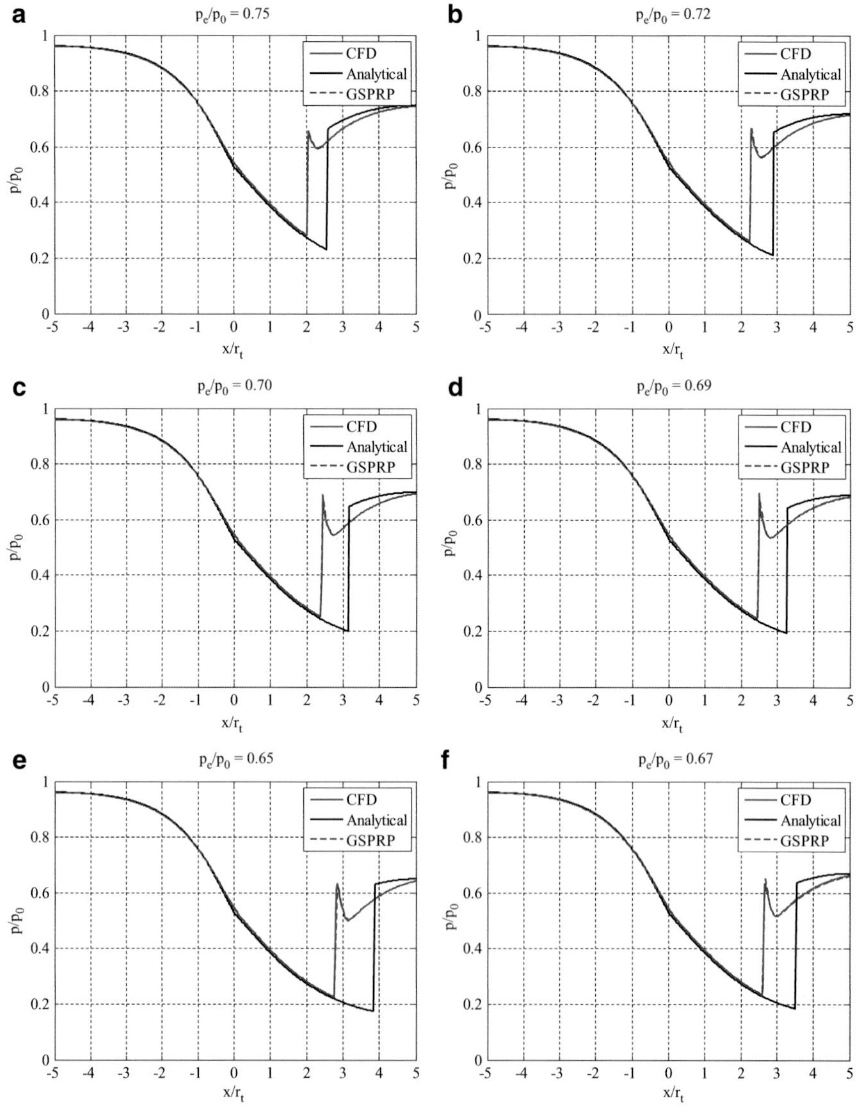

Fig. 12.19 Centerline pressure ratio distributions obtained by the CFD, analytical, and GSPRP models (Leifsson and Koziel 2015c)

12.6.1 Yield Estimation Using Response Features

Let $\boldsymbol{x}^0 = [x_1{}^0 x_2{}^0 \ldots x_n{}^0]^T$ be a nominal design (typically, an optimum design with respect to given performance specifications). It is assumed that due to manufacturing uncertainties, the actual parameters of the fabricated device are $\boldsymbol{x}^0 + d\boldsymbol{x}$, where a random deviation $d\boldsymbol{x} = [dx_1 \; \ldots \; dx_2 dx_n]^T$ is described by a given probability distribution (such as a Gaussian distribution with zero mean and a certain standard deviation, or a uniform distribution with certain lower and upper bounds, e.g., $dx_k \in [-d_{k.max} d_{k.max}]$, $k = 1, \ldots, n$.

We define an auxiliary function $H(\boldsymbol{x})$ as follows (Bandler and Chen 1988):

$$H(\boldsymbol{x}) = \begin{cases} 1 & \text{if } \boldsymbol{f}(\boldsymbol{x}) \text{ satisfies the design specifications} \\ 0 & \text{otherwise} \end{cases}. \tag{12.26}$$

Then, the yield at the nominal design $\boldsymbol{x}^0$ can be estimated as

$$Y(\boldsymbol{x}^0) = \frac{\sum_{j=1}^{N} H(\boldsymbol{x}^0 + d\boldsymbol{x}^j)}{N}, \tag{12.27}$$

where $d\boldsymbol{x}^j$, $j = 1, \ldots, N$, are random vectors sampled according to the assumed probability distribution. Obviously, evaluating (12.27) by means of multiple simulations of the perturbed nominal design may be extremely expensive, particularly because reliable yield estimation requires a large number of samples (typically, a few hundred or more). In the case of small yield values, the number of samples has to be even larger (a few thousand or more) in order to avoid high variance of the estimator.

12.6.1.1 Feature-Based Approximation Model

The yield can be estimated using the so-called feature points. The concept was introduced in Koziel and Szczepanski (2011) in the context of the SPRP technique. Let us consider $|S_{11}|$ responses of a band-pass filter (cf. Fig. 12.20). The plot shows the response at the nominal design (i.e., a typically desired minimax optimum w.r.t. the design specifications marked with horizontal lines, here, $|S_{11}| \leq -20$ dB for 10.55–11.45 GHz, and $|S_{11}| \geq -1$ dB for frequencies lower than 10.3 GHz and higher than 11.7 GHz), as well as a set of so-called feature points, here, represented by –1 and –20 dB levels as well as the peaks of the response in the passband. The location of these points is sufficient to determine whether the response violates or satisfies the given design specifications. In particular, assuming small design perturbations, the feature points corresponding to –1 and –20 dB may move towards lower or higher frequencies violating (in some cases) the specifications regarding passband and/or stopband frequencies; the feature points corresponding to $|S_{11}|$ maxima in the passband may move up leading to violation of the $|S_{11}| \leq -20$ requirement.

The choice of feature points for a given problem is straightforward. Figure 12.20 also shows the response and the corresponding feature points at a perturbed design (which, in this case, violates our specifications). As indicated in Koziel (2012), modeling feature points is easier than constructing response surfaces for entire responses. This is because the dependence of both the frequency and vertical locations of those points on respective designable parameters is much less nonlinear than for the S-parameters modeled (conventionally) as functions of frequency. As a

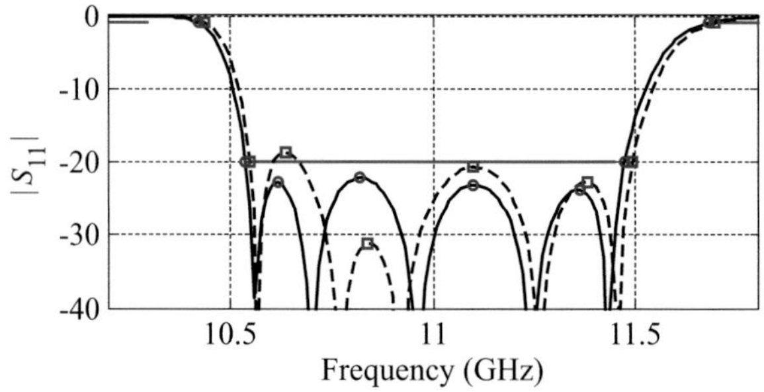

Fig. 12.20 Reflection response of the band-pass filter (—) at the optimum design with respect to given minimax specifications (marked with *horizontal lines*), as well as the response at a perturbed design (- - -) (Koziel and Bandler 2014). *Circles* and *squares* denote feature points for both responses, here, corresponding to the –1 and –20 dB levels as well as the response maxima in the passband. Design specifications are $|S_{11}| \leq -20$ dB for 10.55 to 11.45 GHz, and $|S_{11}| \geq -1$ dB for frequencies lower than 10.3 GHz and higher than 11.7 GHz

result, only a limited number of training samples is necessary for creating such models, particularly if we are only interested in local approximations (i.e., around the nominal design). It should also be emphasized that—unlike in the case of SPRP (Koziel and Szczepanski 2011)—we are not interested in an accurate prediction of the entire response of the structure. The focus is on those critical parts of the response where the design specifications can potentially be violated. This significantly simplifies the modeling process.

In order to construct our model, we consider $2n+1$ evaluations of the original model $\boldsymbol{f}$ at the nominal design, $\boldsymbol{f}(\boldsymbol{x}^0)$, and at the perturbed designs $\boldsymbol{x}^k = [x_{0.1} \ldots x_{0.k} + \text{sign}(k)\cdot\delta_k \ldots x_{0.n}]^T$, $k = -n, \ldots, -1, 1, \ldots, n$, where δ_k may be, e.g., a maximum assumed deviation of the kth parameter from its nominal value. The feature points of the response vector $\boldsymbol{f}(\boldsymbol{x}^k)$ are denoted as $\boldsymbol{p}_k^j = [\omega_k^j r_k^j]^T$, $j = 1, \ldots, K$, where ω and r are the frequency and magnitude components of the respective point, and K is the total number of feature points.

The aim is to predict the position of the feature points corresponding to a perturbed vector $\boldsymbol{x}^0 + d\boldsymbol{x}$, using the available training set $\{\boldsymbol{x}^k; \boldsymbol{p}_{-n}^j, \ldots, \boldsymbol{p}_{-1}^j, \boldsymbol{p}_0^j, \boldsymbol{p}_1^j, \ldots, \boldsymbol{p}_n^j\}$. For any given $d\boldsymbol{x}$, we find a subset X_S of the base set $\{\boldsymbol{x}^k\}$ that defines an area containing $\boldsymbol{x}^0 + d\boldsymbol{x}$. The surrogate model is set up using all the points from X_S, as shown in Fig. 12.21 for $n = 2$. Without loss of generality, we can assume that $X_S = \{\boldsymbol{x}^0, \boldsymbol{x}^1, \ldots, \boldsymbol{x}^n\}$. We define

$$\boldsymbol{p}^j = \boldsymbol{p}_0^j + \beta_1\left(\boldsymbol{p}_1^j - \boldsymbol{p}_0^j\right) + \beta_2\left(\boldsymbol{p}_2^j - \boldsymbol{p}_0^j\right) + \ldots + \beta_n\left(\boldsymbol{p}_n^j - \boldsymbol{p}_0^j\right), \tag{12.28}$$

where $\beta_1, \beta_2, \ldots, \beta_n$, determines a unique representation of $d\boldsymbol{x}$ using vectors $\boldsymbol{v}_i = \boldsymbol{x}^k - \boldsymbol{x}^0$, $i = 1, \ldots, n$. Coefficient β_i can be explicitly found as

$$[\beta_1 \quad \beta_2 \quad \ldots \quad \beta_n]^T = [\boldsymbol{v}_1 \ \ \boldsymbol{v}_2 \ \ \ldots \ \ \boldsymbol{v}_n]^{-1} \cdot \left(\boldsymbol{x} - \boldsymbol{x}^0\right). \tag{12.29}$$

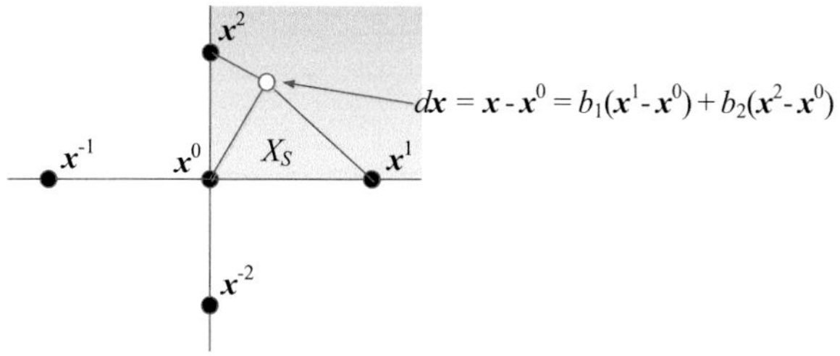

Fig. 12.21 Deviation vector $\boldsymbol{dx}$ and its expansion using star-distributed training vectors $\boldsymbol{x}^0$ and $\boldsymbol{x}^k$, $k = -2, -1, 1, 2$ (denoted as •) (Koziel and Bandler 2014). The shaded area denotes an area defined by a subset X_S of points being the closest to $\boldsymbol{x}^0 + \boldsymbol{dx}$, which is represented as a linear combination of vectors $\boldsymbol{x}^k - \boldsymbol{x}^0$. The feature points at $\boldsymbol{x}^0 + \boldsymbol{dx}$ are calculated using the coefficients of this linear combination and the feature points of $\boldsymbol{f}(\boldsymbol{x}^k)$ for $\boldsymbol{x}^k \in X_S$. Here, we have $\boldsymbol{p}^j = \boldsymbol{p}_0^j + \beta_1(\boldsymbol{p}_1^j - \boldsymbol{p}_0^j) + \beta_2(\boldsymbol{p}_2^j - \boldsymbol{p}_0^j)$

Using (12.19) and (12.20), we can define an approximation model of the feature points as

$$s(\boldsymbol{x}^0 + \boldsymbol{dx}) = \left[\boldsymbol{p}^1(\boldsymbol{x}^0 + \boldsymbol{dx}) \quad \cdots \quad \boldsymbol{p}^K(\boldsymbol{x}^0 + \boldsymbol{dx})\right], \tag{12.30}$$

where $\boldsymbol{p}^j(\boldsymbol{x}^0 + \boldsymbol{dx}) = [\omega^j(\boldsymbol{x}^0 + \boldsymbol{dx})\ r^j(\boldsymbol{x}^0 + \boldsymbol{dx})]^T, j = 1, \ldots, K$. Having $\boldsymbol{s}(\boldsymbol{x}^0 + \boldsymbol{dx})$, one can estimate yield in a way similar to (12.26), (12.27). The fundamental difference is that the satisfaction/violation of the design specification frequencies/levels is verified for the feature points only rather than for the entire responses.

It should be emphasized that using response features for estimating yield rather than constructing, e.g., a linear model of the entire S-parameter response is critical to accuracy. Let us consider a simple, first-order Taylor expansion of the filter model around the nominal design $\boldsymbol{x}^0$

$$\boldsymbol{f}_L(\boldsymbol{x}) = \boldsymbol{f}(\boldsymbol{x}^0) + \boldsymbol{J}_f(\boldsymbol{x}^0) \cdot (\boldsymbol{x} - \boldsymbol{x}^0), \tag{12.31}$$

where $\boldsymbol{J}_f(\boldsymbol{x}^0)$ is an estimated Jacobian of $\boldsymbol{f}$ at the nominal design. The estimate can be obtained using evaluations of $\boldsymbol{f}$ at the perturbed designs $\boldsymbol{x}^k$.

Figure 12.22 shows an S-parameter prediction obtained by evaluating a linear surrogate $\boldsymbol{f}_L$ constructed from the filter responses evaluated for the same training set used for the feature-based model. The lack of accuracy coming from the very sharp responses (as functions of frequency) is reflected in underestimated yield predictions. This indicates that the feature-based yield estimation (although based on the same data set) is fundamentally different from simple linear modeling.

It should be mentioned that a number of sophisticated methods for parametric macromodeling (e.g., Ferranti et al. 2009, 2011) or stochastic macromodeling (e.g., Sumant et al. 2010, 2012) can be found in the literature that allow for avoiding the presence of abnormal responses of the simple linear model (12.22) through, e.g., passivity enforcement. Here, the model (12.22) was only used in order to indicate that the "naive" utilization of the small data set exploited by the feature-based model leads to very poor predictions.

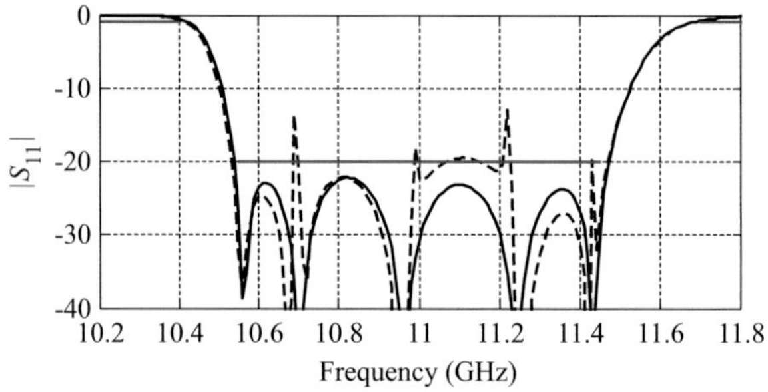

Fig. 12.22 Fifth-order band-pass filter of Section III.*A*: the filter response at the nominal design (—) and the response obtained from a linear model (12.22) constructed using a perturbed design (- - -) (at the selected reference design) (Koziel and Bandler 2014). The spikes that appear due to the linear modeling of sharp responses lead to considerable yield underestimation (cf. Table 12.5)

12.6.1.2 Verification Example: Fifth-Order Waveguide Band-Pass Filter

Consider the X-band waveguide filter with nonsymmetrical irises (Hauth et al. 1993) shown in Fig 12.23. The design variables are $\boldsymbol{x} = [z_1 z_2 z_3 d_1 d_2 d_3 t_1 t_2 t_3]^T$. The filter is simulated in CST (~140,000 tetrahedrons, simulation time about 8 min). The nominal design, $\boldsymbol{x}^0 = [12.0836\ 14.2069\ 14.6875\ 13.9841\ 11.6864\ 10.8076\ 1.5455\ 3.0671\ 2.4557]^T$ mm, is a minimax optimum with respect to the following design specifications: $|S_{11}| \leq -20$ dB for 10.55 GHz $\leq \omega \leq$ 11.45 GHz, and $|S_{11}| \geq -1$ dB for $\omega \leq$ 10.4 GHz and $\omega \geq$ 11.7 GHz. The minimax optimization is understood as minimization of a maximum violation of the aforementioned design specifications within the respective frequency sub-bands.

Yield estimation has been carried out using four scenarios for geometry parameter deviations, including a uniform probability distribution with a maximum deviation equal to 0.01 and 0.02 mm (Cases 1 and 2), and a normal distribution with zero mean and standard deviation 0.01 and 0.02 mm (Cases 3 and 4). The deviations are taken as uncorrelated. The yield has been estimated with the methodology described in Sect. 12.6.1.1, using the eight feature points shown in Fig. 12.20.

For comparison, the yield was also estimated using conventional Monte Carlo analysis with 500 random samples (the number of samples is limited due to the computational cost of the EM simulation). The results are shown in Table 12.5. Figure 12.24 presents a visualization of the yield estimation for Case 2. The agreement between the yield estimation obtained using our proposed method and conventional Monte Carlo analysis is excellent. As a matter of fact, the results obtained using our approach are more reliable than MC: the uncertainty in the latter is relatively large due to the small number of samples used in the process to keep the cost low. Feature-based yield estimation was executed for $N = 5{,}000$.

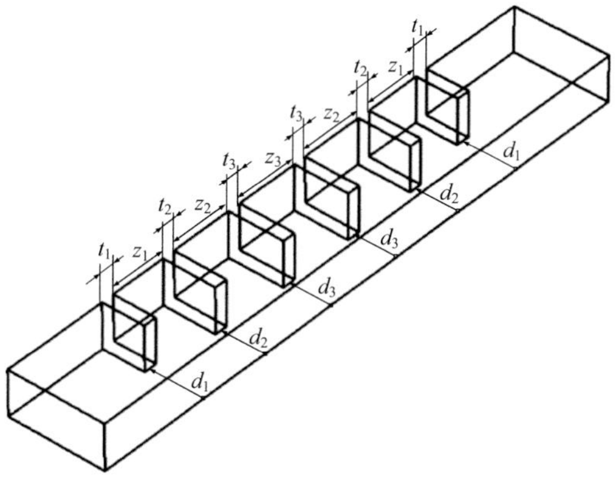

Fig. 12.23 Fifth-order waveguide band-pass filter (Hauth et al. 1993)

Table 12.5 Yield estimation: fifth-order waveguide filter (Koziel and Bandler 2014)

Case	Distribution	Yield estimation method	Estimated yield	CPU cost[a]
1	Uniform (max. dev. 0.01 mm)	Feature points (this work)	0.97	19
		EM-based Monte Carlo	0.97	500
		Linear modeling[b]	0.63	19
2	Uniform (max. dev. 0.02 mm)	Feature points (this work)	0.56	19
		EM-based Monte Carlo	0.55	500
		Linear modeling[b]	0.12	19
3	Gaussian (std. dev. 0.01 mm)	Feature points (this work)	0.69	19
		EM-based Monte Carlo	0.69	500
		Linear modeling[b]	0.19	19
4	Gaussian (std. dev. 0.02 mm)	Feature points (this work)	0.25	19
		EM-based Monte Carlo	0.24	500
		Linear modeling[b]	0.02	19

[a]Estimation cost in number of EM analyses. Feature-based yield estimation utilizes $N = 5{,}000$ random samples

[b]Estimation based on a linear model of the S-parameter response around the nominal design

12.6.2 Tolerance-Aware Design Optimization Using Response Features

As demonstrated in Sect. 12.5.1, the feature-based yield estimation technique ensures very good accuracy at low computational cost. Thus, the feature approximation model (12.21) can be utilized for tolerance-aware design.

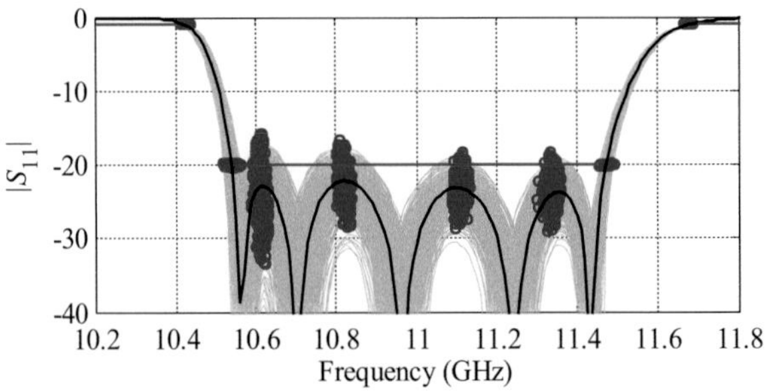

Fig. 12.24 Fifth-order waveguide band-pass filter: yield estimation for Case 2 (Koziel and Bandler 2014). *Gray lines* correspond to 500 EM-simulated random samples for Monte Carlo analysis, *circles* represent corresponding feature points calculated using the approximation model (12.17)–(12.22)

12.6.2.1 Yield Maximization Methodology

Let $Y_s(\boldsymbol{x}^0)$ denote the yield estimated at the nominal design $\boldsymbol{x}^0$ using multiple evaluations of the feature model $\boldsymbol{s}$, as explained in Sect. 12.5.1.2. The yield estimation is based on multiple evaluations of this model, i.e., $\boldsymbol{s}(\boldsymbol{x}^0 + d\boldsymbol{x}^k)$, where $d\boldsymbol{x}^k$, $k = 1, \ldots, N$, are random vectors sampled according to the assumed probability distribution.

One can define a function $Y_s(\boldsymbol{x};\boldsymbol{x}^0)$ realizing a yield estimation for any design $\boldsymbol{x}$ (in practice, $\boldsymbol{x}$ should be in the vicinity of $\boldsymbol{x}^0$), where the yield is calculated using evaluations of $\boldsymbol{s}(\boldsymbol{x} + d\boldsymbol{x}^k)$. However, because $\boldsymbol{s}$ is defined with respect to $\boldsymbol{x}^0$ rather than an arbitrary $\boldsymbol{x}$, the yield estimation has to be performed using the following set of evaluations: $\boldsymbol{s}(\boldsymbol{x}^0 + d\boldsymbol{x}'^k)$, where $d\boldsymbol{x}'^k = d\boldsymbol{x}^k + [\boldsymbol{x} - \boldsymbol{x}^0]$, $k = 1, \ldots, N$. The design which maximizes the yield is then found as

$$\boldsymbol{x}^* = \arg\min_{\boldsymbol{x}} \left\{ -Y_s\left(\boldsymbol{x}; \boldsymbol{x}^0\right) \right\}. \tag{12.32}$$

It should be emphasized that to ensure reliability (in particular, to minimize the yield estimation variance), a large number of samples should be used. In our numerical experiments we set $N = 5{,}000$. Also, the same sample set of perturbations $\{d\boldsymbol{x}^k\}$ should be utilized for all evaluations of $Y_s(\boldsymbol{x};\boldsymbol{x}^0)$ in order to avoid numerical noise related to yield estimation variance.

In practice, the accuracy of a yield estimate using $Y_s(\boldsymbol{x};\boldsymbol{x}^0)$ will degrade as $\boldsymbol{x}$ moves away from $\boldsymbol{x}^0$, so yield optimization should preferably be implemented as an iterative process, namely

$$\boldsymbol{x}^{i+1} = \arg\min_{\boldsymbol{x}} \left\{ -Y_s\left(\boldsymbol{x}; \boldsymbol{x}^i\right) \right\}, \tag{12.33}$$

where $\boldsymbol{x}^i$, $i = 0, 1, \ldots$, are approximations of $\boldsymbol{x}^*$ obtained by optimizing $Y_s(\boldsymbol{x};\boldsymbol{x}^i)$, i.e., a yield estimation function set up similarly to $Y_s(\boldsymbol{x};\boldsymbol{x}^0)$ but centered around $\boldsymbol{x}^i$ rather than $\boldsymbol{x}^0$ and using corresponding perturbations. The cost of each iteration is $2n + 1$ EM simulations, where n is the number of designable parameters. The procedure is terminated when the current iteration does not lead to yield improvement, which is understood as $Y_s(\boldsymbol{x}^{i+1};\boldsymbol{x}^{i+1}) > Y_s(\boldsymbol{x}^i;\boldsymbol{x}^i)$. The feature-based model Y_s (cf. (12.33)) can be optimized using the pattern search algorithm (Koziel 2010g).

Although this section only covers yield optimization starting from certain nominal design (here, minimax optimum with respect to given lower/upper specifications at selected frequencies), it is recommended that when starting from an arbitrary initial design, the yield-driven optimization is split into two separate steps:

1. Conventional minimax-type optimization executed to find the nominal design
2. Yield-driven stage as described in this section

The first step can be realized, for the sake of computational efficiency, using, e.g., space mapping or other type of surrogate-based optimization technique (cf. Chap. 4).

12.6.2.2 Application Example: Yield Optimization of Fifth-Order Waveguide Filter

The yield optimization procedure of the preceding section is applied to the waveguide filter of Sect. 12.5.1.3, assuming a Gaussian probability distribution with a standard deviation equal to 0.02 mm (Case 4). The optimization is started from the nominal design $\boldsymbol{x}^0$ with an estimated yield of 0.25. The optimized design is $\boldsymbol{x}^* = [12.05\ 14.18\ 14.67\ 14.06\ 11.74\ 10.85\ 1.49\ 3.03\ 2.46]^T$ with an estimated yield of 0.46. This design was obtained in four iterations of the yield optimization procedure at a total cost of $4 \times 19 = 76$ filter evaluations. Table 12.6 indicates the yield estimated using both the yield optimization procedure and conventional Monte Carlo analysis. Good agreement between the two estimations is observed. Figure 12.25 shows a visualization of the yield estimate at the optimized design.

Table 12.6 Yield optimization: fifth-order waveguide filter (Koziel and Bandler 2014)

Case[a]	Yield estimation method	Estimated yield	CPU cost[b]
Nominal design (minimax optimum)	Feature points	0.25	19
	EM-based Monte Carlo	0.24	500
Yield-optimized design	Feature points	0.46	19
	EM-based Monte Carlo	0.47	500

[a]Estimation for a uniform distribution with maximum deviations of 0.01 mm
[b]Estimation cost in number of EM analyses

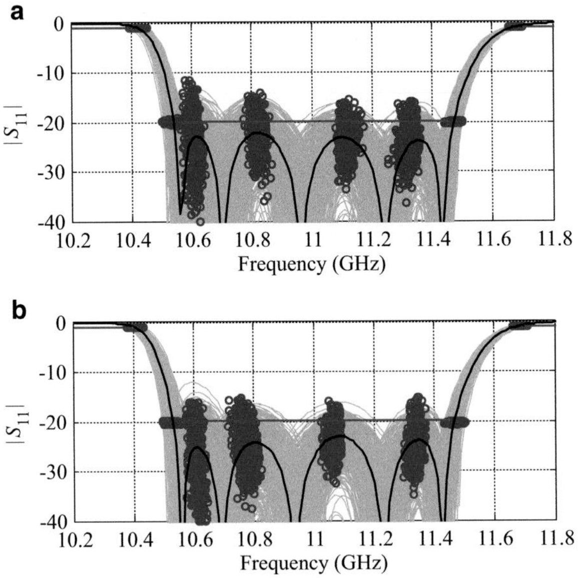

Fig. 12.25 Fifth-order waveguide filter (Koziel and Bandler 2014): yield estimation assuming a Gaussian probability distribution with standard deviations of 0.02 mm (Case 4) at (**a**) the nominal design ($Y = 0.25$), and (**b**) the optimized design ($Y = 0.46$). *Gray lines* correspond to 500 EM-simulated random samples for Monte Carlo analysis, *circles* represent corresponding feature points calculated using the approximation model (12.17)–(12.22)

12.7 Summary and Discussion

This chapter described several physics-based surrogate modeling methodologies and demonstrated their applications to complex engineering systems from various disciplines. It is evident that these physics-based modeling approaches can be reliable, but at the same time highly efficient in computational terms when compared with other surrogate-based modeling techniques, in particular data-driven ones. As a result, physics-based modeling approaches can be good alternatives for certain applications when dealing with modeling problems involving computationally expensive simulations. On the other hand, one should recognize that these modeling approaches are not applicable to all problems or situations, and have certain limitations. For example, methods using space mapping are essentially regression models with limited number of degrees of freedom, which restricts their application to certain types of responses. Methods such as the shape-preserving response prediction technique and feature-based modeling require certain consistency of the low- and high-fidelity model response shapes across the design space so that the sets of characteristic points extracted from these responses are in one-to-one correspondence. In some cases, this limitation is not fundamental, and can be overcome by careful selection of the characteristic points. Physics-based methods, therefore, often require added effort to set up when compared with general-purpose methods. Nevertheless, physics-based surrogate modeling methods, while not being designed for general-purpose applications, can lead to dramatic cost reduction of a simulation-driven design process when used carefully and within their limitations.

Chapter 13
Summary and Discussion

Efficient handling of expensive simulation models, especially in the context of numerical optimization, is of primary importance for contemporary engineering and science. On one hand, simulation-driven design has become ubiquitous, and in many cases indispensable, in numerous engineering disciplines. On the other hand, utilization of computational models faces considerable challenges due to their high evaluation cost. This is especially pronounced for complex systems involving multi-scale and multi-physics models, and only to some extent alleviated by the continuous development of computational resources, both in terms of hardware and software. Clearly, solving design tasks that require a large number of simulations, such as parametric optimization, statistical analysis, or uncertainty quantification, is hindered if the model evaluation time is of concern.

Among various techniques proposed over the years to speed up computational cost of simulation-driven design processes, surrogate-assisted methods seem to be the most promising ones, both in terms of their versatility and potentially high cost reduction rate. Earlier in this book, we briefly outlined conventional optimization methods (Chap. 3) as well as provided introduction to surrogate modeling and surrogate-assisted optimization (Chap. 4). The major topic of this book is surrogate-based optimization by means of response-corrected physics-based models, where the surrogate is constructed by appropriate enhancement of the underlying low-fidelity model (Chap. 5). The latter, however, is often simulation based and it is set up by limiting its evaluation cost at the expense of reduced accuracy.

According to the classification adopted in this book, the main line of distinction between various response correction techniques is whether the low-fidelity model correction is realized using explicit analytical formulas (parametric methods, Chap. 6) or implicitly, by direct exploitation of the system-specific knowledge embedded in the low-fidelity model (non-parametric methods, Chap. 7). As pointed out in Chap. 6, the parametric response correction methods are generic and easy to implement with the surrogate model coefficients computed either directly or by solving appropriate (usually linear) regression problems. On the other hand, because the model parameters are determined for the specific (one or more,

S. Koziel, L. Leifsson, *Simulation-Driven Design by Knowledge-Based Response Correction Techniques*, DOI 10.1007/978-3-319-30115-0_13

depending on a particular method) reference designs, the generalization capability of the respective surrogates may be limited. This is especially the case for models with highly nonlinear responses that change considerably across the design space.

Non-parametric techniques, such as the adaptive response correction (ARC) or shape-preserving response prediction (SPRP), enable better exploitation of the knowledge embedded in the underlying low-fidelity models which typically translates into improved computational efficiency. However, this comes with certain limitations. For example, SPRP (cf. Chap. 7) requires consistency of sets of so-called characteristic points across the design space, which imposes rather strict requirements concerning the shapes of both the low- and high-fidelity model responses. For the same reasons, methods such as SPRP are preferable for design cases where the response shape of the computational models involved is well defined (e.g., narrow-band antennas) or where the ranges of the design parameters are restricted so that the aforementioned consistency assumptions hold (e.g., statistical analysis and uncertainty quantification of microwave filters).

In the book, we also discussed a few methods that are not strictly of the response correction type; however, they are closely related to the non-parametric methods in the sense of a comprehensive exploitation of the problem-specific knowledge embedded in the low-fidelity model. These include adaptively adjusted design specifications (AADS) (Chap. 8) and feature-based optimization (Chap. 9). Both methods are rather specialized: AADS has been designed for handling problems with minimax type of specifications, whereas feature-based optimization is suitable for handling design tasks where the model responses exhibit well-defined response features that can be utilized to determine the solution quality. As demonstrated in Chap. 9, an appropriate reformulation of the optimization problem in terms of the response features allows for considerable reduction of its complexity and, consequently, leads to significant optimization of cost savings.

A majority of the response correction techniques can be enhanced by exploiting derivative information (if available). In Chap. 10, we demonstrated this for several methods, including space mapping, manifold mapping, as well as SPRP. Utilizing response gradients allows for improving all characteristics of the respective method such as their convergence properties as well as reduces the cost of particular stages of the optimization process, e.g., extraction of space mapping parameters, and surrogate model optimization. Clearly, computational benefits are only possible if the derivatives can be evaluated at a low cost (in particular, through adjoints).

Although a majority of the application examples considered in this book refer to single-objective optimization problems, the utilization of response correction methods for handling multi-objective design tasks—within a surrogate-based optimization framework—has also been demonstrated (in Chap. 11). An important stage of the algorithms presented in that part of the book was the refinement of the initial set of Pareto optimal solutions generated using a fast surrogate (typically, a response surface approximation). This is where response correction methods turn out to be helpful, specifically, to create a link between the lower fidelity representation of the system at hand and its high-fidelity model.

Finally, in Chap. 12, the utilization of response correction methods for surrogate modeling purposes has been discussed. As opposed to parametric optimization, where the local alignment between the surrogate and the high-fidelity model is of major concern, the techniques discussed in Chap. 12 aim at constructing models that are accurate in a larger portion of the design space so that—once created—they can be reused multiple times and utilized for handling various design tasks such as optimization or statistical analysis.

Based on the considerations and application examples presented in this book, the following characteristics of response correction techniques can be outlined: simplicity, versatility, and ease of implementation. A certain hierarchy of response correction methods can also be observed. On one hand, we have very simple methods such as additive or multiplicative output space mapping, where the model correction is realized by aligning the models at a single point (typically, the most recent design encountered during the optimization run). Generalization capability of the respective surrogate models is therefore limited, particularly for problems and systems that exhibit highly nonlinear responses. More involved methods utilize multiple designs for model alignment (manifold mapping, multipoint response correction), which increases the surrogate model identification cost but, at the same time, improves the model generalization. Finally, non-parametric techniques such as SPRP and feature-based optimization allow for a full exploitation of the knowledge embedded in the low-fidelity model. Unfortunately, the applicability of these methods is somehow limited by relatively strong assumptions concerning the response structure of the models involved and their consistency across the design space.

Given the large range of available methods, and their various characteristics and limitations, the best choice of a response correction technique for a given problem may not be trivial and usually requires certain engineering experience and problem-specific knowledge. This is one of the many open problems related to a majority of surrogate-assisted optimization methods involving physics-based models, not just the response correction techniques. Nevertheless, the authors believe that this book can be used as a reference, and, to some extent, provide guidance for the right choice of the modeling and optimization methods. It seems that due to the increasing role of computational models in engineering design, the importance of efficient algorithms capable of solving tasks such as parametric optimization or robust (tolerance-aware) design within a reasonable time frame will also be growing. Surrogate-based optimization techniques exploiting response correction methods belong to the most promising approaches with this respect so that their further development is expected to continue for years to come.

References

Abbott, I.H., and Von Doenhoff, A.E., *Theory of Wing Sections*, Dover Publications, 1959.
Abdel-Malek, H.L., and Bandler, J.W. (1978) Yield estimation for efficient design centering assuming arbitrary statistical distributions. *Int. J. Circuit Theory Applicat.*, vol. 6, pp. 289–303.
Abdel-Malek, H.L., Hassan, A.S.O., and Bakr, M.H. (1999) A boundary gradient search technique and its application in design centering. *IEEE Trans. Comput.-Aided Design Integr. Circuits Syst.*, vol. 18, no. 11, pp. 1654–1661.
ADS (Advanced Design System) (2008). Agilent (Keysight) Technologies, Fountaingrove Parkway 1400, Santa Rosa, CA 95403–1799.
Agilent ADS (2011), Agilent Technologies, 1400 Fountaingrove Parkway, Santa Rosa, CA 95403–1799.
Alba, E., and Marti, R., (Eds) (2006) Metaheuristic procedures for training neural networks, Springer.
Alexandrov, N.M., Dennis, J.E., Lewis, R.M., Torczon, V. (1998) A trust region framework for managing use of approximation models in optimization. Struct. Multidisciplinary Optim. 15, pp. 16–23.
Alexandrov, N.M., Lewis, R.M. (2001) An overview of first-order model management for engineering optimization. Optimization and Engineering. 2, 413–430.
Allaire, D., Willcox, K. (2014) A mathematical and computational framework for multifidelity design and analysis with computer models. Int. J. Uncertainty Quantification, 4, pp. 1–20.
Allen, B., Vorus, W.S., and Prestero, T., (2000) Propulsion system performance enhancements on REMUS AUVs. In *Proceedings MTS/IEEE Oceans 2000*, Providence, Rhode Island.
Anderson, J.D. (2007) Modern Compressible Flow, 5th ed., McGraw Hill Inc., New York.
Andrés, E., Salcedo-Sanz, S., Monge, F., Pérez-Bellido, A.M., (2012) Efficient aerodynamic design through evolutionary programming and support vector regression algorithms. *Int. J. Expert Systems with Applications*. 39, 10700–10708.
Angiulli, G., Cacciola, M., Versaci, M. (2007) Microwave devices and antennas modelling by support vector regression machines. IEEE Trans. Magnetics, 43, pp. 1589–1592.
Back, T., Fogel, D.B., Michalewicz Z. (eds). (2000) Evolutionary computation 1: basic algorithms and operators. Taylor & Francis Group.
Bakr, M.H., Bandler, J.W., Biernacki, R.M., Chen, S.H., Madsen, K. (1998) A trust region aggressive space mapping algorithm for EM optimization. IEEE Trans. Microwave Theory Tech., 46, pp. 2412–2425.
Bakr, M.H., Bandler, J.W., Georgieva, N.K., Madsen, K. (1999) A hybrid aggressive space-mapping algorithm for EM optimization. IEEE Trans. Microwave Theory Tech., 47, pp. 2440–2449.

S. Koziel, L. Leifsson, *Simulation-Driven Design by Knowledge-Based Response Correction Techniques*, DOI 10.1007/978-3-319-30115-0

Bakr, M.H., Bandler, J.W., and Georgieva, N. (1999) "Modeling of microwave circuits exploiting space derivative mapping," *IEEE MTT-S Int. Microwave Symp. Dig.*, (Anaheim, CA), pp. 715–718.

Balanis, C.A. (2005) *Antenna theory: analysis and design*. Wiley-IEEE Press.

Bandler, J.W., Liu, P.C., and Tromp, H. (1976a) A nonlinear programming approach to optimal design centering, tolerancing and tuning. *IEEE Trans. Circuits and Systems*, vol. CAS-23, no. 3, pp. 155–165.

Bandler, J.W., Liu, P.C., and Tromp, H. (1976b) Integrated approach to microwave design. *IEEE Trans. Microwave Theory Techn.*, vol. MTT–24, no. 9, pp. 584–591, Sep. 1976.

Bandler, J.W., and Chen, S.H., (1988) Circuit optimization: the state of the art. *IEEE Trans. Microwave Theory Techn.*, vol. 36, no. 2, pp. 424–443.

Bandler, J.W., Biernacki, R.M., Chen, S.H., Grobelny, P.A., Hemmers, R.H., (1994) Space mapping technique for electromagnetic optimization. *IEEE Trans. Microwave Theory Tech.* 42, 2536–2544.

Bandler, J.W., Biernacki, R.M., Chen, S.H., Hemmers, R.H., Madsen, K., (1995) Electromagentic optimization exploiting aggressive space mapping. *IEEE Trans. Microwave Theory Tech.*, 43, 2874–2882.

Bandler, J.W., Mohamed, A.S., Bakr, M.H., Madsen, K., and Søndergaard, J. (2002) "EM-based optimization exploiting partial space mapping and exact sensitivities," *IEEE Trans. Microwave Theory Tech.*, vol. 50, no. 12, pp. 2741-2750.

Bandler, J.W., Cheng, Q.S., Gebre-Mariam, D.H., Madsen, K., Pedersen, F., Søndergaard, J. (2003) EM-based surrogate modeling and design exploiting implicit, frequency and output space mappings. IEEE Int. Microwave Symp. Digest, Philadelphia, PA, pp. 1003–1006.

Bandler, J.W., Cheng, Q.S., Dakroury, S.A., Mohamed, A.S., Bakr, M.H., Madsen, K., Søndergaard, J. (2004a) Space mapping: the state of the art. IEEE Trans. Microwave Theory Tech., 52, pp. 337−361.

Bandler, J.W., Cheng, Q.S., Nikolova, N.K., Ismail, M.A. (2004b) Implicit space mapping optimization exploiting preassigned parameters. IEEE Trans. Microwave Theory Tech., 52, pp. 378–385.

Bandler, J.W., Cheng, Q.S., Hailu, D.M., and Nikolova, N.K. (2004c) "A space-mapping design framework," IEEE Trans. Microwave Theory Tech., vol. 52, no. 11, pp. 2601–2610.

Bandler, J.W., Cheng, Q.S., and Koziel, S. (2006) Simplified space mapping approach to enhancement of microwave device models. *Int. J. RF and Microwave Computer-Aided Eng.*, vol. 16, pp. 518–535.

Beachkofski, B., and Grandhi, R. (2002) Improved distributed hypercube sampling. *American Institute of Aeronautics and Astronautics*. Paper AIAA 2002–1274.

Bekasiewicz, A., Koziel, S., Leifsson, L. (2014a) Low-cost EM-simulation-driven multi-fidelity optimization of antennas. In *International Conference on Computational Science – Procedia Computer Science*, 29, 769–778.

Bekasiewicz, A., Koziel, S., Zieniutycz, W. (2014b) Design Space Reduction for Expedited Multi-Objective Design Optimization of Antennas in Highly-Dimensional Spaces. In *Solving Computationally Extensive Engineering Problems: Methods and Applications*, Springer.

Biernacki, R., Chen, S., Estep, G., Rousset, J., Sifri, J. (2012) Statistical analysis and yield optimization in practical RF and microwave systems. *IEEE MTT-S Int. Microw. Symp. Dig.*, Montreal, pp. 1–3.

Bischof, C., Bücker, H.M., Hovland, P.D., Naumann, U., and Utke, J., (Eds.) (2008) *Advances in Automatic Differentiation, Lecture Notes in Computational Science and Engineering*, Springer.

Bogaerts, W., De Heyn, P., Van Vaerenbergh, T., De Vos, K., Selvaraja, S.K., Claes, T., Dumon, P., Bienstman, P., Van Thourhout, D., Baets, R. (2012) Silicon microring resonators. Laser and Photonics Reviews, 6, pp. 47–73.

Booker, A.J., Dennis, J.E., Frank, P.D., Serafini, D.B., Torczon, V., Trosset, M.W. (1999) A rigorous framework for optimization of expensive functions by surrogates. Structural Optimization. 17, 1–13.

Broyden, C.G. (1965) A class of methods for solving nonlinear simultaneous equations. Math. Comp., 19, pp. 577–593.

Ceperic, V., Baric, A. (2004) Modeling of analog circuits by using support vector regression machines. *Proc. 11th Int. Conf. Electronics, Circuits, Syst.*, Tel-Aviv, Israel, pp. 391–394.

Chakraborty, U., (2008) *Advances in Differential Evolution*. Studies in Computational Intelligence, Springer.

Chen, Z.N. (2008) Wideband microstrip antennas with sandwich substrate. *IET Microw. Ant. Prop.*, vol. 2, pp. 538–546.

Chen, C.A., Cheng, D.K. (1975) Optimum element lengths for Yagi-Uda arrays. *IEEE Trans. Antennas Propag.* 23, 8–15.

Cheng, Q.S., Bandler, J.W., Koziel, S., Bakr, M.H., and Ogurtsov, S. (2010) The state of the art of microwave CAD: EM-based optimization and modeling. *Int. J. RF and Microwave Computer-Aided Eng.*, vol. 20, no. 5, pp. 475–491.

Cheng, D.K., Chen, C.A. (1973) Optimum element spacings for Yagi-Uda arrays. *IEEE Trans. Antennas Propag.* 21, 615–623.

Cheng, Q.S., Bandler, J.W., and Koziel, S. (2008) Combining coarse and fine models for optimal design. *Microwave Magazine*, 9, 79–88.

Cheng, Q.S., Bandler, J.W., and Koziel, S. (2008b) An accurate microstrip hairpin filter design using implicit space mapping. *Microwave Magazine*, vol. 9, no. 1, pp. 79–88.

Cheng, Q.S., Koziel, S., and Bandler, J.W.(2006) Simplified space mapping approach to enhancement of microwave device models. *Int. J. RF and Microwave Computer-Aided Eng.*, 16, 518–535.

Cheng, Q.S., Bandler, J.W., Nikolova, N.K., and Koziel, S. (2011) "Fast space mapping modeling with adjoint sensitivity," *IEEE MTT-S Int. Microwave Symp. Dig.* (Baltimore, MD).

Cheng, Q.S., Bandler, J.W., Nikolova, N.K., and Koziel, S. (2012) "A space mapping schematic for fast EM-based modeling and design," *IEEE MTT-S Int. Microwave Symp. Dig.* (Montreal, QC).

Clerc, M., and Kennedy, J., (2002) The particle swarm – explosion, stability, and convergence in a multidimensional complex space. *IEEE Trans. Evolutionary Computation*, 6, 58–73.

Conn, A.R., Scheinberg, K., Vicente, L.N. (2009) Introduction to Derivative-Free Optimization. MPS-SIAM Series on Optimization.

Conn, A.R., Gould, N.I.M., Toint, P.L. (2000) Trust Region Methods. MPS-SIAM Series on Optimization.

Couckuyt, I., Forrester, A., Gorissen, D., De Turck, F., Dhaene, T. (2012) Blind Kriging: Implementation and performance analysis. Advances in Engineering Software, 49, pp. 1–13.

Couckuyt, I., (2013) Forward and inverse surrogate modeling of computationally expensive problems. *PhD Thesis*, Ghent University.

CST. (2011). CST Microwave Studio, CST AG, Bad Nauheimer Str. 19, D-64289 Darmstadt, Germany.

CST Microwave Studio (2011). CST AG, Bad Nauheimer Str. 19, D-64289 Darmstadt, Germany.

Dalton, C., and Zedan, M.F., (1980) Design of low-drag axisymmetric shapes by the inverse method," *J. Hydronautics*, 15, 48–54.

Deb, K. (2001) Multi-Objective Optimization Using Evolutionary Algorithms. John Wiley & Sons.

De Heyn, P., Kuyken, B., Vermeulen, D., Bogaerts, W., and Van Thourhout, D. (2010) Improved intrinsic Q of silicon-on0insulator microrong resonators using TM-polarized light. *Proc. IEEE Photonics Benelux Chapter Symposium*, 197–200, Delft, Netherlands

Deslandes, D. (2010) Design equations for tapered microstrip-to-substrate integrated waveguide transitions. *IEEE MTT-S Int. Microwave Symp. Dig.*, 704–707.

Deslandes, D., Wu, K. (2001) Integrated microstrip and rectangular waveguide in planar form. *IEEE Microw. Wireless Compon. Lett.*, 11, 68–70.

Deslandes, D., Wu, K. (2005) Analysis and design of current probe transition from grounded coplanar to substrate integrated rectangular waveguides. *IEEE Trans. Microw. Theory Tech.*, 53, 2487-2494.

Devabhaktuni, V.K., Yagoub, M.C.E., and Zhang, Q.J., (2001) A robust algorithm for automatic development of neural-network models for microwave applications. *IEEE Trans. Microwave Theory Tech.*, 49, 2282-2291.

Dielectric resonator filter, Examples (2011) CST Microwave Studio, ver. 2011, CST AG, Bad Nauheimer Str. 19, D-64289 Darmstadt, Germany.

Dorigo M., Gambardella, L.M. (1997) Ant colony system: a cooperative learning approach to the traveling salesman problem. IEEE Transactions on Evolutionary Computation, 1, pp. 53–66.

Dorigo, M., and Stutzle, T., (2004) *Ant colony optimization*. MIT Press, Cambridge.

Director, S.W., Rohrer, R.A. (1969) The generalized adjoint network and network sensitivities. IEEE Trans. Circuit Theory, 16, pp. 318–323.

Echeverria, D., Hemker, P.W. (2005) Space mapping and defect correction. CMAM Int. Mathematical Journal Computational Methods in Applied Mathematics, 5, pp. 107–136.

Echeverría, D., Hemker, P.W. (2008) Manifold mapping: a two-level optimization technique. Computing and Visualization in Science. 11, 193–206.

El Sabbagh, M.A., Bakr, M.H., Nikolova, N.K., (2006) Sensitivity analysis of the scattering parameters of microwave filters using the adjoint network method. Int. J. RF and Microwave Computer-Aided Eng., 16, pp. 596–606.

em™ Version 12.54 (2010), Sonnet Software, Inc., 100 Elwood Davis Road, North Syracuse, NY 13212, USA.

Epstein, B., and Peigin, S., "Optimization of 3D wings based on Navier-Stokes solutions and genetic algorithms," International Journal of Computational Fluid Dynamics, 20, 2006, pp. 75–92.

FEKO (2008), Suite 5.4, EM Software & Systems-S.A. (Pty) Ltd, 32 Techno Lane, Technopark, Stellenbosch, 7600, South Africa.

FLUENT, ver. 14.0, ANSYS Inc., Southpointe, 275 Technology Drive, Canonsburg, PA 15317, 2012.

FLUENT, ver. 15.0, ANSYS Inc., Southpointe, 275 Technology Drive, Canonsburg, PA 15317, 2015.

Ferranti, F., Deschrijver, D., Knockaert, L., and Dhaene, T. (2009) Hybrid algorithm for compact and stable macromodelling of parameterized frequency responses. *IEEE Electron. Lett.*, vol. 45, no. 10, pp. 493–495.

Ferranti, F., Knockaert, L., and Dhaene, T. (2011) Passivity-preserving parametric macromodelling by means of scaled and shifted state-space systems. *IEEE Trans. Microwave Theory Techn.*, vol. 59, no. 10, pp. 2394-2403.

Fonseca, C.M. (1995) Multiobjective genetic algorithms with application to control engineering problems. PhD thesis, Department of Automatic Control and Systems Engineering, University of Sheffield, UK.

Forrester, A.I.J., Sóbester, A., and Keane, A.J. (2007) Multi-Fidelity Optimization via Surrogate Modeling. *Proceedings of the Royal Society A*. 463, 3251-3269.

Forrester, A.I.J., Keane, A.J. (2009) Recent advances in surrogate-based optimization, Prog. in Aerospace Sciences, 45, pp. 50–79.

Geem, Z.W., Kim, J.H., and Loganathan, G.V., (2001) A new heuristic optimization algorithm: harmony search. *Simulation*, 76, 60–68.

Geisser, S. (1993) Predictive Inference. Chapman and Hall.

Giunta, A.A., (1997) Aircraft multidisciplinary design optimization using design of experiments theory and response surface modeling methods. *PhD Thesis*, Virginia Polytechnic Institute and State University.

Giunta, A.A., Eldred, M.S., (2000) Implementation of a trust region model management strategy in the DAKOTA optimization toolkit. *Proc. AIAA/USAF/NASA/ISSMO Symp. Multidisciplinary Analysis and Optimization*, Long Beach, CA, AIAA-2000-4935.

Giunta, A.A., Wojtkiewicz, S.F., Eldred, M.S. (2003) Overview of modern design of experiments methods for computational simulations. American Institute of Aeronautics and Astronautics, paper AIAA 2003—0649.

Glubokov, O., and Koziel, S. (2014a) Substrate integrated waveguide microwave filter tuning through variable-fidelity feature space optimization. *International Review of Progress in Applied Computational Electromagnetics*.

Glubokov, O., and Koziel, S. (2014b) EM-driven tuning of substrate integrated waveguide filters exploiting feature-space surrogates. *IEEE Int. Microwave Symp.*

Glubokov, O., Koziel, S. (2014c) Automated inverse design of bandpass filters with invariable layout through linear approximation of physical dimensions. *European Microwave Conference*, 2014.

Goldberg, D.E., (1989) Genetic algorithms in search, optimization & machine learning. Pearson Education.

Golub, G.H., Van Loan, Ch.F. (1996) Matrix Computations. (3rd ed.), The Johns Hopkins University Press.

Gorissen, D., Crombecq, K., Couckuyt, I., Dhaene, T., Demeester, P. (2010) A surrogate modeling and adaptive sampling toolbox for computer based design. Journal of Machine Learning Research, 11, 2051–2055.

Griewank, A., (2000) *Evaluating Derivatives: Principles and Techniques of Algorithmic Differentiation*. Society for Industrial and Applied Mathematics (SIAM), Philadelphia.

Guan, X., Ma, Z., Cai, P., Anada, T., and Hagiwara, G. (2008) A microstrip dual-band bandpass filter with reduced size and improved stopband characteristics. Microwave and Opt. Tech. Lett., 50, pp. 618–620.

Gunn, S.R. (1998) Support vector machines for classification and regression. Technical Report. School of Electronics and Computer Science, University of Southampton.

Han, S.P., (1977) A globally convergent method for nonlinear programming. *J. Optim. Theory Appl.*, 22, 297–309.

Haraldsson, H.O. (1996) Fluid Dynamics Simulation of Fishing Gear. M.Sc. Thesis. University of Iceland.

Haykin, S. (1998) Neural Networks: A Comprehensive Foundation. 2nd ed. Prentice Hall.

Hauth, W., Keller, R., Papziner, U., Ihmels, R., Sieverding, T., and Arndt, F. (1993) Rigorous CAD of multipost coupled rectangular waveguide components. *Proc. 23rd European Microw. Conf.*, Madrid, Spain, pp. 611–614.

HFSS (2010) Release 13.0, ANSYS, http://www.ansoft.com/products/hf/hfss/.

Hicks, R.M., and Henne, P.A. (1978) *Wing Design by Numerical Optimization*. Journal of Aircraft, Vol. 15, No. 7, pp. 407–412.

Holst, T. L., and Pulliam, T. H., "Transonic wing shape optimization using a genetic algorithm," Fluid Mechanics and its Applications, 73, 2003, pp. 245–252.

Hosder, S., Watson, L.T., Grossman, B., Mason, W.H., Kim, H., (2001) Polynomial response surface approximations for the multidisciplinary design optimization of a high speed civil transport. *Optimization and Engineering*, 2, 431–452.

Hosder, S., Schetz, J.A., Mason, W.H., Grossman, B., and Haftka, R.T. (2010) Computational-Fluid-Dynamics-Based Clean-Wing Aerodynamic Noise Model for Design. *Journal of Aircraft*, 47, 754–762.

Hosder, S. (2012) Stochastic Response Surfaces Based On Non-Intrusive Polynomial Chaos for Uncertainty Quantification. *International Journal of Mathematical Modeling and Numerical Optimization*, Volume 3, No. 1/2, pp. 117–139.

Huang, C.L., Chen, Y.B., and Tasi, C.F. (2008) New compact microstrip stacked slotted resonators bandpass filter with transmission zeros using high-permittivity ceramics substrate. *Microwave Opt. Tech. Lett.*, vol. 50, no. 5, pp. 1377-1379.

Huang, L., Gao, Z., (2012) Wing-body optimization based on multi-fidelity surrogate model. *28th Int. Congress of the Aeronautical Sciences*, Brisbane, Australia.

Hsieh, L.H., and Chang, K. (2003) Tunable microstrip bandpass filters with two transmission zeros. *IEEE Trans. Microwave Theory Tech.*, vol. 51, no. 2, pp. 520–525.

Hsieh, M.Y., and Wang, S.M. (2005) Compact and wideband microstrip bandstop filter. *IEEE Microwave and Wireless Component Letters*, vol. 15, no. 7, pp. 472–474.

ICEM CFD, ver. 14.0, ANSYS Inc., Southpointe, 275 Technology Drive, Canonsburg, PA 15317, 2012.

ICEM CFD, ver. 14, ANSYS Inc., Southpointe, 275 Technology Drive, Canonsburg, PA 15317, 2013.

Jacobs, J.H., Etman, L.F.P., van Keulen, F., Rooda, J.E., (2004) Framework for sequential approximate optimization. *Struct. Multidisc. Optimization*, 27, 384–400.

Jacobs, J.P. (2012) Ba200sian support vector regression with automatic relevance determination kernel for modeling of antenna input characteristics. IEEE Trans. Antennas. Prop., 60, pp. 2114-2118.

Jameson, A., (1988). Aerodynamic design via control theory. *Journal of Scientific Computing*, 3, 233–260.

Jin, Y. (2011) Surrogate-assisted evolutionary computation: recent advances and future challenges. Swarm and Evolutionary Computation, 1, pp. 61–70.

Jones, D., Schonlau, M., Welch, W. (1998) Efficient global optimization of expensive black-box functions. Journal of Global Optimization. 13, pp. 455–492.

Jonsson, E., Hermannsson, E., Juliusson, M., Koziel, S., and Leifsson, L. (2013) Computational Fluid Dynamic Analysis and Shape Optimization of Trawl-Doors. 51st AIAA Aerospace Sciences Meeting including the New Horizons Forum and Aerospace Exposition, Grapevine, Texas, January 7–10.

Jonsson, E., Leifsson, L., and Koziel, S. (2013) Aerodynamic Optimization of Wings by Space Mapping. 51st AIAA Aerospace Sciences Meeting including the New Horizons Forum and Aerospace Exposition, Grapevine,Texas, January 7–10.

Journel, A.G., Huijbregts, Ch.J. (1981) Mining Geostatistics. Academic Press.

Kabir, H., Wang, Y., Yu, M., Zhang, Q.J. (2008). Neural network inverse modeling and applications to microwave filter design. IEEE Trans. Microwave Theory Tech., 56, pp. 867–879.

Khalatpour, A., Amineh, R.K., Cheng, Q.S., Bakr, M.H., Nikolova, N.K., and Bandler, J.W. (2011) "Accelerating input space mapping optimization with adjoint sensitivities," *IEEE Microw. Wireless Comp. Lett.*, vol. 21, no. 6, pp. 280–282.

Kennedy, J. (1997) The particle swarm: social adaptation of knowledge. Proc. 1997 Int. Conf. Evolutionary Computation, Indianapolis, IN, pp. 303−308.

Kennedy, J., Eberhart, R.C., Shi, Y. (2001) *Swarm intelligence*. Academic Press.

Kinsey, D. W., and Barth, T. J. (1984) Description of a Hyperbolic Grid Generation Procedure for Arbitrary Two-Dimensional Bodies," AFWAL TM 84–191-FIMM.

Kirkpatrick, S., Gelatt, C.D., and Vecchi, M.P., (1983) Optimization by simulated annealing. *Science*, 220, 671–680.

Kleijnen, J., (2008) *Design and Analysis of Simulation Experiments. Springer*.

Kleijnen, J.P.C. (2009) Kriging metamodeling in simulation: a review. *European Journal of Operational Research*. 192, 707–716.

Koehler, J.R., Owen, A.B. (1996) Computer experiments. In S. Ghosh and C. R. Rao (Eds.) Handbook of Statistics. Elsevier Science B.V. 13, pp. 261—308.

Kolda, T.G., Lewis, R.M., Torczon, V. (2003). Optimization by direct search: new perspectives on some classical and modern methods. SIAM Rev., 45, pp. 385−482.

Koziel, S., Bandler, J.W., Mohamed, A.S., and Madsen, K. (2005) Enhanced surrogate models for statistical design exploiting space mapping technology. *IEEE MTT-S Int. Microwave Symp. Dig.*, Long Beach, CA, pp. 1609-1612.

Koziel, S., and Bandler, J.W. (2006) Space-mapping-based modeling utilizing parameter extraction with variable weight coefficients and a data base. *IEEE MTT-S Int. Microwave Symp. Dig.*, San Francisco, CA, pp. 1763-1766.

Koziel, S., Bandler, J.W., Madsen, K. (2006) A space mapping framework for engineering optimization: theory and implementation. *IEEE Trans. Microwave Theory Tech.*, 54. 3721-3730.

Koziel, S., Bandler, J.W., Madsen, K. (2006b) Theoretical justification of space-mapping-based modeling utilizing a data base and on-demand parameter extraction. *IEEE Trans. Microwave Theory Tech.*, vol. 54, no. 12, pp. 4316-4322.

Koziel, S., Bandler, J.W. (2007a). Coarse and surrogate model assessment for engineering design optimization with space mapping. IEEE MTT-S Int. Microwave Symp. Dig, Honolulu, HI, pp. 107–110.

Koziel, S., and Bandler, J.W. (2007b) Space-mapping optimization with adaptive surrogate model. IEEE Trans. Microwave Theory Tech., 55, pp. 541–547.

Koziel, S., and Bandler, J.W. (2007c) Microwave device modeling using space-mapping and radial basis functions. *IEEE MTT-S Int. Microwave Symp. Dig.*, Honolulu, Hawaii, pp. 799–802.

Koziel, S., and Bandler, J.W. (2007d) A space-mapping approach to microwave device modeling exploiting fuzzy systems. *IEEE Trans. Microwave Theory Tech.*, vol. 55, no. 12, pp. 2539–2547.

Koziel, S., and Bandler, J.W. (2012) Accurate modeling of microwave devices using kriging-corrected space mapping. International Journal of Numerical Modelling: Electronic Networks, Devices and Fields, vol 25, issue 1, pp. 1–4.

Koziel, S., and Bandler, J.W. (2015) Rapid yield estimation and optimization of microwave structures exploiting feature-based statistical analysis," *IEEE Trans. Microwave Theory Tech.*, 63, pp. 107–114.

Koziel, S., Cheng, Q.S., Bandler, J.W. (2008) Space mapping. IEEE Microwave Magazine, 9, pp. 105–122.

Koziel, S., Bandler, J.W., Madsen, K. (2008b) Quality assessment of coarse models and surrogates for space mapping optimization. Optimization Eng. 9, 375–391.

Koziel, S., Bandler, J.W., Madsen, K. (2009a) Space mapping with adaptive response correction for microwave design optimization. IEEE Trans. Microwave Theory Tech., 57, pp. 478–486.

Koziel, S. (2010a) Shape-preserving response prediction for microwave design optimization. IEEE Trans. Microwave Theory and Tech., 58, pp. 2829–2837.

Koziel, S. (2010b) Adaptively adjusted design specifications for efficient optimization of microwave structures. *Progress in Electromagnetic Research B (PIER B)*, 21, pp. 219–234.

Koziel, S., (2010c) Multi-fidelity multi-grid design optimization of planar microwave structures with Sonnet. *International Review of Progress in Applied Computational Electromagnetics*, Tampere, Finland, 719–724.

Koziel, S. (2010d) Improved microwave circuit design using multipoint-response-correction space mapping and trust regions. *Int. Symp. Antenna Technology and Applied Electromagnetics, ANTEM 2010*, Ottawa, Canada.

Koziel, S. (2010e) Shape-preserving response prediction for microwave circuit modeling. *IEEE MTT-S Int. Microwave Symp. Dig.*, Anaheim, CA, pp. 1660–1663.

Koziel, S. (2010f) Efficient optimization of microwave structures through design specifications adaptation. *IEEE Int. Symp. Antennas Prop.*, Toronto, Canada.

Koziel, S. (2010g) Computationally efficient multi-fidelity multi-grid design optimization of microwave structures. *Applied Computational Electromagnetics Society Journal*, 25, 578–586.

Koziel, S. (2011) Reliable design optimization of microwave structures using multipoint-response-correction space mapping and trust regions. *Int. J. RF and Microwave CAE*, 21, pp. 534–542.

Koziel, S. (2011b) Low-cost modeling of microwave structures using shape-preserving response prediction. *IEEE MTT-S Int. Microwave Symp. Dig.*, Baltimore, MD.

Koziel, S., (2012) Accurate low-cost microwave component models using shape-preserving response prediction. *Int. J. Numerical Modelling: Electronic Devices and Fields*, vol. 25, no. 2, pp. 152–162.

Koziel, S. (2014) "Fast simulation-driven antenna design using response-feature surrogates. *Int. J. RF & Microwave CAE.*

Koziel, S., Bandler, J.W., Cheng, Q.S. (2010a) Robust trust-region space-mapping algorithms for microwave design optimization. IEEE Trans. Microwave Theory and Tech., 58, pp. 2166–2174.

Koziel, S., Cheng, Q.S., Bandler, J.W. (2010b) Implicit space mapping with adaptive selection of preassigned parameters. IET Microwaves, Antennas & Propagation, 4, pp. 361–373.

Koziel, S., Bandler, J.W., Cheng, Q.S. (2010c) Adaptively constrained parameter extraction for robust space mapping optimization of microwave circuits. *IEEE MTT-S Int. Microwave Symp. Dig.*, 205–208.

Koziel, S., Bandler, J.W., and Cheng, Q.S., (2012) "Robust space mapping optimization exploiting EM-based models with adjoint sensitivities," *IEEE MTT-S Int. Microwave Symp. Dig.*, Montreal, QC.

Koziel, S., Ogurtsov, S., and Jacobs, J.P., (2015), "Rapid simulation-based design of covered planar microstrip antenna arrays by means of radiation response surrogates," Loughborough Ant. Prop. Conf., Loughborough, UK, 2–3 Nov. 2015, pp. 1–4.

Koziel, S., and Bekasiewicz, A. (2015) Fast simulation-driven feature-based design optimization of compact dual-band microstrip branch-line coupler. *Int. J. RF and Microwave CAE.*

Koziel, S., Echeverría-Ciaurri, D., Leifsson, L. (2011) Surrogate-based methods, in S. Koziel and X.S. Yang (Eds.) Computational Optimization, Methods and Algorithms, Series: Studies in Computational Intelligence, Springer-Verlag, pp. 33–60.

Koziel, S., and Echeverría Ciaurri, D. (2010) Reliable simulation-driven design optimization of microwave structures using manifold mapping. *Progress in Electromagnetic Research B (PIER B)*, 26, pp. 361–382.

Koziel, S., Yang, X.S., Zhang, Q.J., (Eds.) (2013a) Simulation-driven design optimization and modeling for microwave engineering. Imperial College Press.

Koziel, S., Ogurtsov, S. (2010) Computationally Efficient Simulation-Driven Design of a Printed 2.45 GHz Yagi Antenna. *Microwave and Optical Technology Letters*. 52, 1807–1810.

Koziel, S., Ogurtsov, S., Couckuyt, I., Dhaene, T. (2013b) Variable-fidelity electromagnetic simulations and co-kriging for accurate modeling of antennas. IEEE Trans. Antennas Prop., 61, pp. 1301–1308.

Koziel, S., Ogurtsov, S, Bandler, J.W., and Cheng, Q.S. (2013c) Reliable space mapping optimization integrated with EM-based adjoint sensitivities. IEEE Trans. Microwave Theory Tech., 61, pp. 3493–3502.

Koziel, S., Ogurtsov, S., Cheng, Q.S., and Bandler, J.W. (2014) "Rapid EM-based microwave design optimization exploiting shape-preserving response prediction and adjoint sensitivities," *IET Microwaves, Ant. Prop.*, vol. 8, issue 10, pp. 775–781.

Koziel, S., and Leifsson, L. (2012a) Multi-fidelity airfoil shape optimization with adaptive response prediction.*AIAA/ISSMO Multidisciplinary Analysis and Optimization Conf.*

Koziel, S., and Leifsson, L. (2012b) Knowledge-based airfoil shape optimization using space mapping. *AIAA Applied Aerodynamics Conference.*

Koziel, S., and Leifsson, L. (2012c) Adaptive Response Correction for Surrogate-Based Airfoil Shape Optimization. *30th AIAA Applied Aerodynamics Conference*, New Orleans, Louisiana, June 25–28.

Koziel, S., and Leifsson, L. (2012d) Generalized shape-preserving response prediction for accurate modeling of microwave structures. *IET Microwaves, Ant. Prop.*, vol. 6, pp. 1332–1339.

Koziel, S., and Leifsson, L. (2012e) Response correction techniques for surrogate-based design optimization of microwave structures. *Int. J. RF and Microwave CAE*, 22, 211–223.

Koziel, S., and Leifsson, L. (Eds.) (2013a) Surrogate-Based Modeling and Optimization. Applications in Engineering. Springer.

Koziel, S., and Leifsson, L. (2013b) Multi-level Airfoil Shape Optimization with Automated Low-fidelity Model Selection. *Int. Conf. Comp. Science*, Barcelona, Spain, June 5–7.

Koziel, S., and Leifsson, L. (2013c) Surrogate-Based Aerodynamic Shape Optimization by Variable-Resolution Models. *AIAA Journal*, vol. 51, no. 1, pp. 94–106.

Koziel, S., and Leifsson, L., and Ogurtsov, S. (2013d) "Reliable EM-driven microwave design optimization using manifold mapping and adjoint sensitivity," *Microwave and Optical Technology Letters*, vol. 55, no. 4, pp. 809–813.

Koziel, S., Ogurtsov, S. (2011a) Simulation-driven design in microwave engineering: application case studies, in Yang, X.S., and Koziel, S. (eds), Computational Optimization and Applications in Engineering and Industry, Series: Studies in Computational Intelligence, Springer-Verlag.

Koziel, S., Ogurtsov, S. (2011b) Rapid design optimization of antennas using space mapping and response surface approximation models. Int. J. RF & Microwave CAE, 21, pp. 611–621.

Koziel, S., and Ogurtsov, S. (2011c) Bandwidth enhanced design of dielectric resonator antennas using surrogate-based optimization. *IEEE Int. Symp. Antennas Prop.*, Spokane, WA, July 3–8.

Koziel, S., Ogurtsov, S. (2011d) Bandwidth enhanced design of dielectric resonator antennas using surrogate-based optimization. *IEEE Int. Symp. Antennas Prop.*, Spokane, WA, July 3–8.

Koziel, S., Ogurtsov, S. (2012a) Model management for cost-efficient surrogate-based optimization of antennas using variable-fidelity electromagnetic simulations. IET Microwaves Ant. Prop., 6, pp. 1643–1650.

Koziel, S., and Ogurtsov, S. (2012b) Fast simulation-driven design of microwave structures using improved variable-fidelity optimization technique. Engineering Optimization, 44, pp. 1007–1019.

Koziel, S., and Ogurtsov, S. (2013a) Multi-level microwave design optimization with automated model fidelity adjustment. Int. J. RF and Microwave CAE.

Koziel, S. and Ogurtsov, S. (2013b) Low-cost design of SIW antennas using surrogate-based optimization. IEEE APWC.

Koziel, S., and Ogurtsov, S. (2013c) Multi-objective design of antennas using variable-fidelity simulations and surrogate models. *IEEE Trans. Antennas Prop.*, 61, 5931–5939.

Koziel, S., Ogurtsov, S., Bandler, J.W., and Cheng, Q.S. (2013d) "Reliable space-mapping optimization integrated with EM-based adjoint sensitivities," IEEE Trans. Microwave Theory Tech., vol. 61, no. 10, pp. 3493–3502.

Koziel, S., Yang, X.-S., and Zhang, Q.J. (Eds.) (2013e) *Simulation-driven Design Optimization and Modeling for Microwave Engineering*. London, UK: Imperial College Press.

Koziel, S., and Bandler, J.W. (2014) Rapid Yield Estimation and Optimization of Microwave Structures Exploiting Feature-Based Statistical Analysis. *IEEE Trans. on Microwave Theory and Techniques*, vol. 63, issue 1, pp.107–114.

Koziel, S., Ogurtsov, S. (2014a) Antenna design by simulation-driven optimization. Surrogate-based approach. Springer.

Koziel, S., Ogurtsov, S. (2014b) Fast simulation-driven design of integrated photonic components using surrogate models. IET Microwaves, Antennas Prop.

Koziel, S., and Kurgan, P. (2015) Rapid design of miniaturized branch-line couplers through concurrent cell optimization and surrogate-assisted fine-tuning. IET Microwaves Ant. Prop.

Koziel, S., and Szczepanski, S. (2011) Accurate modeling of microwave structures using shape-preserving response prediction. *IET Microwaves, Antennas & Propagation*, vol. 5, pp. 1116–1122.

Koziel, S., Bekasiewicz, A., and Kurgan, P. (2015) Rapid multi-objective simulation-driven design of compact microwave circuits. *IEEE Microwave Wireless Comp. Letters*

Kuhn, H.W., Tucker, A.W. (1951). Nonlinear programming, in (J. Neyman, Ed.) Proc. Berkeley Symp. Mathematical Statistics Probability, Berkeley: University of California Press, pp 481–492.

Kuo, J. T., Chen, S. P., and Jiang, M. (2003) Parallel-coupled microstrip filters with over-coupled end stages for suppression of spurious responses. *IEEE Microwave and Wireless Comp. Lett.*, 13, 440–442.

Laurenceau, J., and Sagaut, P., (2008) Building efficient response surfaces of aerodynamic functions with kriging and cokriging. *AIAA Journal*, 46, 498–507.

Leary, S., Bhaskar, A., Keane, A. (2003) Optimal orthogonal-array-based latin hypercubes. *Journal of Applied Statistics*. 30, 585–598.

Lee, J.R., Cho, J.H., and Yun, S.W. (2000) New compact bandpass filter using microstrip $\lambda/4$ resonators with open stub inverter. *IEEE Microwave and Guided Wave Letters*, 10, 526–527.

Leifsson, L., Koziel, S., Zhang, Y., and Hosder, S. (2013) Low-Cost Robust Airfoil Optimization by Variable-Fidelity Models and Stochastic Expansions. *51st AIAA Aerospace Sciences*

Meeting including the New Horizons Forum and Aerospace Exposition, Grapevine, Texas, January 7–10.

Leifsson, L., Koziel, S., Hermannsson, E., and Fakhraie, R. (2014) Trawl-Door Design Optimization by Local Surrogate Models. 55th AIAA/ ASMe/ASCE /AHS/SC Structures, Structural Dynamics, and Materials Conference, National Harbor, Maryland, Jan. 13–17.

Leifsson, L., and Koziel, S. (2014) Variable-resolution shape optimization: low-fidelity model selection and scalability. Int. J. Mathematical Modeling and Numerical Optimization.

Leifsson, L., Koziel, S., Kurgan, P. (2014) Automated low-fidelity model setup for surrogate-based aerodynamic optimization. In S. Koziel, L. Leifsson, and X.S. Yang (Eds.) Solving Computationally Extensive Engineering Problems: Methods and Applications, Springer, pp. 87–112.

Leifsson, L., Koziel, S., Hosder, S., and Riggins, D.W. (2014b) Physics-based Multi-fidelity Surrogate Modeling with Entropy-based Availability Method. *AIAA Modeling and Simulation Technologies Conference*, National Harbor, Maryland, Jan. 13–17.

Leifsson, L., and Koziel, S. (2015) Simulation-driven aerodynamic design using variable-fidelity models. Imperial College Press.

Leifsson, L., and Koziel, S. (2015b) Variable-resolution shape optimization: Low-fidelity model selection and scalability. *Int. J. Mathematical Modeling and Numerical Optimization*, vol. 6, no. 1.

Leifsson, L., and Koziel, S. (2015c) Surrogate modeling and optimization using shape-preserving response prediction: A review. *Engineering Optimization*, May 6, pp. 1–21.

Leifsson, L., Koziel, S., and Hosder, S. (2015) Multi-Objective aeroacoustic shape optimization by variable-fidelity models and response surface surrogates. *AIAA Modeling and Simulation Technologies Conference*, Kissimee, Florida, Jan 5–9.

Leoviriakit, K., and Jameson, A. (2005) Multipoint wing planform optimization via control theory, AIAA Paper 2005–0450.

Lepine, J., Guibault, F., Trepanier, J.-Y., and Pepin, F., (2001) Optimized Nonuniform Rational B-Spline Geometrical Representation for Aerodynamic Design of Wings. *AIAA Journal*, 39, 2033–2041.

Leung, T. (2010) A Newton-Krylov Approach To Aerodynamic Shape Optimization in Three Dimensions. PhD Dissertation. Department of Aerospace Science and Engineering, University of Toronto.

Levin, D. (1998) The approximation power of moving least-squares. *Mathematics of Computation*. 67, 1517–1531.

Li, Y.F., and Lan, C.C. (1989) Development of fuzzy algorithms for servo systems. *IEEE Contr. Syst. Mag.*, vol. 9, no. 3, pp. 65–72.

Lim, D., Jin, Y., Ong, Y., Sendhoff, B. (2010) Generalizing surrogate-assisted evolutionary computation. IEEE Trans. Evol. Comp., 14, pp. 329–355.

Lin, Y.F., Chen, C.H., Chen, K.Y., Chen, H.M., and Wong, K.L. (2007) A miniature dual-mode bandpass filter using Al_2O_3 substrate," *IEEE Microw. Wireless Compon. Lett.*, vol. 17, no. 8, pp. 580–582.

Liou, M.S. (1987) "A Generalized Procedure for Constructing an Upwind-Based TVD Scheme," AIAA Paper 87–0355.

Little, B.E., Chu, S.T., Haus, H.A., Foresi, J., and Laine, J.-P. (1997) Microring resonator channel dropping filters. *IEEE J. Lightwave Tech.*, 15, 998–1005.

Liu, J., Han, Z., Song, W., (2012) Comparison of infill sampling criteria in kriging-based aerodynamic optimization. *28th Int. Congress of the Aeronautical Sciences*, Brisbane, Australia.

Liu, B., Zhang, Q., and Gielen, G. (2014) A gaussian process surrogate model assisted evolutionary algorithm for medium scale expensive black box optimization problems. IEEE Trans. Evol. Comp., 18, pp. 180–192.

Lophaven, S.N., Nielsen, H.B., and Søndergaard, J., (2002) DACE: a Matlab kriging toolbox. Technical University of Denmark.

Manchec, A., Quendo, C., Favennec, J.-F., Rius, E., and Person, C. (2006) Synthesis of capacitive-coupled dual-behavior resonator (CCDBR) filters. *IEEE Trans. Microwave Theory Tech.*, vol. 54, no. 6, pp. 2346–2355.

Marheineke, N., Pinnau, R., Reséndiz, E., (2012) Space mapping-focused control techniques for particle dispersions in fluids. *Optimization and Engineering*, 13, 101–120.

Marsden, A.L., Wang, M., Dennis, J.E., Moin, P. (2004) Optimal aeroacoustic shape design using the surrogate management framework. *Optimization and Engineering*. 5, 235–262.

Mavriplis, D. J., (2007) Discrete Adjoint-Based Approach for Optimization Problems on Three-Dimensional Unstructured Meshes. AIAA Journal, Vol. 45, No. 4.

McKay M, Conover W, Beckman R. (1979) A comparison of three methods for selecting values of input variables in the analysis of output from a computer code. *Technometrics*. 21, 239–245.

Meng, J., Xia, L. (2007) Support-vector regression model for millimeter wave transition. *Int. J. Infrared and Millimeter Waves*, 28, pp. 413–421.

Menter, F. (1994) Two-Equation Eddy-Viscosity Turbulence Models for Engineering Applications," *AIAA Journal*, 32, 1598–1605.

Minsky, M.I., and Papert, S.A. (1969) Perceptrons: An Introduction to Computational Geometry. The MIT Press.

Nielsen, H.B., (1999) Damping parameter in Marquardt's method. IMM DTU. Report IMM-REP-1999-05.

Nocedal, J., Wright, S.J. (2000) Numerical Optimization, Springer Series in Operations Research, Springer.

Ogurtsov, S., Koziel, S., Rayas-Sánchez, J.E. (2010) Design optimization of a broadband microstrip-to-SIW transition using surrogate modeling and adaptive design specifications. *International Review of Progress in Applied Computational Electromagnetics*, Tampere, Finland.

Ogurtsov, S., Koziel, S. (2011) Design of microstrip to substrate integrated waveguide transitions with enhanced bandwidth using protruding vias and EM-driven optimization. *International Review of Progress in Applied Computational Electromagnetics*, Williamsburg, VA, USA, pp. 91–96.

O'Hagan, A. (1978) Curve fitting and optimal design for predictions. *Journal of the Royal Statistical Society B*. 40, 1–42.

Palacios, F., Colonno, M.R., Aranake, A.C., Campos, A., Copeland, S.R., Economon, T.D., Lonkar, A.K., Lukaczyk, T.W., Taylor, T.W.R., Alonso, J.J. (2013) Stanford University unstructures (SU^2): an open-source integrated computational environment for multi-physics simulation and design. In 51th AIAA Aerospace Sciences Meeting, Grapevine, TX, USA.

Palmer, K., Tsui, K.-L. (2001) A minimum bias latin hypercube design. *IIE Transactions*. 33, 793–808.

Papadimitriou, D.I., and Giannakoglou, K.C., (2008) Aerodynamic shape optimization using first and second order adjoint and direct approaches. *Arch. Comput. Methods Eng.*, 15, 447–488.

Passino, K.M., and Yurkovich, S. (1998) Fuzzy Control. Addison Wesley Longman Inc., Menlo Park, CA, USA.

Pérez, V.M., Renaud, J.E., Watson, L.T., (2002) Interior point sequential approximate optimization methodology. *Proc. 9th AIAA/ISSMO Symp. Multidisciplinary Analysis and Optimization*, Atlanta, GA, AIAA-2002-5505.

Petosa, A. (2007) Dielectric Resonator Antenna Handbook, Artech House.

Pironneau, O. (1984) Optimal Shape Design for Elliptic Systems. *Springer-Verlag*, New York.

Prabhu, A. M., Tsay, A., Han, Z., and Van, V. (2009) Ultracompact SOI microring add–drop filter with wide bandwidth and wide FSR. *IEEE Photonics Tech. Lett.*, 21, 651–653.

Price K., Storn R. and Lampinen J., (2005). *Differential Evolution: A Practical Approach to Global Optimization*, Springer.

Priess, M., Koziel, S., and Slawig, T., (2011) Surrogate-based optimization of climate model parameters using response correction. *J. Comp. Science.*, 2, 335–344.

Queipo, N.V., Haftka, R.T., Shyy, W., Goel, T., Vaidynathan, R., Tucker, P.K. (2005) Surrogate-based analysis and optimization. Progress in Aerospace Sciences, 41, pp. 1–28.

Rasmussen, C.E., Williams, C.K.I. (2006) Gaussian Processes for Machine Learning. MIT Press, Cambridge, Massachussets.

Rayas-Sanchez, J.E. (2004) EM-based optimization of microwave circuits using artificial neural networks: the state-of-the-art. *IEEE Trans. Microwave Theory Tech.* 52, 420–435.

Redhe, M., Nilsson, L. (2004) Optimization of the new Saab 9–3 exposed to impact load using a space mapping technique. *Structural and Multidisciplinary Optimization.* 27, 411–420.

Robinson, T.D., Eldred, M.S., Willcox, K.E., and Haimes, R., (2008) Surrogate-Based Optimization Using Multifidelity Models with Variable Parameterization and Corrected Space Mapping. *AIAA Journal*, 46, 2316–2326.

Rojo-Alvarez, J.L., Camps-Valls, G., Martinez-Ramon, M., Soria-Olivas, E., Navia-Vazquez, A., Figueiras-Vidal, A.R. (2005) Support vector machines framework for linear signal processing. *Signal Processing*, 85, pp. 2316–2326.

Roux, W.J., Stander, N., Haftka, R.T., (1998) Response surface approximations for structural optimization. *Int. J. Numerical Methods in Engineering*, 42, 517–534.

Salleh, M.K.M., Pringent, G., Pigaglio, O., and Crampagne, R. (2008) Quarter-wavelength side-coupled ring resonator for bandpass filters. IEEE Trans. Microwave Theory Tech., 56, pp. 156–162.

Sans, M., Selga, J., Rodriguez, A., Bonache, J., Boria, V.E., Martin, F. (2014) Design of planar wideband bandpass filters from specifications using a two-step aggressive space mapping (ASM) optimization algorithm. IEEE Trans. Microwave Theory Tech., 62, pp. 3341–3350.

Santner, T.J., Williams, B., Notz, W. (2003) The Design and Analysis of Computer Experiments. Springer-Verlag.

Scotti, G., Tommasino, P., and Trifiletti, A. (2005) MMIC yield optimization by design centering and off-chip controllers. *IET Proceedings - Circuits, Devices and Systems*, vol. 152, no. 1, pp. 54–60, Feb. 2005.

Shaker, G.S.A., Bakr, M.H., Sangary, N., Safavi-Naeini, S. (2009) Accelerated antenna design methodology exploiting parameterized Cauchy models. J. Progress in Electromagnetic Research (PIER B), 18, pp. 279–309.

Simpson, T.W., Maurey, T.M., Korte, J.J., and Mistree, F. (2001) Kriging models for global approximation in simulation-based multidisciplinary design optimization. *AIAA Journal*, vol. 39, no. 12, pp. 2233–2241.

Simpson, T.W., Peplinski, J., Koch, P.N., Allen, J.K. (2001) Metamodels for computer-based engineering design: survey and recommendations. Engineering with Computers, 17, pp. 129−150.

SMA Edge Mount P.C. Board Receptacles (2013), Catalog, Applied Engineering Products, New Haven, CT, USA.

Smola, A.J., Schölkopf, B. (2004) A tutorial on support vector regression. Statistics and Computing, 14, pp. 199−222.

Sobester, A., Forrester, A.I.J. (2015) Aircraft aerodynamic design: geometry and optimization. John Wiley & Sons.

Sobieszczanski-Sobieski, J., Haftka, R.T., (1997). Multidisciplinary aerospace design optimization: survey of recent developments. *Structural Optimization*, 14, 1–23.

Soliman, E.A., Bakr, M.H., and Nikolova, N.K. (2004) "An adjoint variable method for sensitivity calculations of multiport devices," *IEEE Trans. Microwave Theory Tech.*, vol. 52, no. 2, pp. 589–599.

Søndergaard, J. (2003) Optimization using surrogate models – by the space mapping technique. *Ph.D. Thesis, Informatics and Mathematical Modelling, Technical University of Denmark*, Lyngby.

Sonnet EM (2012). Sonnet Software, Elwood Davis Road 100, North Syracuse, NY 13212.

Spence, T. G., and Werner, D. H. (2006) A novel miniature broadband/multiband antenna based on an end-loaded planar open-sleeve dipole. *IEEE Trans. Antennas Propag.*, 54, 3614–3620.

Star-CCM+ (2015) CD-adapco Group, 60 Broadhollow Road, Melville, NY 11747, USA.

Storn R., Price, K. (1997) Differential evolution – a simple and efficient heuristic for global optimization over continuous spaces. *Journal of Global Optimization*, 11, pp. 341–359.
Styblinski, M.A., Oplaski, L.J. (1986) Algorithms and software tools for IC yield optimization based on fundamental fabrication parameters. IEEE Trans. Comput.-Aided Design Integr. Circuits Syst., 5, pp. 79–89.
Sumant, P.S., Wu, H., Cangellaris, A.C., and Aluru, N.R. (2010) A sparse grid based collocation method for model order reduction of finite element approximations of passive electromagnetic devices under uncertainty. *IEEE MTT-S Microwave Symp. Dig.*, pp. 1652–1655.
Sumant, P.S., Wu, H., Cangellaris, A.C., and Aluru, N.R. (2012) Reduced-order models of finite element approximations of electromagnetic devices exhibiting statistical variability. *IEEE Trans. Antennas Prop.*, vol. 60, no. 1, pp. 301–309.
Swidzinski, J.F., and Chang, K. Nonlinear statistical modeling and yield estimation technique for use in Monte Carlo simulations. IEEE Trans. Microwave Theory Techn, 48(12): 2316–2324.
Tan, K.C., Khor, E.F., Lee, T.H. (2005) Multiobjective evolutionary algorithms and applications. Springer-Verlag.
Tannehill, J.A., Anderson, D.A., and Pletcher, R.H., (1997) *Computational fluid mechanics and heat transfer*, 2nd edition, Taylor & Francis.
Tesfahunegn, Y.A., Koziel, S., and Leifsson, L. (2015) Surrogate-based airfoil design with multi-level optimization and adjoint sensitivities. *53rd AIAA Aerospace Sciences Meeting, Science and Technology Forum*, Kissimee, Florida, Jan 5–9.
Toal, D.J.J., and Keane, A.J. (2011) Efficient Multipoint Aerodynamic Design Optimization via Cokriging. *Journal of Aircraft*. 48, 1685–1695.
Toropov, V.V. (1989) Simulation approach to structural optimization. Structural Optimization, 1, pp. 37–46.
Toropov, V.V., Filatov, A.A., Polynkin, A.A., (1993). Multiparameter structural optimization using FEM and multipoint explicit approximations. *Struct. Optim.*, 6, 7–14.
Toivanen, J.I, Rahola, J., Makinen, R.A.E., Jarvenpaa, S., and Yla-Oijala, P. (2010) "Gradient-based antenna shape optimization using spline curves," *Annual Review of Progress in Applied Comp. Electromagnetics*, Tampere, Finland, pp. 908–913.
Tu, S., Cheng, Q.S., Zhang, Y., Bandler, J.W., Nikolova, N.K. (2013). Space mapping optimization of handset antennas exploiting thin-wire models, IEEE Trans. Ant. Prop., 61, 3797–3807.
Wang, L.-X., and Mendel, J.M. (1992) Generating fuzzy rules by learning from examples. *IEEE Trans. Systems, Man, Cybernetics*, vol. 22, no. 6, pp. 1414–1427, Nov./Dec. 1992.
Wild, S.M., Regis, R.G., Shoemaker, C.A. (2008) ORBIT: Optimization by radial basis function interpolation in trust-regions. SIAM J. Sci. Comput., 30, pp. 3197–3219.
Wu, K. (2009) Substrate Integrated Circuits (SiCs) – A new paradigm for future GHz and THz electronic and photonic systems. *IEEE Circuits and Systems Society Newsletter*, 3.
Xia, L., Xu, R.M., Yan, B. (2007) Ltcc interconnect modeling by support vector regression. *Progress In Electromagnetics Research*, 69, pp. 67–75.
Yamamoto, I., (2007) Research and development of past, present, and future AUV technologies," in *Proc. Int. Mater-class AUV Technol. Polar Sci. – Soc. Underwater Technol.*, 17–26.
Yang, X.S., (2005) Engineering optimization via nature-inspired virtual bee algorithms. IWINAC 2005, *Lecture Notes in Computer Science*, 3562, 317–323.
Yang, X.S., (2008) *Nature-inspired metaheuristic algorithms*. Luniver Press.
Yang, X.S. (2010) Engineering optimization: an introduction with metaheuristic applications. Wiley.
Yang, Y., Hu, S.M., Chen, R.S., (2005) A combination of FDTD and least-squares support vector machines for analysis of microwave integrated circuits. *Microwave Opt. Technol. Lett.*, 44, 296–299.
Ye, K.Q., (1998) Orthogonal column latin hypercubes and their application in computer experiments. *Journal of the American Statistical Association*. 93, 1430–1439.

Tesfahunegn, Y.A., Koziel, S., Leifsson, L., and Bekasiewicz, A. (2015) "Surrogate-based airfoil design with space mapping and adjoint sensitivity," *Int. Conf. Comp. Science*, Reykjavik, Iceland, June 1–3.

Zadeh, L.A. (1965) Fuzzy sets. *Inform. Contr.*, vol. 8, no. 3, pp. 338–353.

Zhang, K., and Han, Z., (2013) Support vector regression-based multidisciplinary design optimization in aircraft conceptual design. *AIAA Aerospace Sciences Meeting*. AIAA paper 2013–1160.

Zhang, Y., Hosder, S., Leifsson, L., and Koziel, S., (2012) Robust Airfoil Optimization Under Inherent and Model-Form Uncertainties Using Stochastic Expansions. AIAA-Paper 2012-0056, *50th AIAA Aerospace Sciences Meeting Including the New Horizon Forum and Aerospace Exposition*, Nashville, TN, June 9–12, 212.

Printed in the United States
By Bookmasters